RENEWALS 458-4574
DATE DUE

**WITHDRAWN**
UTSA LIBRARIES

# Social Science Knowledge and Economic Development

*Economics, Cognition, and Society*

This series provides a forum for theoretical and empirical investigations of social phenomena. It promotes works that focus on the interactions among cognitive processes, individual behavior, and social outcomes. It is especially open to interdisciplinary books that are genuinely integrative.

Editor: Timur Kuran
Editorial Board: Tyler Cowen   Avner Greif
                 Diego Gambetta   Viktor Vanberg

**Titles in the Series**

Ulrich Witt, Editor. *Explaining Process and Change: Approaches to Evolutionary Economics*

Young Back Choi. *Paradigms and Conventions: Uncertainty, Decision Making, and Entrepreneurship*

Geoffrey M. Hodgson. *Economics and Evolution: Bringing Life Back into Economics*

Richard W. England, Editor. *Evolutionary Concepts in Contemporary Economics*

W. Brian Arthur. *Increasing Returns and Path Dependence in the Economy*

Janet Tai Landa. *Trust, Ethnicity, and Identity: Beyond the New Institutional Economics of Ethnic Trading Networks, Contract Law, and Gift-Exchange*

Mark Irving Lichbach. *The Rebel s Dilemma*

Karl-Dieter Opp, Peter Voss, and Christiane Gern. *Origins of a Spontaneous Revolution: East Germany, 1989*

Mark Irving Lichbach. *The Cooperator s Dilemma*

Richard A. Easterlin. *Growth Triumphant: The Twenty-first Century in Historical Perspective*

Daniel B. Klein, Editor. *Reputation: Studies in the Voluntary Elicitation of Good Conduct*

Eirik G. Furubotn and Rudolf Richter. *Institutions and Economic Theory: The Contribution of the New Institutional Economics*

Lee J. Alston, Gary D. Libecap, and Bernardo Mueller. *Titles, Conflict, and Land Use: The Development of Property Rights and Land Reform on the Brazilian Amazon Frontier*

Rosemary L. Hopcroft. *Regions, Institutions, and Agrarian Change in European History*

E. L. Jones. *Growth Recurring: Economic Change in World History*

Julian L. Simon. *The Great Breakthrough and Its Cause*

David George. *Preference Pollution: How Markets Create the Desires We Dislike*

Alexander J. Field. *Altruistically Inclined? The Behavioral Sciences, Evolutionary Theory, and the Origins of Reciprocity*

David T. Beito, Peter Gordon, and Alexander Tabarrok, Editors. *The Voluntary City: Choice, Community, and Civil Society*

Randall G. Holcombe. *From Liberty to Democracy: The Transformation of American Government*

Omar Azfar and Charles A. Cadwell, Editors. *Market-Augmenting Government: The Institutional Foundations for Prosperity*

Stephen Knack, Editor. *Democracy, Governance, and Growth*

Phillip J. Nelson and Kenneth V. Greene. *Signaling Goodness: Social Rules and Public Choice*

Vernon W. Ruttan. *Social Science Knowledge and Economic Development: An Institutional Design Perspective*

# Social Science Knowledge and Economic Development

## An Institutional Design Perspective

Vernon W. Ruttan

UNIVERSITY OF MICHIGAN PRESS
*Ann Arbor*

To the social scientists who serve on the staffs of the national and international development agencies. They have become increasingly effective in bringing social science knowledge to bear on the design and implementation of development policies and programs.

Copyright © by the University of Michigan 2003
All rights reserved
Published in the United States of America by
The University of Michigan Press
Manufactured in the United States of America
⊗ Printed on acid-free paper

2006  2005  2004  2003    4  3  2  1

No part of this publication may be reproduced, stored in a retrieval system, or transmitted in any form or by any means, electronic, mechanical, or otherwise, without the written permission of the publisher.

*A CIP catalog record for this book is available from the British Library.*

Library of Congress Cataloging-in-Publication Data

Ruttan, Vernon W.
    Social science knowledge and economic development / Vernon W. Ruttan.
      p.  cm. — (Economics, cognition, and society)
    Includes bibliographical references and index.
    ISBN 0-472-11355-0 (Cloth : alk. paper)
    1. Economic development.  2. Applied sociology.  3. Technological innovations—Economic aspects—Developing countries.  4. Organizational change—Economic aspects—Developing countries.  I. Title.  II. Series.
HD75 .R87   2004
338.9—dc21                                                              2003012783

We contract the boundaries of our subjects, and of our sub-subjects, to make them more manageable; and we are enabled to do this because our academic specialization is in fact happening in the "real world." But it is not all that is happening in the world.

—Sir John Hicks (1969)

# Contents

| | | |
|---|---|---|
| | List of Figures | xi |
| | List of Tables | xiii |
| | List of Boxes | xiii |
| | Preface | xv |

PART I

CHAPTER 1 **Induced Institutional Innovation**   3
*What Is Institutional Innovation?*   4
*Demand for Institutional Innovation*   9
*The Supply of Institutional Innovation*   16
*Toward a More Complete Model of
   Induced Innovation*   22

PART II

CHAPTER 2 **Cultural Endowments and Economic Development**   33
*Cultural Endowments in
   Development Economics*   33
*Why Anthropology?*   42
*Deconstructing Development*   52
*Constructing Culture*   59
*Perspective*   63

CHAPTER 3 **The Sociology of Development
   and Underdevelopment**   68
*Why Sociology?*   69
*What Happened to Modernization Theory?*   71
*Alternative Sociologies*   79
*Rational Choice, Social Norms,
   and Development*   87
*Perspective*   97

| | | |
|---|---|---|
| CHAPTER 4 | What Happened to Political Development? | 100 |
| | Political Development in Development Economics | 101 |
| | Political Science and Political Development | 106 |
| | The Political Basis of Economic Development | 112 |
| | The Economic Foundations of Political Development | 117 |
| | Political Power and Political Development | 119 |
| | Institutional Design | 125 |
| | Perspective | 131 |
| CHAPTER 5 | Growth Economics and Development Economics | 135 |
| | Classical and Schumpeterian Growth | 136 |
| | Modern Growth Theory | 139 |
| | Dialogue with Data | 150 |
| | Growth Economics as Development Economics | 154 |
| | A More Comprehensive Growth Economics? | 163 |
| | Perspective | 165 |
| | A Postscript on Method | 167 |

PART III

| | | |
|---|---|---|
| CHAPTER 6 | Technology Adoption, Diffusion, and Transfer | 171 |
| | The Convergence of Traditions | 171 |
| | The Diffusion of Diffusion Research | 184 |
| | International Technology Transfer | 190 |
| | Resistance to Technology | 193 |
| | The Divergence of Traditions | 197 |
| CHAPTER 7 | Social Capital and Institutional Renovation | 200 |
| | The Kombi-Naam Cultural Endowment | 202 |
| | Understanding the Kombi-Naam | 203 |
| | Renovating the Kombi-Naam as a Development Organization | 205 |
| | Groupements Naam and Technical Innovation | 207 |
| | The Continuity and Replicability of the Groupements Naam | 209 |
| | Conclusions | 213 |
| CHAPTER 8 | Religion, Culture, and Nation | 215 |
| | Fundamentalist Religion and Economic Development | 215 |
| | Culture and Development | 226 |
| | Nationalism and Nation Building | 240 |
| | Perspective | 248 |

| | | Contents ix |
|---|---|---|
| CHAPTER 9 | Why Foreign Economic Assistance? | 251 |
| | *Donor Self-Interest* | 252 |
| | *Ethical Considerations* | 255 |
| | *Lessons from Experience* | 260 |
| | *A Foreign Economic Assistance Future?* | 266 |
| | *In Conclusion* | 268 |
| PART IV | | |
| CHAPTER 10 | Postscript | 271 |
| | Appendix: Definitions of Culture | 273 |
| | Bibliography | 277 |
| | Indexes | 327 |

# Figures

| | | |
|---|---|---|
| 1.1. | Interrelationships between changes in resource endowments, cultural endowments, technology, and institutions | 23 |
| 1.2. | Macro- and microlevel propositions: effects of religious doctrine on economic organization | 28 |
| 2.1. | A Marxist model. The forces of production and the relations of production together make up the economic base or mode of production. | 46 |
| 2.2. | The cultural materialism model | 47 |
| 3.1. | Linear and interactive models of the relations between advances in scientific and technical knowledge | 86 |
| 5.1. | Real per capita gross domestic product in the United States, 1870–1994 | 152 |
| 6.1. | The epidemic model | 176 |
| 6.2. | Cumulative number of diffusion research publications by year | 178 |
| 6.3. | Number of diffusion publications by rural sociologists by year | 179 |
| 6.4. | Number of horses and cars in the United States, 1850–2000 | 188 |

## Tables

| | | |
|---|---|---|
| 1.1. | Factor Shares of Rice Output per Hectare, 1976 Wet Season | 12 |
| 1.2. | Comparison between the Imputed Value of Harvesters' Share and Imputed Cost of *Gamma* Labor | 14 |
| 6.1. | Major Diffusion Research Traditions | 172 |

## Boxes

| | | |
|---|---|---|
| 2.1. | Postmodernism | 54 |
| 3.1. | Rational Choice | 88 |
| 4.1 | Institutional Design Principles | 127 |
| 5.1. | The Keynesian (Harrod-Domar) Model | 140 |
| 5.2. | The Neoclassical (Solow-Swan) Growth Model | 142 |
| 5.3. | Endogenous Growth Theory (Romer-Lucas) | 145 |

# Preface

In the first several decades after World War II there was a dramatic increase in interest in the development of non-Western societies in all of the social sciences. Prior to World War II interest in non-Western societies by social scientists was dominated by ethnographic studies of primitive isolates or studies of native cultures by applied anthropologists in the service of colonial administration. The emergence of new states committed to rapid economic growth contributed to the rapid emergence of a new field of inquiry into the process of development in each of the social sciences.

In the literature of the 1950s there was a presumption that economic growth would be strongly influenced by cultural endowments and institutional development. This resulted in the rapid emergence of disciplinary subfields such as development sociology, political development, and development administration. Historians, geographers, legal scholars, and historians also began to explore the contribution that their particular disciplinary skills and insights might offer to the understanding of development process or development practice.

The early expectation that these several streams of inquiry might merge to enrich each other has only recently begun to materialize. In some fields, such as political development, early enthusiasm was replaced by serious doubt about the viability of the political development research agenda. In other fields, interest in development studies faded along with the decline of financial support by foundations and official development assistance agencies as the Cold War wound down. Interest by economists in the contribution of other social science disciplines declined as economic approaches to policy analysis and planning became more formalized.

The rapid pace of economic development in a number of developing countries during the 1960s and 1970s seemed to confirm the perspective that economic development could be left to the economists. During the 1980s, however, much of the optimism about the pace of development in both Western and non-Western economies eroded. By the early

1990s institutional and cultural explanations for stagnation or distorted development again emerged as important themes in both popular and professional literature.

My own earlier work on the role of institutional innovation in economic development (and with colleagues Yujiro Hayami and Hans Binswanger) largely involved the extension of neoclassical microeconomics in attempts to understand (1) how changes in resource endowments and technical change induced institutional innovations and (2) how institutional change induced changes in resource endowment and technology. My work departed from the mainstream "new institutional economics" in that it drew more heavily on the work of Theodore W. Schultz and Leonid Hurwicz than on that of Ronald Coase for its inspiration. In this book I go beyond my earlier concerns to explore what development economists can (or should) learn from other social scientists, and from macroeconomic growth theory, in their attempts to design more effective development policies and institutions. A consistent theme in this book is that advances in social science knowledge represent a powerful source of economic development. Theoretical inquiry carried on apart from a continuing dialogue with data is arid.

This book originated in a program of summer reading that began in the mid-1980s. The initial result was a series of articles published mainly in the 1990s. In the spring of 2000 I was able to interest the University of Michigan Press in publishing a book, drawing on these earlier articles, under the title *Social Science Knowledge and Economic Development*. The Press informed me, however, that they were not interested in simply publishing a collection of my old articles. But they would be interested in a revised, updated, and integrated treatment. The title of the book was chosen very deliberately. It is about the contribution of social science knowledge to the project of economic development.

I would like to thank Professor Henry J. Bruton, the University of Michigan Press reviewer, for his helpful comments on the proposed book. He has continued to offer useful comments and suggestions on subsequent drafts. Christopher Clague, Greta Friedemann-Sanchez, Robert T. Holt, Timur Kuran, Douglass C. North, and Stephen Gudeman have read and made recommendations on earlier drafts of the manuscript. Heidi Van Schooten has typed, retyped, and edited portions of the manuscript.

I have used earlier drafts of these chapters in an interdisciplinary graduate seminar offered by the MacArthur Program at the University of Minnesota. The book has benefited from my students' comments. I anticipate that the book may be of interest to behavioral economists and to other social scientists. It should also be of interest to social science professionals in development assistance agencies.

PART I

CHAPTER 1

# Induced Institutional Innovation

The central premise of this book is that the demand for social science knowledge is derived from the demand for institutional change.[1] If this view is correct then any claim by the social science disciplines and related professions for substantial public support depends on a credible promise that advances in social science knowledge represent an efficient source of institutional innovation. Another way of articulating the same point is that advances in social science knowledge reduce the cost of institutional change. I employ the term *cost* in its broadest sense, to include the psychological, the social, and the political as well as the economic costs of institutional change.

The interpretation of technical and institutional change as endogenous rather than exogenous to the economic system is a relatively new development in economic thought. In work published in the 1970s Yujiro Hayami, Hans Binswanger, and I extended the theory of induced technical change and tested it against the history of agricultural development in the United States and Japan (Hayami and Ruttan 1971; Binswanger and Ruttan 1978). It is now generally accepted that the theory of induced technical change provides very substantial insight into the process of agricultural development for a wide range of developed and developing countries. Industrial economists and economic historians have drawn extensively on the theory of induced technical change in attempting to interpret differential patterns of productivity growth among firms, industries, and countries and over time (Thirtle and Ruttan 1987, 68–72). In the late 1980s growth theory was revitalized by the introduction of new models of endogenous economic growth (Romer 1986; Lucas 1988; chap. 5, this vol.).

---

1. In this chapter I draw heavily on Ruttan and Hayami (1984). I am indebted to Henry J. Bruton, James F. Oemke, and Christopher K. Clague for comments on an earlier draft of this chapter.

The demonstration that technical change can be treated as largely endogenous to the development process does not imply that the progress of either agricultural or industrial technology can be left to an "invisible hand" that drives technology along an "efficient" path determined by relative resource endowments. The capacity to advance knowledge in science and technology is itself a product of institutional innovation—"the great invention of the nineteenth century was the invention of the method of invention" (Whitehead 1925, 96).

In the case of agriculture, for example, in both the United States and Japan, much of the technical change that has led to growth of output per hectare was, until very recently, produced by public sector institutions. These institutions—state (or prefectoral) and federal (or national) agricultural experiment stations—obtained their resources in the political marketplace and allocated their resources through bureaucratic mechanisms (Hayami and Ruttan 1985, 206–52; Ruttan 2001b, 179–234). The success of the theory of induced technical change gives rise, therefore, to the need for a more careful consideration of the sources of institutional innovation and design.

In this chapter I elaborate a theory of institutional innovation in which shifts in the demand for institutional innovation are induced by changes in relative resource endowments and by technical change. I also consider the impact of advances in social science knowledge and of cultural endowments on the supply of institutional change. After examining the forces that act to shift the supply and demand of institutional innovation, the elements of a more general model of institutional change are presented.

### What Is Institutional Innovation?

There is considerable disagreement regarding the concept of institution. Institutions are commonly defined as the organizations and rules of a society that facilitate coordination among people by helping them form expectations that each person can reasonably hold in dealing with others. They reflect the conventions and ideologies that have evolved in different societies regarding the behavior of individuals and groups relative to their own behavior and the behavior of others. In the area of economic relations they have a crucial role in establishing expectations about the rights to use resources in economic activities and about the partitioning of the income streams resulting from economic activity: "institutions provide *assurance* respecting the actions of others, and give order and stability to expectations in the complex and uncertain world of economic relations."[2]

In my work I find that an inclusive definition that includes organizations is most useful, which is consistent with the view expressed by both Commons (1950, 24) and Knight (1952, 51). The more inclusive definition is employed in order to be able to consider changes in the rules or conventions that govern behavior (1) within economic units such as firms and bureaucracies, (2) among economic units as in the cases of the rules that govern market relationships, and (3) between economic units and their environment, as in the case of the relationship between a firm and a regulatory agency. It includes policy, mechanism, and system innovations.[3] The distinction that I make between institutions and cultural endowments is that institutions are the formal rules and arrangements that govern behavior among and within organizations, while cultural endowments are the informal codes and norms that influence individual and group behavior. The state is a useful example. It is an organization. It is also a source of formal rules and relationships within itself and between itself and private agents (Aoki 2001, 151–79).

In order to perform the essential role of forming reasonable expectations in dealings among people, institutions must be stable for an extended time period. But institutions, like technology, must also change if development is to occur. Anticipation of the latent gains to be realized by overcoming the disequilibria resulting from changes in factor endowments, product demand, and technical change represents powerful inducements to institutional innovation (North and Thomas 1970; Schultz 1975; Binswanger and Deininger 1997). Institutions that have been efficient in generating growth in the past may, over time, come to protect the vested interests of some of their members in maintaining the status quo and thus become obstacles to further economic development.[4] The growing disequilibrium in resource allocation due

---

2. See Runge (1981b, xv). Formal analysis of the role of institutions in providing assurance of stability in economic relationships emerged from dissatisfaction with the implications of the assumption of strict dominance of individual strategy in modern welfare economics (Sen 1967; Runge 1981a). North argues that shared ideological and ethical perspectives provide assurance that is lacking in models built on the dominance of individual strategies (1981, 45–58).

3. In his more recent work, North employs a definition of *institution* that is similar to anthropologists' use of the term *culture* (1991, 1994) and to my use of the term *cultural endowment* (fig. 1.1). For the evolution of the term *culture* see the appendix.

4. The role of special interest "distributional coalitions" in slowing society's capacity to adopt new technology and reallocate resources in response to changing conditions is a central theme in Olson (1982, 74).

to institutional constraints generated by economic growth creates incentives for political entrepreneurs or leaders to organize collective action to bring about institutional change.

This perspective on the sources of demand for institutional change is similar, in some respects, to the traditional Marxist view.[5] Marx considered technological change to be the primary source of institutional change. The induced innovation perspective is somewhat more complex in that it considers that changes in cultural endowments, factor endowments, and product demand are also important sources of institutional change. The definition of institutional change employed in this book is not limited to the dramatic or revolutionary changes of the type anticipated by Marx. Institutions such as property rights and markets are more typically altered through the accumulation of secondary or incremental institutional changes such as modifications in contractual relations or shifts in the boundaries between market and nonmarket activities (Davis and North 1971, 9).

There is a supply dimension as well as a demand dimension in institutional change. Advances in social science knowledge represent an increasingly important source of shifts in the supply of institutional innovations. Collective action leading to the implementation of institutional innovations involves struggles among various vested interest groups. Clearly, the process is much more complex than the clear-cut, two-class conflict between property owners and the property-less as assumed by Marx. The supply of institutional innovations is strongly influenced by the cost of achieving social consensus (or of suppressing opposition). The cost of implementing an institutional innovation depends on the distribution of both economic and political resources. It also depends critically on cultural tradition and ideology (such as nationalism) that make certain institutional arrangements more easily accepted than others.

Advances in knowledge in the social sciences (and in related professions such as law, administration, planning, and social service) can reduce the cost of institutional change in a manner somewhat similar

---

5. "At a certain stage of their development, the material forces of production in society come in conflict with existing relations of production, or—what is but a legal expression for the same thing—with the property relations within which they had been at work before. From forms of development of the forces of production these relations turn into their fetters. Then comes the period of social revolution. With the change of the economic foundation the entire immense superstructure is more or less rapidly transformed" (Marx 1913, 11–12). For a discussion of the role of technology in Marxian thought see Rosenberg (1982, 34–54).

to how advances in the natural sciences reduce the cost of technical change. Advances in game theory have, during the last several decades, enabled economists and political scientists to bring an increasingly powerful set of tools to bear on the understanding of the processes of institutional change (Schotter 1981; Ostrom 1990; Platteau 2000; Aoki 2001). Still, I continue to find the application of standard neoclassical microeconomic theory to interpret the sources of the supply and demand of institutional change exceedingly useful.

Insistence that important advances in the understanding of the processes of institutional innovation and diffusion can be achieved by treating institutional change as endogenous to the economic system represents a clear departure from the tradition of modern analytical economics.[6] This does not mean that modern analytical economics must be abandoned. On the contrary, the scope of modern analytical economics is expanded by treating institutional change as endogenous.[7]

There is general agreement that institutional change has evolved and continues to evolve in response to long-term changes such as the pressure of population against land resources or the rise in the price of labor relative to capital. Disequilibrium between institutional and economic rents represents a powerful source of institutional change. But there has been substantial disagreement within the social sciences about the role of purposeful or rational design in institutional innova-

---

6. The orthodox view was expressed by Samuelson (1948, 221–22): "The auxiliary [institutional] constraints imposed upon the variables are not themselves the proper subject of welfare economics but must be taken as given." Contrast this with the statement by Schotter (1981, 61): "We view welfare economics as a study . . . that ranks the system of rules which dictate social behavior." There are now five fairly well-defined political economy traditions that have attempted to break out of the constraints imposed by traditional welfare economics and treat institutional change as endogenous. These include (1) the theory of property rights, (2) the theory of economic regulation, (3) the theory of interest group rent seeking, (4) the liberal-pluralist theories of government, and (5) the neo-Marxian theories of the state. In the property rights theories the government plays a relatively passive role; the economic theory of regulation focuses on the electoral process; the rent-seeking and liberal-pluralist theories concentrate on both electoral and bureaucratic choice processes; and the theory of the state attempts to incorporate electoral, legislative choice, and, bureaucratic choice processes. For a review and criticism see Rausser, Lichtenberg, and Lattimore (1982).

7. My use of the neoclassical microeconomic approach to interpret the process of institutional change is closer in spirit to that of Hicks (1969) and North and Thomas (1973) than to North's more recent work (1994). It is similar to that employed by Gary Becker in analyzing the institutions of the family (1981, 1993). An important difference is that my work focuses on the effects of long-term changes in the external environment that must be treated as exogenous by the agents who act to bring about institutional change.

tion.[8] Those holding an "organic" perspective argue that the fact that the institutions of civilization have been created by human action "does not mean that man must also be able to alter them at will" (Hayek 1978b, 3).[9] The organic view of the sources of institutional change is reinforced by a theory of "unintended consequences" that runs through the work of Adam Smith, Max Weber, and Friedrich Hayek (Lal 1998).[10] In contrast, the constructivist or design perspective holds that advances in social science knowledge can play an important role in the rational design of institutional reform and institutional innovation.

Much of my work with Yujiro Hayami on induced institutional innovation reflects an organic perspective. In other work, on the development of agricultural research institutions, for example, I have employed a constructivist or design perspective (Ruttan 1982). I reject any demand to choose between the organic and constructivist perspectives. They should be viewed as complements rather than as alternatives. I also reject the ideological implication, advanced by some proponents of the organic approach, that the unintended consequences of institutional change preclude the possibility of a rational or analytical approach to institutional reform and design. I next employ an organic approach to interpret a series of institutional changes in land and labor relationships, then discuss the supply of institutional innovation from a constructivist or design perspective.

---

8. Schotter (1981, 3–4) notes that in economics there have been, historically, two distinct interpretations of the sources of institutional change—organic and collectivist. He identifies the organic view with the work of Hayek and the collectivist view with the work of Commons. Hayek (1978b, 3–22) uses the term *constructivist* rather than *collectivist*. The collectivist perspective, as employed by Schotter, is similar in concept to the designer perspective as employed by Hurwicz (1998).

9. Hayek was apparently referring to a statement by Karl Marx: "Men make their own history, but they do not make it as they please; they do not make it under circumstances chosen by themselves, but under circumstances directly formed, given and transmitted from the past" (Marx 1936, 15).

10. In an earlier work Hayek argued that it was misleading to divide all phenomena into "natural" and "artificial." He addresses a threefold classification: (1) phenomena that are natural in the sense that they are wholly independent of human action; (2) those unintended patterns and regularities in human society that are due to human action but not to human design; (3) those patterns and regularities that are the deliberate product of human design. He regarded the explanation of the unintended patterns and regularities, which he termed "spontaneous order," as the proper task of social theory. He was, and remained, skeptical of constructivism because of the inability of social theory to anticipate unintended consequences (1967, 96–105).

## Demand for Institutional Innovation

In some cases the demand for institutional innovation can be satisfied by the development of new forms of property rights, more efficient market institutions, or even evolutionary changes arising out of direct contracting by individuals at the level of the community or the firm. In other cases, where externalities are involved, substantial political resources may have to be brought to bear to organize nonmarket institutions in order to provide for the supply of public goods. Drawing from agricultural history, this section illustrates how changes in factor endowments, technical change, and growth in product demand have induced changes in property rights and contractual arrangements.

The agricultural revolution that occurred in England between the fifteenth and the nineteenth centuries involved a substantial increase in the productivity of land and labor. It was accompanied by the enclosure of open fields and the replacement of a system of small peasant cultivators, who held their land from manorial lords, by one in which large farmers used hired labor to farm the land they leased from the landlords. The First Enclosure Movement, in the fifteenth and sixteenth centuries, resulted in the conversion of open arable fields and commons to private pasture in areas suitable for grazing. It was induced by expansion in the export demand for wool. The Second Enclosure Movement, in the eighteenth century, involved conversion of communally managed arable land into privately operated units. It is now generally agreed that demand for changes in land tenure arrangements was largely induced by the growing disequilibrium between the fixed institutional rent that landlords received under copyhold tenures (with lifetime contracts) and the higher economic rents expected from adoption of new technology that became more profitable as a consequence of higher grain prices and lower wages. When the land was enclosed there was a redistribution of income from farmers to landowners, and the disequilibrium was reduced or eliminated.[11]

In nineteenth-century Thailand, the opening of the nation to international trade and the reduction in shipping rates to Europe following the completion of the Suez Canal resulted in a sharp increase in the demand for rice. The land available for rice production, which had

---

11. There has been a continuing debate among students of English agricultural history about whether the higher rents that landowners received after enclosure was because enclosed farming was more efficient than open field farming or because enclosures redistributed income from farmers to landowners. See Chambers and Mingay (1966), Dahlman (1980), and Allen (1982).

been abundant, became scarcer. Investment in land development for rice production for export became profitable. The rise in the profitability of rice production for export induced a demand for the reform of property rights in both land and man. Traditional rights in human property (corvée and slavery) were replaced by more precise private property rights in land (fee-simple titles) (Feeney 1982, 2002).

In Japan, at the beginning of the feudal Tokugawa period (1603–1867), peasants' rights to cropland had been limited to the right to till the soil with the obligation to pay a feudal land tax in kind. As the population grew, commercialization progressed, and irrigation and technology were developed to make intensive farming more profitable. Some peasants divided their holdings into smaller units and leased them out to former servants or extended family members. Some accumulated land through mortgaging arrangements that made other peasants de facto tenants. As a result of the accumulation of illegal leasing and mortgaging practices, peasants' property rights in land approximated those of a fee-simple title by the end of the Tokugawa period. These rights were readily converted to the modern private property system in the succeeding Meiji period (Hayami and Kikuchi 1981, 28).

Research conducted by Yujiro Hayami and Masao Kikuchi in the Philippines during the late 1970s has enabled us to examine a contemporary example of the interrelated effects of changes in resource endowments and technical change on the demand for institutional change in land tenure and labor relations (Kikuchi and Hayami 1980; Hayami and Kikuchi 1981, 2000). The case is particularly interesting because the institutional innovations occurred as a result of private contracting among individuals—what Hayek termed "spontaneous order" and in more recent literature has been referred to as "Coasian bargains" (Hayek 1978b; Olson 2000). The Philippine study is unique in that it is based on a rigorous analysis of microeconomic data from a single village over several decades.[12]

## Land Tenure and Labor Relations in a Philippine Village

Between 1956 and 1976, rice production per hectare in the study village rose dramatically, from 2.5 to 6.7 metric tons per hectare per year. This was due to two technical innovations. In 1958, the national irrigation

---

12. For additional case studies using the framework employed in this chapter see Feeney (1988).

system was extended to the village. This permitted double-cropping to replace single-cropping, thereby more than doubling the annual production per hectare of rice land. The second major technical change was the introduction in the late 1960s of modern high-yielding rice varieties. The diffusion of modern varieties was accompanied by increased use of fertilizer and pesticides and by the adoption of improved cultural practices such as straight-row planting and intensive weeding.

Population growth in the village was rapid. Between 1966 and 1976 the number of households rose from 66 to 109 and the population rose from 383 to 464, while cultivated area remained virtually constant. The number of landless laborer households increased from 20 to 54. In 1976, half of the households in the village had no land to cultivate, not even rented land. The average farm size declined from 2.3 hectares to 2.0 hectares.

The land is farmed primarily by tenants. In 1976, only 1.7 hectares of the 108 hectares of cropland in the village were owned by village residents. Traditionally, share tenancy was the most common form of tenure. In both 1956 and 1966, 70 percent of the land was farmed under share tenure arrangements. In 1963, a new agricultural land reform code was passed that was designed to break the political power of the traditional landed elite and to provide greater incentives to peasant producers of basic food crops.[13] A major feature of the new legislation was an arrangement that permitted tenants to initiate a shift from share tenure to leasehold, with rent under the leasehold set at 25 percent of the average yield for the previous three years. Implementation of the code between the mid-1960s and the mid-1970s resulted in a decline in the percentage of land farmed under share tenure to 30 percent.

The shift from share tenure to lease tenure was not, however, the only change in tenure relationships that occurred between 1966 and 1976. There was a sharp increase in the number of plots farmed under subtenancy arrangements. The number increased from one in 1956, to

---

13. Although the passage and implementation of the Land Reform Code of 1963 was exogenous to the economy of the village, the land reform of the 1960s has been interpreted as the result of efforts by an emerging industrial elite to simultaneously break the political power of the more conservative landowning elite and to provide incentives to peasant producers to respond to the rapid growth in demand for marketable surpluses of wage goods, primarily rice and maize, needed to sustain rapid urban industrial development. Thus, the Land Reform Code can be viewed as an institutional innovation designed to facilitate realization of the opportunities for economic growth that could be realized through rapid urban industrial development. See Ruttan (1969).

TABLE 1.1. Factor Shares of Rice Output per Hectare, 1976 Wet Season

| | Number of Plots | Area (ha) | Rice Output[b] | Current Inputs[b] | Land-owner | Sub-tenancy[b] | Factor Shares[a] | | | |
|---|---|---|---|---|---|---|---|---|---|---|
| | | | | | | | Total | Labor | Capital[c] | Operators' Surplus |
| Leasehold land | 44 | 67.7 | 2,889 (100.0) | 657 (22.7) | 567 (19.6) | 0 (0) | 567 (19.6) | 918 (31.8) | 337 (11.7) | 410 (14.2) |
| Share tenancy land | 30 | 29.7 | 2,749 (100.0) | 697 (25.3) | 698 (25.4) | 0 (0) | 698 (25.4) | 850 (30.9) | 288 (10.5) | 216 (7.9) |
| Subtenancy land | 16 | 9.1 | 3,447 (100.0) | 801 (23.2) | 504 (14.6) | 801[d] (23.2) | 1,305 (37.8) | 1,008 (29.3) | 346 (10.1) | −13 (−0.4) |

*Source*: Yujiro Hayami and Masao Kikuchi, *Asian Village Economy at the Crossroads: An Economic Approach to Institutional Change* (Tokyo: University of Tokyo Press, 1981, and Minneapolis: University of Minnesota Press, 1982), 111–13.
[a]Percentage shares are shown in parentheses.
[b]In kilograms per hectare.
[c]Sum of irrigation fee and paid and/or imputed rentals of carabao, tractor, and other machines.
[d]Rents to subleasors; in the case of pledged plots are imputed by applying the interest rate of 40 percent crop season (a mode in the interest rate distribution in the village).

five in 1966, then sixteen in 1976. Subtenancy is illegal under the land reform code. The subtenancy arrangements are usually made without the consent of the landowner. All cases of subtenancy were on land farmed under a leasehold arrangement. The most common subtenancy arrangement was a fifty-fifty sharing of costs and output.

It was hypothesized that an incentive for the emergence of the subtenancy institution was that the rent paid to landlords under the leasehold arrangement was below the equilibrium rent—the level that would reflect both the higher yields of rice obtained with the new technology and the lower wage rates implied by the increase in population pressure against the land.

To test this hypothesis, market prices were used to compute the value of the unpaid factor inputs (family labor and capital) for different tenure arrangements during the 1976 wet season. The results indicate that the share-to-land was lowest and the operators' surplus was highest for the land under leasehold tenancy. In contrast, the share-to-land was highest and no surplus was left for the operator who cultivated the land under the subtenancy arrangement (table 1.1). Indeed, the share-to-land when the land was farmed under subtenancy was very close to the sum of the share-to-land plus the operators' surplus under the other tenure arrangement.

The results are consistent with the hypothesis. A substantial portion of the economic rent was captured by the leasehold tenants in the form of operators' surplus. On the land farmed under a subtenancy arrangement, the rent was shared between the leaseholder and the landlord.

A second institutional change, induced by higher yields and the increase in population pressure, has been the emergence of a new pattern of employer-labor relationship between farm operators and landless workers. According to the traditional system called *hunusan*, laborers who participated in the harvesting and threshing activity received a one-sixth share of the paddy (rough rice) harvest. By 1976, most of the farmers (83 percent) adopted a system called *gamma*, in which participation in the harvesting operation was limited to workers who had performed the weeding operation without receiving wages.

The emergence of the *gamma* system can be interpreted as an institutional innovation designed to reduce the wage rate for harvesting to a level equal to the marginal productivity of labor. In the 1950s, when the rice yield per hectare was low and labor was less abundant, the one-sixth share may have approximated an equilibrium wage level. With the higher yields and the more abundant supply of labor, the one-sixth share became larger than the marginal product of labor in the harvesting operation.[14]

To test the hypothesis that the *gamma* system was rapidly adopted primarily because it represented an institutional innovation that permitted farm operators to equate the harvesters' share of output to the marginal productivity of labor, imputed wage costs were compared with the actual harvesters' shares (table 1.2). The results indicate that a substantial gap existed between the imputed wage for the harvesters' labor alone and the actual harvesters' shares. This gap was eliminated if the imputed wages for harvesting and weeding labor were added. Those results are consistent with the hypothesis that the changes in institutional arrangements governing the use of production factors were induced when disequilibria between the marginal returns and the marginal costs of factor inputs occurred as a result of changes in factor endowments and technical change. Institutional change, therefore, was

**TABLE 1.2. Comparison between the Imputed Value of Harvesters' Share and Imputed Cost of *Gamma* Labor**

|  | Based on Employers' Data | Based on Employers' Data |
|---|---|---|
| No. of working days of *gamma* labor (days/ha)[a] | | |
| Weeding | 20.9 | 18.3 |
| Harvesting/threshing | 33.6 | 33.6 |
| Imputed cost of *gamma* labor (P/ha)[b] | | |
| Weeding | 167.2 | 146.4 |
| Harvesting/threshing | 369.6 | 369.6 |
| Total | 536.8 | 516.0 |
| Actual share of harvesters | | |
| (1) In kind (kg/ha)[c] | 504.0 | 549.0 |
| (2) Imputed value (P/ha)[d] | 504.0 | 549.0 |
| (2) − (1) | −32.8 | 33.0 |

*Source:* Yujiro Hayami and Masao Kikuchi, *Asian Village Economy at the Crossroads: An Economic Approach to Institutional Change* (Tokyo: University of Tokyo Press, 1981, and Minneapolis: University of Minnesota Press, 1982), 121.

[a]Includes labor of family members who worked as *gamma* laborers.
[b]Imputation using market wage rates (daily wage = P8.0 for weeding, P11.0 for harvesting).
[c]One-sixth of output per hectare.
[d]Imputation using market prices (1 kg = P1).

---

14. Real wages for agricultural labor declined significantly between the mid-1950s and the mid-1960s in the Philippines. See Khan (1977). Thus, while we cannot be certain that the labor market was in equilibrium in the 1950s, it is clear that the degree of disequilibrium widened, as a result of both higher yields and lower wage rates, prior to the introduction and diffusion of the gamma system.

directed toward the establishment of a new equilibrium in factor markets.[15]

It is important to recognize that subtenancy and *gamma* contracts were the institutional innovations arrived at by voluntary agreements among farm operators, tenants, and laborers. The land reform laws gave leasehold tenants strong protection of their tenancy rights. It gave them the right to continue tilling the soil at an institutional rent that was lower than the economic rent. But the laws prohibited tenants from renting their land to someone else who might utilize it more efficiently, when they became elderly or found more profitable off-farm employment, for example. Subtenancy reduced such inefficiency due to the institutional rigidity in the land rental market resulting from the land reform programs. Likewise, the *gamma* system counteracted the institutional rigidity in the labor market associated with the institutional wage rate based on the traditional harvest share.

It might appear that these institutional innovations increased efficiency at the expense of equity. But, if the subtenancy system had not been developed, the route would have been closed for some of the landless laborers to become farm operators and use their skills more profitably. If the wage rate for harvesting work had been raised in the absence of the implicit *gamma* contract, it might have encouraged mechanization in threshing, thereby reducing both employment and labor earnings.

In the case reviewed here the induced innovation process leading toward the establishment of equilibrium in factor markets occurred very rapidly in spite of the fact that many of the transactions—between landlords, tenants, and laborers—were less than fully monetized. Informal contractual arrangements or agreements were utilized. The subleasing and the *gamma* labor contract evolved without the mobilization of substantial political activity or bureaucratic effort. Indeed, the subleasing arrangement evolved in spite of legal prohibition. Where substantial political and bureaucratic resources must be mobilized to bring about technical or institutional change, the changes occur much more slowly, as in the cases of the English enclosure move-

---

15. A second round of technical and institutional changes occurred in the 1990s. Nonfarm employment opportunities expanded as a result of better transport to the metropolitan Manila area and the location of a small metal-craft firm in the village. Higher wage rates have induced the substitution of small portable threshing machines for manual rice threshing. The labor share for harvesting has declined, and a new form of labor contract, referred to as *new hunusan,* has emerged. As a result of the new nonfarm employment opportunities the incomes of landless labor households have risen (Hayami and Kikuchi 2000).

ments and the Thai and Japanese property rights cases referred to previously.

The examples of institutional change advanced in this section, such as the enclosure in England and the evolution of private property rights in land in Japan and Thailand, have contributed to the development of a more efficient market system. Institutional changes of this type are profitable for society only if the costs involved in the assignment and protection of rights are smaller than the gains from better resource allocation. If those costs are very high, it may be necessary to design nonmarket institutions in order to achieve more efficient resource allocation.[16]

### The Supply of Institutional Innovation

The disequilibria in economic relationships associated with economic growth, such as technical change leading to the generation of new income streams and changes in relative factor endowments, have been identified as important sources of demand for institutional change. But the sources of supply of institutional innovation are less well understood (Olson 1968; Ostrom 1990). The factors that reduce the cost of institutional innovation have received only limited attention by economists or by other social scientists.

In the Philippines village case discussed earlier, innovations in tenure and labor market institutions were supplied, in response to the changes in demand generated by changing factor endowments and new income streams, through the individual and joint decisions of owner-cultivators, tenants, and laborers. But even at this level it was necessary for gains to the innovators to be large enough to offset the risk of ignoring the land reform prohibitions against subleasing and the transaction costs involved in changing traditional harvest-sharing arrangements. While mobilization of substantial political resources was not required to introduce and extend the new land and labor market institutions, the distribution of political resources within the village did influence the initiation and diffusion of the institutional innovations.

The supply of major institutional innovations necessarily involves

---

16. Demsetz (1964) has pointed out that the relative costs of using market and political institutions is rarely given explicit consideration in the literature on market failure. An appropriate way of interpreting the public goods vs. private goods issue is to ask whether the costs of providing a market are too high relative to the cost of nonmarket alternatives. A similar point is made by Hurwicz (1972b).

the mobilization of substantial political resources by political entrepreneurs and innovators. It is useful to think in terms of a supply schedule of institutional innovation that is determined by the marginal cost schedule facing political entrepreneurs as they attempt to design new institutions and resolve the conflicts among various vested interest groups (or suppress the opposition when necessary). It was hypothesized that institutional innovations will be supplied if the expected return from the innovation that accrues to the political entrepreneurs exceeds the marginal cost of mobilizing the resources necessary to design and introduce the innovation. To the extent that the private return to the political entrepreneurs is different from the social return, the institutional innovation will not be supplied at a socially optimum level.[17]

Thus, the supply of institutional innovation depends critically on the distribution of economic and political resources among interest groups in a society. If the power balance is such that the political entrepreneurs' efforts to introduce an institutional innovation with a high rate of social return are adequately rewarded by greater prestige and stronger political support, a socially desirable institutional innovation may occur. However, if the institutional innovation is expected to result in a loss to a dominant political bloc, the innovation may not be forthcoming even if it is expected to produce a large net gain to society as a whole. And socially undesirable institutional innovations may occur if the returns to the entrepreneur or the interest group exceed the gains to society (Tullock 1967; Krueger 1974; Tollison 1982).

The failure of many developing countries to institutionalize the agricultural research capacity needed to take advantage of the large gains from relatively modest investments in technical change may be due, in part, to the divergence between social returns and the private returns to political entrepreneurs. In the mid-1920s, for example, agricultural development in Argentina appeared to be proceeding along a path roughly comparable to that of the United States. Mechanization of crop production lagged slightly behind that in the United States. Grain yields per hectare averaged slightly higher than in the United States. In contrast to the United States, however, output and yields in Argentina remained relatively stagnant between the mid-1920s and the mid-1970s. It was not until the late 1970s that Argentina began to real-

---

17. See, for example, Frohlich, Oppenheimer, and Young (1971). For a review and extension of concepts of political entrepreneurship see Guttman (1982).

ize significant gains in agricultural productivity. Part of this lag in Argentine agricultural development was due to the disruption of export markets in the 1930s and 1940s. Students of Argentine development have pointed to the political dominance of the landed aristocracy, to the rising tensions between urban and rural interests, and to inappropriate domestic policies toward agriculture (de Janvry 1973; Smith 1969, 1974; Cavallo and Mundlak 1982). In the Argentine case, it would seem that the bias in the distribution of political and economic resources imposed exceptionally costly delays in the institutional innovations needed to take advantage of the relatively inexpensive sources of growth that technical change in agriculture could have made available.

Cultural endowments, including religion and ideology, exert a strong influence on the supply of institutional innovation. They make some forms of institutional change less costly to establish and impose severe costs on others. For example, the traditional moral obligation in the Japanese village community to cooperate in joint communal infrastructure maintenance has made it less costly to implement rural development programs than in societies where such traditions do not prevail. These activities had their origin in the feudal organization of rural communities in the pre-Meiji period. But practices such as maintenance of village and agricultural roads and of irrigation and drainage ditches through joint activities in which all families contribute labor were still practiced in well over half of the hamlets in Japan as recently as 1970 (Ishikawa 1981). The traditional patterns of cooperation have represented an important form of social capital on which to erect modern forms of cooperative marketing and joint farming activities. Similar cultural resources are not available in South Asian villages where, for example, the caste structure inhibits cooperation and encourages specialization (Lal 1998; chap. 8, this vol.).

Likewise, the adoption of new ideology may reduce the cost to political entrepreneurs of mobilizing collective action for institutional change. For example, the Jeffersonian concept of agrarian democracy provided ideological support for the series of land ordinances culminating in the Homestead Act of 1862, which established the legal framework designed to encourage an owner-operator system of agriculture in the American West (Cochrane 1979, 41–47, 179–88). Strong nationalist sentiment in Meiji Japan, reflected in slogans such as "A Wealthy Nation and Strong Army" (*Fukoku Kyohei*), helped mobilize the resources needed for the establishment of vocational schools and agricultural and industrial experiment stations (Hayami and Kikuchi

1981). In China, communist ideology, reinforced by the lessons learned during the guerrilla period in Yenan, inspired the mobilization of communal resources to build irrigation systems and other forms of social overhead capital (Schran 1975). Thus, ideology can be a critical resource for political entrepreneurs and an important factor affecting the supply of institutional innovations.

Advances in social sciences that improve knowledge relevant to the design of institutional innovations that are capable of generating new income streams or that reduce the cost of conflict resolution act to shift the supply of institutional change to the right. Throughout history, improvements in institutional performance have occurred primarily through the slow accumulation of successful precedent or as by-products of expertise and experience. Institutional change was generated through the process of trial and error much in the same manner that technical change was generated prior to the invention of the research university, the agricultural experiment station, or the industrial research laboratory. With the institutionalization of research in the social sciences and related professions, the process of institutional innovation has begun to proceed much more deliberately; it has become increasingly possible to substitute social science knowledge and analytical skill for the more expensive process of learning by trial and error.

The research that led to advances in our understanding of the production and consumption of rural households in less developed countries represents an important example of the contribution of advances in social science knowledge to the design of more efficient institutions (Schultz 1964; Nerlove 1974; Binswanger, Evenson, Florencio, and White 1980). In a number of countries this research has led to the abandonment of policies that viewed peasant households as unresponsive to economic incentives. And it has led to the design of policies and institutions to make more productive technologies available to peasant producers and to the design of more efficient price policies for factors and products.

I next present a case study of the contribution of advances in social science knowledge to the design of a contemporary institutional innovation—an emissions trading system designed to reduce the transaction costs of controlling sulfur dioxide ($SO_2$) emissions, an important industrial pollutant. Advances in economic knowledge led to an understanding of the very large cost reductions that could be achieved by designing a "constructed market" to replace the "command and control" approach to the management of $SO_2$ emissions.

## Constructed Markets for Emissions Trading

The concept behind the design of a constructed market for the control of $SO_2$ pollutants is fairly simple. It is based on the realization that the behavioral sources of the pollution problem can often be traced to poorly defined property rights in open access natural resources such as air and water.[18] A system of property rights and tradable permits for the management of pollution was first proposed in the late 1960s by Crocker (1966) and Dales (1968a, 1968b). The suggested institutional innovation did not emerge from its inventors in a fully operational form. Their proposals were followed by a large theoretical and empirical literature by resource and environmental economists (Bohm 1985). Design and implementation involved an extended process of "learning by doing" and "learning by using."

Proposals by Presidents Johnson and Nixon to replace the command and control approach by effluent fees or taxes on pollutants were dismissed as impractical and characterized by environmental activists as a "license to pollute." Beginning in the mid-1980s, however, a series of events conspired to make a more market-oriented approach to reducing $SO_2$ emissions politically feasible (Taylor 1989, 28–34; Hahn and Stavins 1991; Stavins 1998). One was the predilection of President George H. W. Bush in favor of a market-oriented approach to environmental policy. Another was the enthusiasm of Environmental Protection Agency administrator William Reilly and a number of key staff members in the executive office of the president for validating Bush's desire to be known as "the environmental president." There was also bipartisan support in key congressional committees for a variety of market-based approaches to environmental policy.

Within the environmental community the Environmental Defense Fund (EDF) began to differentiate itself from the rest of the environmental community by advocating market-based approaches as early as the mid-1980s. In 1989 EDF staff began to work closely with the White House staff in drafting an early version of proposed legislation. The credibility of the effort was enhanced by the fact that EPA administrator Reilly, formerly president of the Conservation Foundation, was a card-carrying environmentalist. Executives of several major corporations, influenced by subtle lobbying by the EDF, commented favorably on the emissions trading proposals.

---

18. This section draws heavily on Ruttan (2001b, 511–16). For a retrospection on the use of tradable permits see Tietenberg (2002).

The design of the $SO_2$ emissions trading system advanced in the Clean Air Act of 1990 drew on earlier EPA experience. The EPA began experimenting with emissions trading permits in 1974. The early programs included the elimination of lead in gasoline, the phaseout of chlorofluorocarbons and halons in refrigeration, and the reduction of water pollution from nonpoint sources. The early programs had a mixed record. They were typically grafted onto existing command-and-control programs. The difficulty of converting from command-and-control programs encountered substantial transaction costs. These experiences did, however provide important lessons for the design of more market-oriented trading programs in the 1990s.

The Clean Air Act created a national market for $SO_2$ allowances for coal-burning electrical utilities. The commodity exchanged in the $SO_2$ emissions trading program is a property right to emit $SO_2$ that was created by the EPA and allocated to individual firms. A firm can make allowances that had been issued to it available to be traded to other firms by reducing its own emissions of the pollutant below its own baseline level. In 1995, the program's first year, 110 of the nation's dirtiest coal-burning plants were included in the program. The affected plants were allowed to emit 2.5 pounds of $SO_2$ for each million British thermal units (Btu) of energy that they generated. During Phase II, to begin in 2000, almost all coal-burning plants were scheduled to be included, and allowances for each plant were to be reduced to 1.2 pounds per million Btu. Utilities that overcomply by reducing their emissions more than required may sell their excess allowances. Utilities that find it more difficult, or expensive, to meet the requirements may purchase allowances from other utilities.

The evidence available at the time this book was completed suggests that emissions trading has been even more cost effective than originally anticipated. Prior to initiation of the program the utility industry had complained that reducing $SO_2$ in amounts sufficient to meet the projected target (down from about 19 million tons in 1980 to 8.95 million tons in 2000) might cost as much as $1,500 per ton. By the late 1990s allowances were being sold in the $100–150 range. The decline in the cost of abatement has been due in part to technical changes in coal mining and deregulation of rail transport that have lowered the cost of low sulfur coal to Midwestern power producers. It has also been due to technical changes in fuel blending and $SO_2$ scrubbing that was induced by the introduction of performance-based allowance trading. As a result, benefits have substantially exceeded early estimates (Joskow, Schmalensee, and Bailey 1998).

The successful experience with $SO_2$ emissions trading illustrates a very important principle in designing new property rights institutions to manage formerly open access resources. In a now classic paper, Coase (1960) argued that when only a few decision makers are involved in the generation of externalities, the two parties, if left to themselves, will voluntarily negotiate new institutional mechanisms—rules and payments—that result in a reduction of the externalities to an acceptable level. However important the Coase theorem might be for understanding the small institutional innovations in the Philippine village case presented earlier, it has little relevance to most contemporary large-scale externality problems. The important externality problems that concern society today—such as $SO_2$ pollution, ozone pollution, or the greenhouse gases responsible for global climate change—typically involve large numbers of polluters and even larger numbers of persons affected by the externalities. In contrast to the evolution of a natural market, the government must establish the conditions necessary for a constructed market to function. In the $SO_2$ case it was necessary for an outside principal, the U.S. Congress, to define the size (or the boundaries) of the resource, in this case the maximum tons of $SO_2$ emissions, and to establish the trading rules. The social science effort involved in the design and implementation of the institutional arrangements to confront such problems requires the mobilization of large economic and political resources.

**Toward a More Complete Model of Induced Innovation**

The elements of a pattern (or structural) model that maps the general equilibrium relationships among changes in resource endowments, cultural endowments, technologies, and institutions are presented in figure 1.1.[19] The model goes beyond the conventional general equilibrium model in which resource endowments, technologies, institutions,

---

19. Fusfeld used the terms *pattern* or *Gestalt model* to describe a form of analysis that links the elements of a general pattern together by logical connections. The recursive multicausal relationships of the pattern model imply that the model is always open—"it can never include all of the relevant variables and relationships necessary for a full understanding of the phenomenon under investigation" (1980, 33). Ostrom uses the term *framework* rather than *pattern model*. "The framework for analyzing problems of institutional choice illustrates the complex configuration of variables when individuals . . . attempt to fashion rules to improve their individual and joint outcomes. The reason for presenting this complex array of variables as a framework rather than a model is precisely because one cannot encompass the degree of complexity within a single model" (1990, 214).

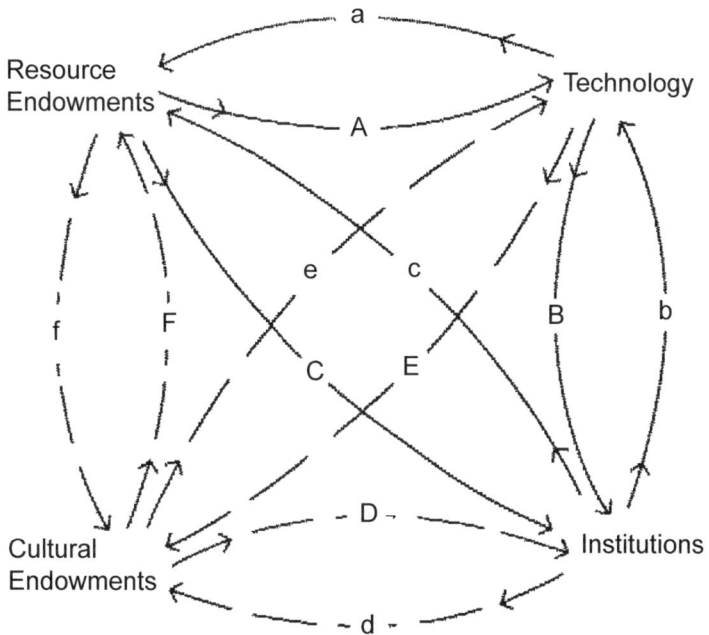

Fig. 1.1. Interrelationships between changes in resource endowments, cultural endowments, technology, and institutions. (From Yujiro Hayami and Vernon Ruttan, *Agricultural Development: An International Perspective*, rev. ed. [Baltimore: Johns Hopkins University Press, 1985], 111.)

and culture (conventionally designated as tastes) are given.[20] It represents an attempt to articulate a more holistic analytical framework for understanding the interrelationships between changes in resource endowments, technology, institutions, and cultural endowments during the process of economic development. In the study of long-term social and economic change, the relationships among the variables must be treated as recursive and dynamic (Harsanyi 1960). The formal microeconomic models that are employed to analyze the supply and demand for technical and institutional change can be thought of as nested within the general equilibrium framework of figure 1.1.

---

20. In economics the concept of cultural endowments has traditionally been subsumed under the concept of tastes, which are regarded as given, that is, not subject to economic analysis (Stigler and Becker 1977). I use the term *cultural endowments* to capture those dimensions of culture that have been transmitted from the past. Contemporary changes in institutions, for example, can be expected to harden into the next generation's cultural endowments.

An important advantage of the pattern model outlined in figure 1.1 is that it avoids the necessity of choosing between a materialist conception of human action, in which agents mechanically respond to changes in resource endowments, and an idealist conception of human action, in which agents respond only to subjective changes in cultural endowments (such as religion or ideology). Another advantage of this model is that it helps to identify areas of ignorance. Our capacity to model and test the relationships between resource endowments and technical change is relatively strong. Our capacity to model and test the relationships between cultural endowments and either technical or institutional change is relatively weak. The model is also useful in identifying the model components that enter into other attempts to account for secular economic and social change. Failure to analyze historical change in a general equilibrium context tends to result in a unidimensional perspective on the relationships bearing on technical and institutional change.[21]

For example, historians working within the Marxist tradition often tend to view technical change as dominating both institutional and cultural change. In his book *Oriental Despotism,* Wittfogel (mistakenly) views the irrigation technology used in wet rice cultivation in East Asia as determining political organization (1957). In terms of figure 1.1, his primary emphasis was on the impact of changes in resources and technology on institutions (C) and (B).

A serious misunderstanding can also be observed in the neo-Marxist critiques of the green revolution in rice production in Asia (Cleaver 1972; Hayami and Ruttan 1985, 336–45). These criticisms have focused attention almost entirely on the impact of technical change on labor and land tenure relations. Both the radical and populist critics have emphasized relation (B). But they have tended to ignore relationships (A) and (C).[22] This has led to repeated failure to identify effectively the

---

21. Induced innovation theory should be viewed as a diagnostic tool. Accurate prediction is not an appropriate test of the theory. If, for example, an increase in population pressure against land resources fails to induce the expected innovation in property rights institutions, the appropriate response is to augment the model. Thus in my own work I employ induced innovation theory not to predict the effects of changes in resource endowments, technology, institutions, and culture but rather as a guide to a dialogue with data. For an exceedingly useful discussion of the role of cultural constraints associated with traditional property rights in land on induced technical and institutional change, particularly in Africa, see Platteau (2000, 73–188).

22. A major limitation of the Marxian model is the emphatic rejection of a causal link between demographic change and technical and institutional change (North 1981, 60–61). This blindness to the role of demographic factors, and to the impact of relative resource endowments, originated in the debates between Marx and Malthus. An attempt to correct this deficiency represents the major innovation of the cultural materialism school of anthropology. See Harris (1979) and chapter 2, this volume.

separate consequences of population growth and technical change on the growth and distribution of income. The analytical power of the more complete induced innovation model was illustrated in the work by Hayami and Kikuchi, discussed earlier, on the impact of both technical change and population growth on changes in land tenure and labor market relationships in the Philippines.

Economists such as Coase (1960) and Alchian and Demsetz (1973) identify a primary function of property rights as guiding incentives to achieve greater internalization of externalities. They consider that the clear specification of property rights reduces transaction costs in the face of growing competition for the use of scarce resources as a result of population growth and/or growth in product demand. North and Thomas, building on the Alchian-Demsetz paradigm, attempted to explain the economic growth of Western Europe between 900 and 1700 primarily in terms of changes in property institutions.[23] During the eleventh and thirteenth centuries the pressure of population against increasingly scarce land resources induced innovations in property rights that in turn created profitable opportunities for the generation and adoption of labor-intensive technical changes in agriculture. The population decline in the fourteenth and fifteenth centuries was viewed as a primary factor leading to the demise of feudalism and the rise of the national state (line C). These institutional changes in turn opened up new possibilities for economies of scale in nonagricultural production and in trade (line b).

Olson (1968, 1982) has emphasized the proliferation of institutions as a source of economic decline. He also regards broad-based encompassing organizations as having incentives to generate growth and redistribute incomes to their members with little excess burden. For example, a broadly based coalition that encompasses the majority of agricultural producers is more likely to exert political pressure for growth-oriented policies that will enable its members to obtain a larger share of a larger national product than a smaller organization that represents the interests of the producers of a single commodity. Small organizations representing narrow interest groups are more likely to pursue the interests of their members at the expense of the welfare of other producers and the general public. In contrast, an even more broadly based farmer-labor coalition would be more concerned with promoting economic growth than an organization representing a sin-

---

23. See North and Thomas (1970, 1–17a; 1973). For a critical perspective on the North-Thomas model see Field (1981). Field is critical of North and Thomas for treating institutional change as endogenous.

gle sector. But large groups, in Olson's view, are inherently unstable because rational individuals will not incur the costs of contributing to the realization of the large group program—they have strong incentives to act as free riders. As a result, organizational space in a stable society will be increasingly occupied by special-interest distributional coalitions. These distributional coalitions make political life more divisive. They slow down the adoption of new technologies (line b) and limit the capacity to reallocate resources (line c). The effect is to slow down economic growth or in some cases initiate a period of economic decline.[24]

The relationships in the lower lefthand corner of figure 1.1 (dashed lines) have until recently received relatively little attention from economists. The classic analysis by Weber (1958) of the impact of the Protestant Reformation, particularly Calvinism, on the emergence of capitalism in *The Protestant Ethic and the Spirit of Capitalism* is an important exception (line D). The analysis by Greif (1994) of how the differential impact of the collectivist cultural endowments of Maghrebi traders and the individualistic cultural endowments of Genoese traders influenced the development of commercial institutions in the Mediterranean region in the eleventh and twelfth centuries is a second example.[25] Political scientist Ronald Inglehart employs a model in which cultural endowments (value changes) respond to changes in resource endowments (line f). Materialist values are stronger in poor societies that are resource constrained, while wealthy societies are characterized by postmaterialist (or postmodern) values (Inglehart 1997).

The effect of resource endowments on the international diffusion of institutions has been explored by Acemoglu, Johnson, and Robinson

---

24. For a dramatic example see Eggertsson (1996, 2003). Eggertsson poses the question of why Iceland, until well into the nineteenth century, neglected to exploit its rich offshore fishing resources. His answer was that the country was stuck in "a pernicious equilibrium trap that had an external and internal component. The internal component was related to the economic self-interest of landlords and farmers who feared that the development of high productivity fisheries would weaken the institutions that tied labor to the land . . . The external element was the policy of the Danish Crown of isolating the country from foreign trade and taxing Icelanders by selling monopoly rights to trade with Iceland" (Eggertsson 1996, 21). These institutional constraints were broken only after a subsistence crisis in the latter eighteenth and early nineteenth centuries.

25. In England the cultural changes associated with the civil war, the beheading of the king, the restoration of the monarchy, and the Glorious Revolution (1688–89) led to a series of institutional innovations in property rights, public finance, and governance that enhanced the capacity of the English state to raise revenue and project its power in international relations (line D) (North and Weingast 1989; Weingast 1997).

(2001). They found, where the disease environment was not favorable to settlement, that European colonizers established extractive states (such as Britain in the Gold Coast and Belgium in Congo). Where the disease environment was favorable, the European colonizers established settler colonies. Where extractive states were established, legal institutions were adopted that favored the extraction and transfer of resources to the metropolitan country and, after independence, to the new ruling elites. In settler colonies, in contrast, legal institutions that favored the rule of law and encouraged investment were established. Those differences in legal culture and institutions explain substantial differences in contemporary per capita income (lines c, f, and d).

A potential criticism of the pattern model depicted in figure 1.1 is that it does not stipulate the mechanisms through which changes in resource endowments, for example, induce changes in technology. However, it is not too difficult to visualize the mechanisms that mediate the relationships among changes in resource endowments, technical change, and institutional change. The market represents a "master mechanism" for translating the uncoordinated behavior of individuals into system-level coordination (Headstrom and Swedberg 1998, 3). It is somewhat more difficult, however, to describe the mechanisms that link institutional change and changes in cultural endowments in terms of the neoclassical model (other than as metaphor). Another potential criticism of the pattern model of figure 1.1 is that it is "overdetermined." Identification problems become intractable since every variable in the system is subject to influences arising from changes in every other variable (Resnick and Wolff 1987). But because changes in the different relationships in the model occur at different rates, the identification problem, while difficult, is tractable.

Coleman, a leading social theorist of the last generation, advanced what he termed a macro-micro-micro-macro model (1986; 1990b, 1–23). In figure 1.2 the Coleman model is used to interpret the Weber thesis on the relationship between the Protestant ethic and the spirit of capitalism. Protestant theology inculcated a change in social values among its adherents (line 1); individuals internalized new value orientations (rationalism, antitraditionalism, asceticism) toward economic behavior (line 2); the new value orientations resulted in the actions by individuals and groups that induced the development of the economic institutions of capitalism (line 3). Coleman argues that Weber's own interpretation was incomplete because he did not address the critical theoretical problem—how individual actions combined to produce the

## 28  Social Science Knowledge and Economic Development

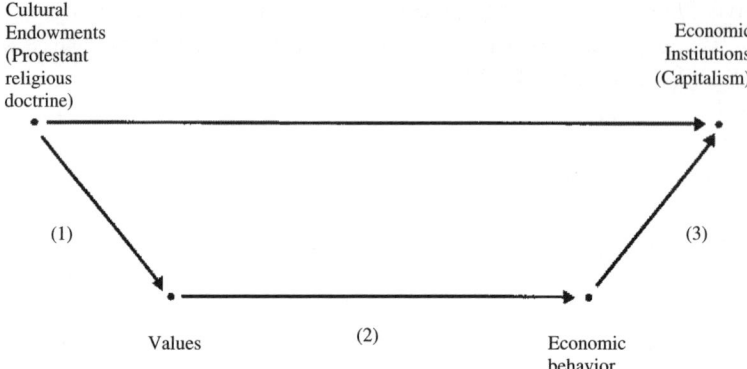

Fig. 1.2. Macro- and microlevel propositions: effects of religious doctrine on economic organization. (Adapted from James S. Coleman, *Foundations of Social Theory* [Cambridge: Harvard University Press, 1990], 8. Reprinted by permission of the publisher. Copyright © 1990 by the President and Fellows of Harvard College.)

unanticipated behavior of groups of individuals that brought about the economic institutions of capitalism.[26] "What is necessary to account for the growth or occurrence of any social organization, whether capitalist organization or something else, is how the structure of organizations come into being, how persons who come to occupy each of the positions in the organization are motivated to do so, and how this interdependent system of incentives is sustainable" (Coleman 1990b, 9). Coleman's challenge to the social sciences research community has seldom been met.[27] This is an issue that I will return to several times in subsequent chapters.

---

26. Weber's thesis has also been criticized for not attempting to explain the social forces that led to the Protestant Reformation. The assertion by Douglas (1986, 36)—"Religion does not explain. Religion has to be explained"—is at least half correct. Thus at a deeper level it may be possible to explain the emergence of Protestantism in terms of the economic changes associated with late medieval urban development or even the early financial reforms of the Catholic Church (Harsanyi 1960, 143; Lal 1998; chap. 8, this vol.).

27. The issue of whether adherence to the extreme methodological individualism implied by Coleman's stipulation is necessary to understand social behavior is a source of continuing debate in the social sciences. One response is to argue that, while desirable, it imposes a demand on social science research that cannot be met in practice. But even if it could be met, would it be sufficient? The answer to this question appears to be negative. Variables that are not attached exclusively to individuals, such as culture and technology, are essential to efforts to understand the behavior of economic systems and, more broadly, of social systems (Arrow 1994; Satz and Ferejohn 1994).

## Perspective

What are the implications of this theory of induced institutional innovation for the research on the contribution of social science knowledge to economic development? In my research with Hayami and Binswanger on the direction and rate of technical change, we were able to advance significantly our knowledge by treating technical change as largely endogenous—as induced primarily by changes in relative resource endowments and the growth of demand. We were also able to interpret the advances in knowledge about the role of changes in the economic environment on the rate and direction of technical change for the design of research systems and the allocation of research resources (Ruttan 1982, 2001b).

This theory of induced institutional change has advanced our understanding of the process of institutional change. It suggests that substantial new insights have been obtained by treating institutional change as an economic response to changes in resource endowments and to technical change. But, as in the case of technical change, my concern goes beyond advancing our understanding of the process of institutional innovation. It is essential for the social sciences to advance our understanding of the historical processes of social and economic development. But that is not sufficient! If social science knowledge is to be valued by society it must also advance the knowledge to successfully intervene in the process of development—to reduce the cost of the trial and error that accompany the historic organic processes of institutional innovation. My interest in this book is to explore the contributions of the several social sciences to the design of policies, mechanisms, and systems that can become efficient sources of development.[28]

I have also insisted on the significance of cultural endowments, including the factors that economists typically conceal under the rubric of tastes and that political scientists include under ideology, for economic development. In an article published in the mid-1980s Yujiro Hayami and I insisted that until our colleagues in the other social sciences provide us with more helpful analytical tools, we would be forced to adhere to a strategy that focused primarily on the interactions

---

28. For a criticism and assessment of that strategy see the articles assembled in Koppel (1995). For a more positive assessment see Runge (1999). One reason that explanations of institutional change in terms of changes in economic forces have been productive is that the economic system is a major channel through which exogenous changes or differences in natural environments act upon the social system (Harsanyi 1960, 141).

between resource endowments, technical change, and institutional change (Ruttan and Hayami 1984). This approach had the clear advantage of allowing us to explore how far a strategy based on the rather straightforward extension of standard neoclassical microeconomic theory could take us in advancing out understanding of institutional change.[29]

In spite of the fact that this strategy has yielded very substantial insight into the process of institutional change I do not regard it as a very satisfactory conclusion. Beginning in the mid- 1980s I initiated a program of research and writing designed to explore in greater depth what development economists should learn from scholars in the other nomothetic social sciences—anthropology, sociology, and political science—working in the field of development. This book grew out of that effort.

In Part II, I review the evolution of thought about development in anthropology, sociology, political science, and growth economics and attempt to assess the potential contributions from these fields for research and policy in the field of development economics. These reviews are necessarily highly selective. No attempt is made to assess or draw inferences from the broader work in these four nomothetic disciplines. In Part III, I draw on contributions to knowledge from the several social sciences in addressing the state of knowledge on the adoption, diffusion, and transfer of technology; in the renovation and transformation of traditional institutions as agents of development; the role of religion, culture, and nationalism; and for insight into the economic and ethical issues associated with foreign economic assistance. Chapter 10 concludes by reemphasizing the contribution of advances in social science knowledge to economic development.

---

29. I have discussed some of the issues involved in working across disciplines in Ruttan (2002). For an exceedingly useful introduction to a number of important economic concepts with applications in the other social sciences see Sandler (2001).

PART II

CHAPTER 2

# Cultural Endowments and Economic Development

What role does culture play in the process of economic development? Economists have been reluctant to incorporate cultural endowments, or cultural change, in their models of economic development. Anthropologists have been critical of economic theory for avoiding the role of culture. But they have provided very little insight into how culture should be taken into account (Bardhan 1989; Harrison and Huntington 2000). In the previous chapter I presented a model of the dialectical relationships among changes in cultural endowments, resource endowments, technology, and institutions (fig. 1.1). In this chapter an attempt will be made to explore in greater depth the relationships between cultural endowments and economic development.[1]

I first review the efforts of a number of development economists to draw on the anthropological literature to understand the role of cultural endowments and cultural change in the process of economic development. I then explore the potential value of a deeper mining of the anthropological literature for economists who would like to incorporate the role of cultural endowments, and of cultural change, into the analysis of the processes of economic development or the design of development institutions.

## Cultural Endowments in Development Economics

Economists have made substantial contributions to our understanding of the sources of technical and institutional change. But, until recently, they have given relatively little attention to the role of cultural endowments or cultural change in economic development. To the extent that cultural endowments are considered at all by economists, they have

---
1. I am indebted to Robert C. Hunt, Stephen Gudeman, Frank Miller, Robert E. Rhoades, Lia M. Ruttan, and Lore M. Ruttan for helpful comments on an earlier version of this chapter (Ruttan 1988).

generally been subsumed under the concept of tastes. And tastes, even more than technology and institutions, have traditionally been regarded as not subject to economic analysis.[2]

At an intuitive level I have little difficulty in accepting the view that cultural endowments, including religion and ideology, exert at least some influence on the supply of institutional innovation (chap. 8, this vol.). Cultural endowments can make some forms of institutional change less costly to establish and can impose severe costs on others. It has been argued, for example, that the traditional moral obligation in the Japanese village community to cooperate in communal infrastructure maintenance has made it less costly to implement rural development programs than in societies lacking such traditions (Ishakawa 1981). In northern Italy, dense networks of social institutions ranging from choral societies to soccer clubs, some with traditions running back to the late Middle Ages, have contributed to the effectiveness of recent reforms in regional governance (Putnam 1993). The traditional patterns of cooperation have represented an important cultural resource on which to erect modern forms of cooperative marketing and joint farming activities. In China, communist ideology, reinforced by the lessons learned during the guerrilla period in Yenan, inspired the mobilization of communal resources to build irrigation systems and other forms of physical and social capital (Schran 1975, 345–72). Similar cultural endowments have not been available in South Asian villages where, for example, the caste structure has inhibited cooperation and encouraged specialization (Lal 1988; chaps. 4 and 8, this vol.) or in many areas in sub-Saharan Africa where the egalitarian values of lineage-based family structure have often been the source of powerful sanctions against individual achievement (Platteau and Hayami 1998; Platteau 2000, 189–240).

---

2. Stigler and Becker (1977) note that the traditional view in economics is that tastes represent the unchallengeable axioms of man's behavior and that economic analysis "is abandoned at this point to whoever studies and explains tastes (psychologists? anthropologists? phrenologists? sociobiologists?)" (76). They then argue for an alternative approach based on the view "that tastes neither change capriciously nor differ importantly between people" (76). Their argument is that this reformulation permits the economist to explain any differences or changes in behavior through differences in prices or income. Thus "the (relative) consumption of music appreciation rises with exposure not because tastes shift in favor of music, but because its shadow price falls as skill and experience in the appreciation of music are acquired with exposure" (79). Feminist economists were particularly critical of the Stigler-Becker analysis. See, for example, the chapters in Cook, Roberts, and Waylen (2000) and von Weizsacker (1971, 345–72). In more recent work Becker (1996) treats changes in tastes as endogenous. For an excellent review of more recent economic thought on the endogeneity of tastes and preferences see Bowles (1998).

In the development literature, cultural endowments have traditionally been viewed as obstacles to technical or institutional change. Kusum Nair (1979) insisted that the differential response to the green revolution seed-fertilizer technology among regions in India could be explained, at least in part, on cultural grounds. George Foster has argued that indigenous innovation in peasant societies is often blocked by an "Image of Limited Good"—"peasants view their social, economic and natural universe—their total environment—as one in which all of the desired things in life such as land, wealth, health, friendship and love, manliness and honor, respect and status, power and influence, security and safety exist in a finite quality and are always in short supply" (1965, 296). It has been argued that the relative economic decline of Britain over the last century was due in part to cultural changes associated with the gentrification of bourgeois culture: "the rooting of pseudoaristocratic attitudes and values in upper-middle-class educated opinion shaped an unfavorable context for economic endeavor" (Weiner 1981).

The first postwar generation of development economists gave a prominent role, at least at the rhetorical level, to the role of cultural endowments in constraining or facilitating economic growth. They accepted the body of scholarship in history, anthropology, sociology, and political science that insisted that cultural endowments exerted a major impact on behavior and hence on the response in traditional societies to the opportunities associated with the modernization of community life and the possibilities of national economic development (Hagen 1980; Rogers 1969). Without attempting to be exhaustive, let me refer to the work of Bert F. Hoselitz, Everett E. Hagen, Irma Adelman and Cynthia Taft Morris, Gunnar Myrdal, and Peter T. Bauer. I emphasize the work of Hoselitz because of his interdisciplinary entrepreneurship; Hagen because of his attempt to develop a unified theory of social change; Adelman and Morris because of their effort to quantify the role of sociocultural variables; Myrdal because of his effort to take cultural variables explicitly into account in development policy reform; and Bauer because of the influence his work has had on recent development assistance policy.

## Hoselitz

Bert F. Hoselitz played a particularly important entrepreneurial role in the 1950s in urging economists to give greater consideration to the role of cultural factors in economic development. His activities included the organization of the Center on Economic Development and Cultural Change at the University of Chicago and the founding of the journal *Economic Development and Cultural Change*. He authored and edited a number of influential publications dealing with noneconomic barriers and aids to economic development (1952, 1960). Among the noneconomic factors he identified were: (1) the emergence of cultural minorities or classes that serve as the spearhead for both technical and institutional change; (2) a social and political system that encourages a high degree of social mobility; (3) a social and cultural environment that facilitates the development of institutions capable of generating the technical and institutional knowledge necessary to operate a modern society; and (4) the weakening of commitment to traditional methods of production and institutions (1952).

This last consideration was particularly important in Hoselitz's view since traditional value systems "offer special resistance to change . . . their change is facilitated if the material economic environment in which they can flourish is destroyed or weakened. . . . Economic development plans which combine industrialization with an extension of traditional or near traditional forms of agriculture are thus creating a dilemma which in the long run may present serious repercussions in the speed or facility with which ultimate objectives can be reached" (1952, 15).

### Hagen

The most ambitious attempt to incorporate cultural variables into the analysis of economic development was that of Everett E. Hagen. In an important book published in 1962 Hagen argued that advances in the fields of anthropology, sociology, psychology, and economics had reached the point where a synthesis could be achieved to form a unified theory of society and social change. He drew on the literature from these several social science fields to analyze the development history of England, Japan, Colombia, Indonesia, Burma, and the Sioux nation.

Hagen's analysis led him to place primary emphasis on personality formation. He argued that the interrelations between personality formation and social structure are such that social change could not occur without prior or concurrent personality change (1962, 80). Such factors as political development, nationalization, religious change, urbanization, infrastructure development, and commercial innovation are "primarily incidents in the process of change but not initial causal factors in change" (250).

Traditional societies were characterized by an authoritarian personality. "The image of the world . . . includes a perception of uncontrollable forces. . . . Each individual finds his place in the authoritarian hierarchy of human relationships" (83–84). In his historical studies, Hagen gave particular attention to the emergence of personality characteristics conducive to innovation. In a retrospective review in 1980, Hagen argued that a disproportionate share of entrepreneurs were drawn from social groups that were excluded from traditional elite roles (215–31). Hagen's work received enthusiastic reviews. But in retrospect it must be seen as the culmination of an effort to enrich the theory of development by drawing on anthropology, sociology, and psychology, rather than as the foundation for further advances.

### Adelman and Morris

An early effort by economists to obtain quantitative estimates of association between sociocultural variables and economic development was made by Irma Adelman and Cynthia Taft Morris (1965, 1967, 1973). Their approach was to use factor analysis techniques to compress a large set of indicators into groups of closely associated sociocultural, political, and economic development indicators to find a small set of underlying latent variables that could be aggregated into an index of socioeconomic development (1965, 1967). In a 1973 study they employ the same data to search for factors that account for differences in income distribution across countries. Among the indicators selected to reflect change in sociocultural endowments were the size of the traditional agricultural sector, the extent of dualism, the character of basic social organization, the extent of social mobility, and the degree of ethnic homogeneity. The analysis is performed first for a set of seventy-four countries and then for three subsets classified by level of development. An attempt was made to differentiate between long-run and short-run patterns of association by performing the analysis first without and then within a group of economic variables that can be interpreted as responsive to short-run policy interventions. Adelman and Morris emphasize that "relationships found between levels of economic development and differences in social and political structure are neither caused nor causal. Rather they reflect the interaction of an organic system of institutional and behavioral change which underlies the process of economic development" (1967, 172).

But they do draw some fairly firm conclusions. During the earliest stage of development, cultural and social constraints are a burden on economic growth. The sociocultural environment must be transformed in order to enlarge the scope for economic activity (1967, 202). Furthermore, their research "suggests that one may look at the entire process of national modernization as the progressive differentiation of the social, economic, and political spheres from each other and the development of specialized institutions and attitudes within each sphere. More specifically, the process of economic development in underdeveloped countries consists basically of the separation of the economic sphere, first from the complex of social organization and the norms that govern it, and, subsequently and to a lesser extent, from the political environment by which it is constrained" (266).

Adelman and Morris suggest that the appropriate policy mix will differ depending on the level of development. At low levels of development, the growth of the market sector and the narrowing of dualism among sectors should have high priority. At an intermediate level, social tensions increase as income distribution becomes more unequal. In this stage, political development that is capable of reducing stress among social classes becomes particularly important in creating a favorable socioeconomic environment for economic growth.[3]

## Myrdal

One of the most ambitious attempts by an economist to employ cultural variables to interpret economic behavior, assess the prospects for growth, and prescribe economic policy was the massive study of South Asian development by Nobel laureate Gunnar Myrdal, *Asian Drama: An Inquiry into the Poverty of Nations* (1968).[4] Myrdal contrasts modernization ideals, which represent the official ideology of a Westernized elite, with the traditional values of the rest of society. The official creed, held by the politically alert, articulate, and active part of the population, particularly by the intellectuals, emphasized the values that, in the West, were a product of the Enlightenment: rationality, equity, efficiency, diligence, honesty, innovation, national independence, democracy, and social discipline (53–69).

---

3. The Adelman and Morris work has largely been ignored by development economists. An important recent exception is a reevaluation by Temple and Johnson (1998). They show that forecasts of long-run economic growth rates that incorporate the Adelman-Morris index of socioeconomic development have been more accurate than the conventional projections of economic growth rates made in the early 1960s.

4. Myrdal began to take an active interest in development thought, economics, and planning in the early 1950s. See particularly Myrdal (1957).

Although Myrdal regards casual speculation about the impact of personality, culture, and religion as unscientific, his research leads him to the view that the people of South Asia "have lived for a long time under conditions very different from those in the Western world and this has left its mark upon their bodies and minds. Religion has, then, become the emotional container of this whole way of life and work and by its sanction has rendered it rigid and resistant to change" (1968, 112). Popular religion "sanctifies a whole system of life and work, attitudes and institutions, that contribute to the resistance of that system to . . . changes along the lines of the modernization ideal" (109).[5] But this weight of social and political inertia must be overcome, in Myrdal's view, by planned development.

But planning and plan implementation in South Asia are inhibited by political limitations that Myrdal labels as the "soft state." Policies decided on are not enforced. The authorities are reluctant to place obligations on people. "Planning for development requires a readiness to place obligations on people in all social strata to a much greater extent than is done in any of the South Asian countries. . . . Under present South Asian conditions development cannot be achieved without much more social discipline than the prevailing interpretation of democracy in the region permits" (67). At times Myrdal comes close to implying that economic development in South Asia can only be achieved by an authoritarian socialist regime—but without Stalin or Mao! In a retrospective view, published in 1984, Myrdal still regarded the failure of the "soft state" to achieve internal reforms as a major obstacle to development in South Asia (151–65).

## Bauer

The role of cultural endowments in economic development has also been a consistent theme in the work of Peter Bauer. Bauer has insisted that successful development in poor countries has not been the result of "the forced mobilization of their resources. Nor was it the result of forcible modernization of attitudes and behavior, nor of large-scale state-sponsored industrialization, nor of any other form of big push. And it was not brought about by the achievement of political independence, . . . or by any other form of political or cultural revolution" (1984b, 30).

---

5. For a similar but more rigorous interpretation of the constraints of Hindu religion on economic development in India see Lal (1988); chap. 8, this vol.

Bauer does insist that economic achievement and progress depend largely on human aptitude and attitudes, on social and political institutions and arrangements, on historical experience, and, to a lesser extent, on external contact, market opportunities, and natural resources (1972). Cultural endowments, reflected in differences among ethnic groups, have been particularly emphasized by Bauer. He has repeatedly drawn on his early studies in Southeast Asia.

> Many rubber estates kept records of the daily output of each tapper, and distinguished between the output of the Chinese and Indian workers. The output of the Chinese was usually more than double that of the Indians, with all of them using the same equipment of tapping knife, latex cup, and bucket. . . . The pronounced differences between Chinese and Indians could not be attributed to the special characteristics often possessed by migrants, as both groups were recent immigrants. The great majority of both Indians and Chinese were uneducated coolies, so that the differences in their performance could not be explained in terms of differences in human capital. . . . I was to encounter similar phenomena in West Africa, in the Levant, in India, and elsewhere . . . differences in economic performance among different cultural groups is a feature of much of economic history.[6]

Bauer's perspective has not been reinforced by new investigations. It has retained its currency through frequent repetition. Myrdal and Bauer share remarkably similar views on the role of cultural constraints on economic development, but this does not lead them to similar views on development policy. Myrdal's enthusiasm for strong state intervention is countered by Bauer's faith in market forces.

In spite of the wide attention that each of the five bodies of work reviewed here has received, they have not been incorporated into mainstream economics or economic development thought. Professional opinion in economics has not dealt kindly with the reputations of those development economists who made serious efforts to incorporate cultural variables into development theory or into the analysis of the

---

6. Bauer (1984b, 32, 33). Bauer has not regarded consistency as a virtue. In his book with Yamey (Bauer and Yamey 1957) he notes that economists are not qualified to pronounce on cultural factors (59) but then goes on to comment on the role of the extended family in impeding economic progress (64–67) and on differences in entrepreneurship among ethnic groups (102–12).

development process. Their work has typically been favorably reviewed and then ignored. Their work has often had wider currency outside than within the field of economics. There was no rush by other scholars or by graduate students to refine or test either their theories or their results.

A premature obituary to the cultural endowments school was pronounced by Albert O. Hirschman (1965). He grouped the several cultural barriers referred to in the literature as (1) obstacles that turn into assets, (2) obstacles whose elimination turns out to be unnecessary, and (3) obstacles whose elimination is postponable. The publication in 1964 of *Transforming Traditional Agriculture* by T. W. Schultz, which shifted attention from peasant culture as an obstacle to development and set forth the "poor but efficient" view of the peasant cultivator in traditional societies, was even more influential (though not referred to by Hirschman) in turning the attention of development economists away from the issue of cultural factors in development. But it was the rapid adoption of green revolution agricultural technology by peasant producers throughout Asia that gave plausibility to Hirschman's and Schultz's skepticism about cultural constraints.

Experience has taught us that when peasants refuse to adopt the practices recommended by agronomists and economists it may be the experts rather than the peasants who are wrong. But in spite of the failure of research on the economic implications of cultural endowments to find a secure place in economic development literature, the conviction that "culture matters" remains pervasive in the underworld of development thought and practice.[7] The fact that the scholars and practitioners of development are forced to deal with cultural endowments at an intuitive level rather than in analytical terms should be regarded as a deficiency in professional capacity rather than as evidence that culture does not matter.

In our book *Agricultural Development,* Yujiro Hayami and I, while insisting on the potential significance of cultural endowments, argued that until our colleagues in the other social sciences are able to provide us with more helpful analytical tools, economists are forced to adhere to a strategy of exploring how far modest extensions of microeconomic

---

7. For an example of renewed interest in the role of cultural endowments in economic development see the articles in Harrison and Huntington (2000) and the studies by Eggertsson on the history of the institutions governing property rights and labor relations in agriculture and fisheries in Iceland (1992, 1996, 2003). Among development economists Robert Chambers has been particularly vigorous in stressing the importance of anthropological insight.

theory can take us in analyzing both the sources and impact of technical and institutional change (1985, 114). Although I continue to believe that one should first try to understand economic phenomena primarily in economic terms, it may be time to reassess what the advances in other social sciences might be able to contribute to development economics. I begin with the field of anthropology.

### Why Anthropology?

There are a number of reasons that development economists might look to the field of anthropology for guidance in attempting to understand the sources and impact of cultural endowments on economic development. One is that anthropology has traditionally embraced a broad conception of culture (see the appendix). The term *culture* was used by the early anthropologists, such as Franz Boas, "to designate the totality of human social behavior that was independent of the genetic constitution and biological characteristics of organisms" (Kroeber and Parsons 1958, 582). In this view, culture comprised the totality of inherited artifacts, material goods, technical processes, and mental constructs. Over time, however, distinct traditions of physical and cultural anthropology emerged. Cultural anthropology focused on the evolution and diffusion of custom, social organization, values, and ideology (Goodenough 1964, 36; Singer 1968, 540). Since the now classic work of Raymond Firth in the 1950s, it has become common within anthropology to make a distinction between organization and structure (Firth 1951; Bennett 1976).

A second reason for looking to anthropology is the large body of ethnographic studies that have become available since the early 1950s. It is the insistence on descriptive realism that makes the use of these ethnographic studies so potentially attractive to economists. The descriptive detail in ethnographic studies has often made it possible for economists to draw on research by anthropologists for analytical and policy purposes.[8] But relatively few economists have been willing to

---

8. See, e.g., how Schultz (1964, 41–44) uses the ethnographic studies of a Guatemalan Indian economy by Tax (1953) and of an Indian village by Hopper (1965) in formulating his "poor but efficient" hypothesis. In my own work at the International Rice Research Institute in the Philippines in the mid-1960s I found the work of Howard Conklin (1957) on swidden agriculture and Clifford Geertz (1963) on agricultural involution exceedingly useful. When I first met Geertz I complemented him, somewhat facetiously, on being the author of the best work on farm management in Indonesia. He was not amused! For a critique of the Geertz agricultural involution thesis see White (1983).

make the investment in time needed to generate the information or evaluate the available information necessary to assure a reasonably adequate understanding of even economic relations at the village or community level. Two important exceptions are the community studies in Laguna (Philippines) by Hayami and Kikuchi (1981, 2000) and the Palanpur (India) studies by Bliss and Stern (1982) and Lanjouw and Stern (1998).[9]

Ethnographic studies are now available for many peasant and urban communities as well as the isolated indigenous communities that were the traditional focus of anthropological research. There are two major obstacles to drawing on anthropology for an understanding of the relations between cultural endowments and technical and institutional change. The first is that ethnographic studies, as a result of a commitment to learning primarily through fieldwork, have often avoided embodying their interpretation either in a historical or a contemporary political and economic context (Marcus and Fischer 1986, 77–110). A professional commitment to avoid intervention has also often deterred anthropologists from directing research to problems that are of significance to the communities being studied.[10]

The second obstacle is the intellectual fragmentation within the discipline of anthropology. In addition to the classical subfields of archaeology, physical anthropology, cultural anthropology, and linguistics the economist who attempts to "read anthropology" is confronted by a bewildering array of postmodern anthropologies (Marcus and Fischer 1986, 16). There has also emerged within anthropology a body of literature that questions the concept of culture (Clifford 1988; Abu-Lughod 1991).[11]

In spite of its fractionated appearance, it is possible to make a separation, perhaps somewhat oversimplified, between the several "materi-

---

9. See also Hunt's (2000) very useful review and evaluation of studies by anthropologists that have attempted to test the Boserup hypothesis that labor productivity declines with intensification. Hunt demonstrated that the Boserup hypothesis was rejected by the few studies that were designed with sufficient rigor to actually test the hypothesis.

10. See, for example, the exhaustive review of agrarian ecology by Netting (1974). Netting is critical of the neglect of agrarian societies by anthropologists. But in his review of several hundred studies he does not once mention the value of the studies to the communities in which the studies were conducted.

11. Shweder (2000b) notes that it is ironic that an "anticulture" or a "postculture" position has emerged in anthropology just as economists and political scientists are again beginning to regard culture as a legitimate topic of investigation.

alist" and "interpretive" schools of anthropology.[12] The materialists' perspective interprets differences in social life and behavior as arising out of universal physiological, economic, and political concerns. Idealists argue for a deeper study of the meaning of life and for the interpretation of behavior in terms that are considered significant to the society being studied. Both represent divergent streams within post-1950s U.S.-style cultural anthropology.[13]

### Materialist Perspectives

Because of the pervasive role played by resource endowments and self-interest in economic analysis, the materialist approaches seem, at first instance, more congenial to economists. There is a strong (fossilized) Marxist tradition of historical materialism in anthropology (White 1949; Sahlins 1976). In this tradition, culture (superstructure) is viewed as so largely determined by the forces and relations of production (fig. 2.1) that it offers little in the way of insight or additional analytical power. Little weight could be given to cultural differences in a world that seemed to be inevitably moving toward a single integrated economic and political system and in which culture constituted primarily a source of resistance that has to be taken into account in planning for change (Marcus and Fischer 1986, 86; Fukuyama 1989).

The cultural materialist perspective, which has been articulated most forcefully by Marvin Harris, draws on the ecological perspectives of White (1949) and Steward (1955) to produce a richer and less ideological materialism than the traditional Marxist approach (Harris 1968, 1979, 1999).[14] The cultural materialist tradition puts Malthus back on the stage from which he was banished by Marx. Marx held that "technology discloses man's mode of dealing with nature, the

---

12. In the 1950s the theoretical tool kit available to anthropologists "consisted of three major and somewhat exhausted paradigms—British structural-functionalism (descended from A. R. Radcliffe-Brown and Bronislaw Malinowski), American cultural and psychological anthropology (descended from Margaret Mead and Ruth Benedict), and American evolutionist anthropology (centered around Leslie White and Julian Steward and having strong links to archeology)" (Ortner 1994, 374).

13. The structuralist approach in anthropology, as advanced in the work of Claude Lévi-Strauss, holds that the diverse customs and beliefs of different societies represent surface manifestations of deeper motivations and structures. The evolution of kinship structure and language provide the model institutions that Lévi-Strauss drew on in developing his essentialist approach to understanding cultural change. See particularly Lévi-Strauss (1966b). For a useful introduction see De George and De George (1972).

14. For a review and evaluation of the ecological approach in anthropology see Netting (1982).

process of production by which he sustains his life and thereby also lays bare the mode of formation of his social relations, and of the mental concepts that flow from them" (1936, 406n). Harris insists that the modes of production and reproduction determine (probabilistically) domestic and political economic organization and behavior, which in turn determine (probabilistically) the superstructure (fig. 2.2).

The cultural materialist framework and research agenda is similar, in some respects, to the induced institutional innovation framework and agenda (chap. 1, this vol.). Objectively determinable behavioral components include (1) an infrastructure, defined to include the ecosystem and the modes of production and reproduction; (2) a structure that includes the elements of domestic and political economy; and (3) a superstructure that includes both objectively determinable (etic) and culturally specific (emic) components.[15] Harris's superstructure is highly congruent with the cultural endowments category in figure 1.1, and his structure component is largely congruent with the institutions component.

> Cultural materialists give highest priority to the effort to formulate and test theories in which infrastructural variables are the primary causal factors.... Cultural materialists give still less priority to exploring the possibility that the solution to sociocultural puzzles lies primarily within the behavioral superstructures.... In other words, cultural materialism asserts the strategic priority of etic and behavioral conditions over emic and mental conditions and processes, and of infrastructural over structural and superstructural conditions and processes; but it does not deny the possibility that emic, mental, superstructural and structural components may achieve a degree of autonomy from the etic behavioral infrastructure. Rather, it merely postpones and delays that possibility in order to guarantee the fullest exploration of the determining influences exerted by the etic behavioral infrastructure. (Harris 1980, 56)

---

15. The use of the terms *emic* to designate culturally specific and *etic* to designate universal models of interpretation is based on the analogy with the terms *phonemic* and *phonetic* in linguistics. Emic statements refer to logical empirical systems whose phenomenal distinctions, or "things," are built up out of contrasts and discriminations that are "significant, meaningful, real, accurate, or in some other fashion regarded as appropriate by the actors themselves" (Harris 1968, 571). Etic statements depend on phenomenal distinctions judged appropriate by the community of scientific observers. "Etic statements are verified when independent observers using similar operations agree that a given event has occurred or about the determinants of the behavior of classes of people" (575).

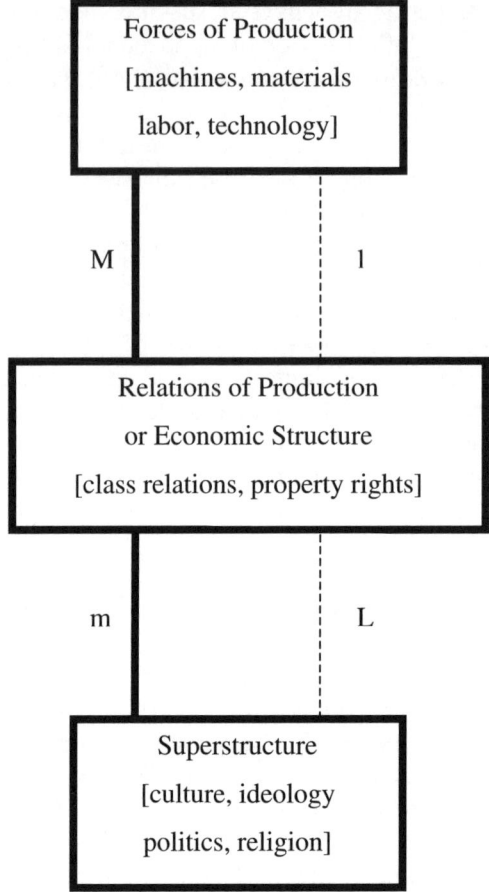

Fig. 2.1. A Marxist model. The forces of production and the relations of production together make up the economic base or mode of production. The arrows labeled M and m draw on Marx, while the arrows labeled L and l draw on Lenin. (Vernon M. Ruttan, 1988, "Cultural Endowments and Economic Development: What Can We Learn from Anthropology?" *Economic Development and Cultural Change* 36 [3]: S258.)

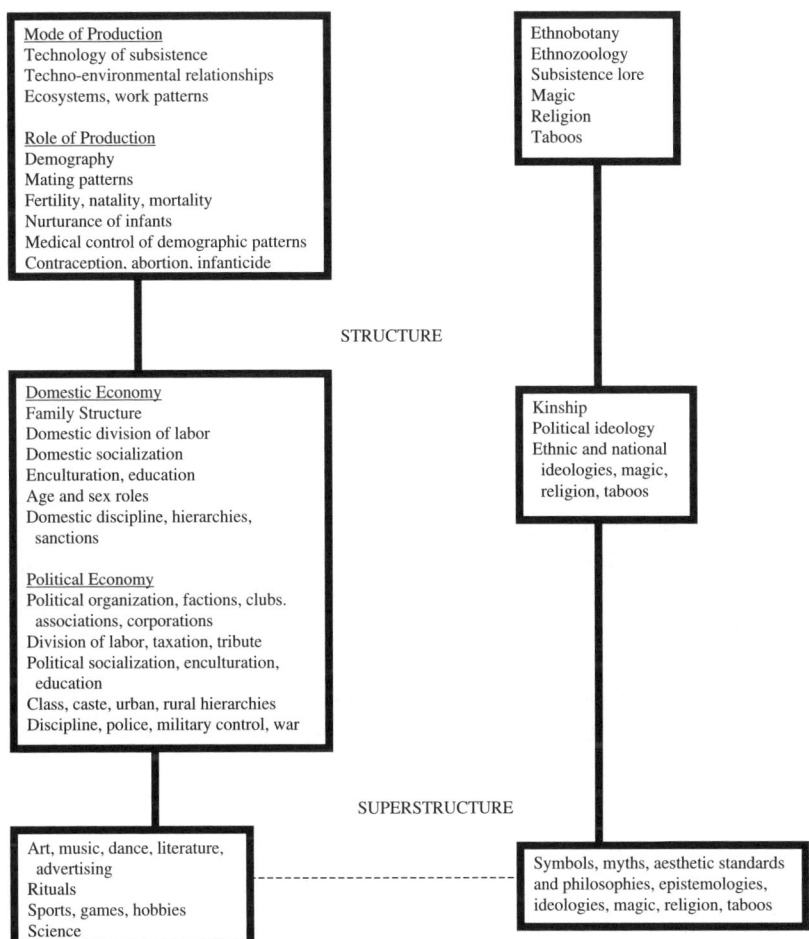

Fig. 2.2. The cultural materialism model. (Adapted from Marvin Harris, *Cultural Materialism: The Struggle for a Science Culture* [New York: Random House, 1980], 51–54.)

When one examines the research studies conducted within a cultural materialist perspective, the conceptual similarity with studies conducted within the induced institutional innovation framework is further reinforced. A useful example is the attempt by Harris to understand why, in the state of Kerala in southwestern India, the mortality rate of male calves is much higher than that of female calves, while in the northern state of Uttar Pradesh the mortality rate of female calves is much higher than that of male calves (1980, 32–34, 248–53).[16] In both areas farmers indicated a strong personal commitment to Hindu prohibitions against the slaughter of domestic cattle. They insisted that they would never kill or starve one of their cattle. Yet economic factors were, in both provinces, powerful predictors of cattle sex ratios. In Kerala cattle were valued primarily for milk rather than traction; in Uttar Pradesh cattle were valued primarily for traction rather than milk. This is precisely the modification in cultural behavior that would have been predicted using the microeconomic analysis employed in studies drawing on the induced institutional innovation perspective. Compare, for example, the interpretation by Hayami and Kikuchi of the changes in labor relations on Laguna rice farms associated with the introductions of higher-yielding rice varieties (chap. 1, this vol.).

In other studies, Harris advances a materialist interpretation of cannibalism and the biblical prohibition against pork consumption. These and other curiosities represent test cases for the materialist hypotheses. While advancing a materialist interpretation of the evolution of cultural forms, Harris insists that he is not a cultural relativist: some cultures really are better than others! For example, modern forms of contraception are better than traditional forms of induced abortion (1977).

The materialist interpretations offered by Harris have been criticized from widely different perspectives—as tautological, as functionalist, and as "vulgar economic determinism" (Plattner 1989b, 39). Gans has argued, in defense of Harris, that "if he can explain such bizarre, and apparently functionless culture traits by the principle of cultural materialism, then surely he can account for the main run of cultural development, the economic rationality of which is at least plausible on the surface" (1985, 81).

What help can development economists draw from the materialist approaches in anthropology for extending the induced innovation model to include cultural endowments? It seems clear that ethno-

---

16. See also Harris's interpretation and the discussion of the cultural prohibitions on beef consumption and cattle slaughter in India (Harris 1966, 57–59; 1971, 199–201; Heston 1971, 199–200).

graphic studies drawing on a materialist perspective can be quite useful to economists who are attempting to utilize the tools of microeconomic analysis to understand the impact of resource endowments and technology on differences in institutional performance and on institutional change. The ethnographic studies would be even more useful, and more credible, if their authors were more fully informed in modern microeconomic theory and the methodology used in the empirical testing of hypotheses generated from the use of microeconomic theory. Harris did not, at least in his citations, indicate any familiarity with modern neoclassical microeconomics. Familiarity with farm management and production economics literature and with the "household economics" literature would be particularly useful (Binswanger et al. 1980). Research carried out within the materialist agenda could provide development economists with substantial insight into the production and consumption decisions of peasant households. It would be particularly valuable, for example, to be able to compare the economic benefits to peasant families of production and consumption decisions based on emic (culturally based) and etic (formal analytic) decision criteria (Finkler 1979).[17]

Neither the current research output nor the research agenda of the cultural materialist school, now largely out of fashion, is likely to provide much information on the questions that I have attempted to raise in this chapter—what guidance can development economists obtain from anthropology in attempting to understand how differences or changes in cultural endowments affect behavior leading to technical and institutional change? This same point has been stated in a somewhat different manner by Gans who notes that in the cultural materialist strategy the "existence of human society and its fundamental institutions is simply taken for granted and hypotheses are formulated to explain certain of its features as adaptations to infrastructural conditions" (1985, 85).

## Interpretive Anthropology

From the 1920s until well into the 1960s there was a continuing struggle to resolve the conflicts between the Boas "cultural anthropology" school, which focused its attention on the identification of "culture patterns," and the Radcliffe-Brown "social anthropology" school, which

---

17. During the 1980s and 1990s economic anthropologists have generated a large body of literature that has drawn more directly on formal analytic and quantitative methods taken from economics (Binswanger et al. 1980; Plattner 1989a; Gudeman 1998).

emphasized social structure. The main difference between the two schools is that the cultural pattern approach subordinated social structures to culture, while the structure approach subordinated culture to social structure. Singer notes that "the structural theory considers an 'explanation' achieved when it has shown how each part contributes functionally to the existence and continuity of a particular type of social structure while the (cultural) pattern theory's desideratum for 'explanations' is to show how each part fits into an overall configuration or stylistic pattern of the culture" (1968, 533). During the 1960s and 1970s, efforts emerged, drawing on a wide range of philosophical and social science traditions, to direct anthropological theory and ethnographic research to "elucidate how different cultural constructions of reality affect social action" (Marcus and Fischer 1986, 25).

Marshall Sahlins has been among the most outspoken critics of materialist interpretations of cultural development. In his later work Sahlins insists that material forces play no independent role in the formation of culture—that resource endowments and the entire natural world are as much cultural constructions as ideas and values are.[18] "Anthropology can no longer be content with the idea that custom is merely fetishized utility" (Sahlins 1976, 76). He dismisses the conceptual basis of materialist anthropology: "The material forces in production contain no cultural order, but merely a set of physical possibilities and constraints selectively organized by the cultural system" (207). He suggests, somewhat more pungently, that materialist theory assumes that "manure is thicker than blood" (25).

In a more positive tone Sahlins argues, "The real issue posed for anthropology by all . . . practical reason is the existence of culture. The utility theories have gone through many changes in custom, but always play out the same denouement: the elimination of culture as a distinctive object of the discipline. One sees through the variety of these theories two main types. . . . One type is naturalistic or ecological . . . while the second is utilitarian . . . invoking the familiar means-ends calculus of the rational human subject" (1976, 101). But, he insists, neither type of theory has been able to explain fully the anthropological discovery that the creation of meaning is the distinguishing quality of humans (101).

Beginning in the early 1970s, Clifford Geertz, drawing on the

---

18. For articulation of Sahlins's perspective on the role of cultural endowments and his argument for a larger role of history in anthropological research see Sahlins (1985, 1993, 1995). In his early work Sahlins was more receptive to the role of resource and environmental change on cultural evolution (1960). In the late 1960s Sahlins "abruptly abandoned the evolutionist position to which he had adhered for the better part of two decades" (Kuper 1999, 164). For his manifesto embracing cultural determinism see Sahlins (1976).

organic vision of culture of the later Wittgenstein, began to advance an alternative to the materialist and structural interpretation of cultures. Geertz employed, in his own work, a semiotic (symbolic) concept of culture and an interpretive approach, based on "thick description" in the study of culture.[19] Thick description is employed to capture the symbolic elements of culture. The objective is to aid the ethnologist in interpreting the conceptual world in which his informants live so that it is intelligible to people from outside their culture and does not do violence to the self-perceptions of the people being studied (Keohane 1988, 42). "The aim of the anthropologist is the enlargement of the universe of human discourse" (Geertz 1973, 14).

Geertz does not minimize the difficulties involved in conducting research and of validating knowledge when employing interpretive theory. The method is clinical rather than experimental. Its product is diagnostic rather than predictive. The interpretive anthropologist approaches "broader interpretation and more abstract analysis from extended acquaintance with extremely small matters." The broader significance of interpretive analysis emerges because "anthropologists don't study villages; they study in villages" (Geertz 1973, 22). The objective is to use small facts to speak to large issues!

A useful illustration is Geertz's analysis of the formation and impact of ideology (1964). He argues that formal ideologies first emerge and begin to guide political action at the point when a political system begins to free itself from the dominance of received tradition—"from the direct and detailed guidance of religious and philosophical canons on the one hand and from the unreflective precepts of conventional moralism" (64). Geertz asserts further that it is the ability of "ideologies to render otherwise incomprehensible social situations meaningful, so to construct them as to make it possible to act purposefully within them, that accounts . . . for the intensity with which, once accepted, they are held" (64). The objective of Geertz's later work was to move anthropology closer to the humanities and away from his own earlier materialist interpretation of cultural change. The effect was to encourage a form of "cultural fundamentalism" that characterizes the postmodernist turn in anthropology (Kuper 1999, 75–121).

In the interpretive reaction to materialist approaches, there does not

---

19. For the most accessible exposition see his chapter, "Thick Description: Toward an Interpretive Theory of Culture" (Geertz 1973, 3–30). The classic example of Geertz's interpretive approach is his discussion of the cultural significance of the institution of cock fighting in Bali (1972). I interpret Geertz's earlier work, particularly *Agricultural Involution,* as falling solidly within the materialist tradition. For additional perspective on the role of interpretive theory in the social sciences see Rabinow and Sullivan (1987) and Hatch (1997).

seem to be any way to connect the process of cultural change to historical changes in resource endowments technology or institutions. Just as intellectual history runs the danger of losing its authority when not linked to institutional history, cognitive and symbolic anthropology need to maintain a continuing dialogue with the broader trajectories of technical and institutional change. At this stage, the interpretations offered by interpretive anthropologists often strike me as excessively personal and idiosyncratic—thick description sometimes seems to come out of thin air (Connor 1984, 271). They provide little insight into the process of institutional change or guidance for institutional design. I doubt that thick description, in the Geertz tradition, would provide insight into the differences (or changes) in cow demography between Uttar Pradesh and Kerala as analyzed by Harris (1980, 32–34, 348–53).

More recently, scholars working within the interpretive approach have argued that Geertz has not been sufficiently ambitious. He did not give sufficient weight to the larger context that "informed the interpreter's interpretation" (Hatch 1997, 309). By placing an exclusive emphasis on meaning he neglected the possibility of explaining, for example, the historical sources of Balinese commitment of large resources to court pageantry. Placing the cultural forms studied by anthropologists working in the interpretive tradition within a broader context of political and economic development could open up the possibility of theoretical developments that might bridge the gap between materialist and interpretive anthropology (Sahlins 1993; Ortner 1994).

**Deconstructing Development**

The relationship between anthropology and development has been and remains troubled. One source of the troubled relationship with the idea of development has centered around the concept of cultural evolution. A number of early anthropologists—Edward B. Tylor, Herbert Spencer, and Lewis Henry Morgan—conceptualized development in terms of cultural evolution. Franz Boas, a dominant voice in anthropology from the 1920s through the 1940s, was a vigorous critic of both physical and cultural evolution. Cultural relativism was a major theme in the work of his students such as Margaret Mead and Ruth Benedict.[20] A second, arguably more important source of the continuing tension extends back to the early history of British social anthropology

---

20. For a vigorous defense of the evolutionary perspective in cultural anthropology see Sahlins (1960).

and to the arguments about the role of applied anthropology in the service of colonial administration (Malinowski 1930; Asad 1973; Firth 1981; Stocking 1984). The history of the association of applied anthropology with colonial administration has contributed to the reluctance of anthropologists to participate as fully as sociologists, political scientists, and economists in the "development project."

The reluctance of anthropologists to engage in research in support of development or to contribute to development administration and policy has deep cultural roots within the anthropology profession (Firth 1981; Klitgaard 1991; Gow 2002). The early commitment to fieldwork among "primitive" isolates as a unifying methodological instrument also contributed to a view of social structure as essentially static. External intervention, in the form of commercial penetration or military force, or even to enhance well-being, was regarded as subversive of traditional institutions and culture. Anthropologists have often viewed themselves as collectors, interpreters, and defenders of "endangered cultures."[21]

Since the early 1980s, anthropologists, traveling under the rubric of applied anthropology (or, more recently, development anthropology), have rapidly colonized international and national development agencies (Cernea 1996). Employment opportunities were opened up by "New Directions" policies in agencies such as the U.S. Agency for International Development (USAID) and the World Bank in new program areas such as rural development, health and nutrition, family planning, and poverty reduction (Ruttan 1996).[22] Insistence that projects "had to be socially relevant, culturally appropriate, and involve their direct beneficiaries . . . created unprecedented demand for the anthropologists' skills" (Escobar 1991, 663). The expanded role of anthropologists in development planning and evaluation opened up a new and contentious debate about the role of anthropology "at the service of power" (663).

---

21. Geertz notes that the view that primitive cultures "are radical different from us, morally superior to us, and need only to be protected, presumably by us, from our greed and cruelty . . . is not much in favor these days" (2000, 117). This comment was made in the context of a review of Clastres (1972, 1977). Clifford presents a more sophisticated interpretation in a comment on the work of Marcel Griaule, the great French ethnographer: "It is as though the Dogon had recognized the need for a kind of cultural ambassador, a qualified representative who would dramatize and defend their culture in the colonial world and beyond" (1988, 87).

22. At the World Bank the new opportunities for anthropologists and sociologists were advanced by the entrepreneurial skills of the Romanian rural sociologist Michael Cernea. For examples of research sponsored by the World Bank see Cernea (1991a).

The Colombian anthropologist Arturo Escobar has been a prominent voice in the critical assessment of development anthropology. He has drawn substantial inspiration from the work of the postmodernist French philosophers Michel Foucault and Jean-François Lyotard who sought to liberate the human mind from what they considered the parochialism of modern rational humanism. In their view, modern rationality has become a coercive source of power exercised through the formation and accumulation of knowledge. The effect of the socially constructed knowledge derived from theoretical and empirical inquiry in the social sciences has been that the emancipatory ideals of the Enlightenment have become a coercive force leading to new forms of domination. Foucault believed that all global (or general) theories, including the Marxist mode of production theory, economic development theory, sociological modernization theory, or world systems theory, were reductionist, universalistic, coercive, and even totalitarian (Peet and Hartwick 1999, 129–32).

> **Box 2.1. Postmodernism**
>
> Postmodernism was imported into the social sciences from the humanities. Its penetration in the social sciences traces to the writings of a group of French intellectuals during the 1960s particularly Michel Foucault, Jacques Derrida, Jacques Lacan, and Jean-François Lyotard (Epstein 2000, 215). The philosophical perspective employed by practitioners of postmodernist theory is illustrated by the critique by Jacques Derrida of the work of Claude Lévi-Strauss. Derrida's critique focused on the internal contradictions of Lévi-Strauss's anthropological interpretation of South American Indian culture. Derrida's "deconstruction" of the Lévi-Strauss interpretation was based on his reading and criticisms of the apparent contradictions in the Lévi-Strauss text. Derrida felt no need for training in anthropology or field experience in South America in order to deconstruct Lévi-Strauss's text (Miller 1993, 218).
>
> Postmodernism has not yet achieved, or sought, the status of a single coherent body of thought. Its practitioners have appropriated, transformed, and transcended ideas from "French structuralism, romanticism, Phenomenology, nihilism, populism, existentialism, hermeneutics, Western Marxism, Critical Theory and anarchism," and it has important quarrels with each (Rosenau 1992, 13). Despite its diversity, however, two general

orientations—skeptical postmodernism and affirmative postmodernism—can be distinguished.

Postmodernists, particularly the skeptical postmodernists, take a nihilistic approach to traditional social science criteria for evaluating the results of intellectual inquiry. They reject the tests of both coherence and correspondence. Theory must be liberated from the tyranny of data and observation! This rejection of modern approaches for evaluating knowledge reflects an epistemological view that there is no objective reality that the social sciences can comprehend. Affirmative postmodernists are somewhat less extreme. They imply that knowledge of the "local, the decentered, the marginal and the excluded" should be "privileged" relative to that of the center. But they share an opposition to the application of formal tests to confirm or reject social science knowledge.

The diffusion of postmodernist "cultural studies" from the humanities, particularly literary criticism, has differed among the several social sciences. Postmodernist anthropologists contend that modern anthropology creates the very phenomena it seeks to study, and therefore it cannot be expected, at least since the late nineteenth century, to discover truth about cultural endowments. "Traditional culture is increasingly recognized to be more of an invention constructed for contemporary purpose than a stable heritage handed on from the past" (Hanson 1989, 890–99). Anthropology in this view is "persuasive fiction or poetry" rather than social science.

Postmodernism has penetrated less deeply into sociology than anthropology. It has, however, generated substantial debate over the problem of representation in sociology. The response by some postmodernist sociologists has been to reject the attempt at formal theory construction. Others have advocated transforming sociological analysis into storytelling— to eliminate the distinction between sociology and literature (Richardson 1988, 200–204). The "strong" postmodernist agenda has, however, been rejected by most sociologists who, while accepting the view that perceptions of reality are mediated by language and culture, reject the implication that all claims to knowledge have equal status. They continue to insist that it is worth advancing social knowledge even if that knowledge will always be subject to refinement or revision.

The impact of postmodernist thought has had a much more

limited impact in political science, as a discipline, than on either anthropology or sociology. In economics, postmodernist thought, primarily in the work of Donald (Deirdre) McCloskey, has barely penetrated the most revisionist fringes of the discipline (Osteen and Woodmansee 1999). But the political system, particularly the traditional left-right alignment, has been strongly influenced by the diffusion of postmodernist social issues such as environmental concerns; racial, ethnic, and women's rights; and gay and lesbian emancipation, in addition to traditional concerns about income distribution and economic growth in the more highly developed Western economies (Inglehart 1997).

The postmodernist perspective toward development rejects Western technology and modern social, political, and economic development thought. It tends to be supportive of populist, fundamentalist, and nationalistic social movements. It calls for the defense of the primitive, sacred, and traditional. It is hard to avoid viewing postmodernism, at least in part, as a form of self-indulgent intellectual luxury practiced in societies in which scarcity and poverty are no longer central concerns. By the mid-1990s it was losing intellectual credibility in Europe, particularly in France (Epstein 2000, 218). While it remains an important intellectual movement in the United States it has never achieved a substantial following in the poor countries of the world (Rosenau 1992).

In this box I draw primarily on Jean-François Lyotard, *The Postmodern Condition: A Report on Knowledge* (Minneapolis: University of Minnesota Press, 1984; French ed., 1979); Michele Lamont, "How to Become a Dominant French Philosopher: The Case of Jacques Derrida," *American Journal of Sociology* 93 (1987): 584–622; Laurel Richardson, "The Collective Story: Post Modernism and the Writing of Sociology," *Sociological Focus* 21 (1988): 199–207; Allan Hanson, "The Making of the Maori: Cultural Invention and Its Logic," *American Anthropologist* 91 (1989): 890–902; Ronald Inglehart, *Modernization and Postmodernization: Cultural, Economic and Political Change in Forty-three Societies* (Princeton: Princeton University Press, 1997); Pauline Marie Rosenau, *Postmodernism and the Social Sciences: Insights, Inroads and Intrusions* (Princeton: Princeton University Press, 1992); Christopher L. Miller, "Literary Studies

> in African Literature: The Challenge of Intercultural Literacy," in *Africa and the Disciplines: The Contributions of Research in Africa to the Social Sciences and Humanities,* ed. Robert H. Bates, V. Y. Mudimbe, and Jean O'Barr, 213–31 (Chicago: University of Chicago Press, 1993); Martha Woodmansee and Mark Osteen, eds., *The New Economic Criticism: Studies at the Intersection of Literature and Economics* (London: Routledge, 1999); Barbara Epstein, "Postmodernism and the Left," in *The Sokal Hoax: The Sham That Shook the Academy,* ed. Lingua Franca, 214–29 (Lincoln: University of Nebraska Press, 2000), reprinted from *New Politics* (winter 1997): 130–44; Perry Anderson, *The Origins of Post Modernity* (London: Verso, 1998); George Snedeker, "Defending the Enlightenment: Jürgen Habermas and the Theory of Communicative Reason," *Dialectical Anthropology* 25 (2000): 239–50.

When Escobar extended postmodernist thought to interpret the effects of Western intervention in the development of poor countries, in a work written while still a graduate student at Berkeley, he saw advances in social science knowledge about development as a source of power leading to Western control over economic and political development in the Third World (1984, 1988). When Escobar imposed the Foucault-Lyotard model of the link between power and knowledge on post–World War II development thought and practice, what he saw was a reinstatement of international control in American terms—the transfer of "modern scientific and technical knowledge" and a global "war on poverty" as a replacement for European colonialist assumptions of "the white man's burden" (Escobar 1995, 3; Peet and Hartwick 1999, 157).

Escobar has directed even more vigorous criticism toward the theory and practice of development anthropology. Development anthropologists function, in his view, by applying a conventional concept of culture to a broad range of "development situations." They see themselves as culture brokers or translators working on behalf of the poor (1991). In practice, however, development anthropologists are more effective in translating the development agencies' points of view than in interpreting "the natives' point of view" in terms that influence program and project design. They are institutionally conditioned to fail to respect the widespread resistance to development in many parts of the Third World. They are unable to escape the dominant "top down, ethnocentric and technocratic" approach to development (1995, 44).

Finally, Escobar insists that most development anthropologists are not even good anthropologists. "They have failed to take into account, perhaps even to notice, the significant changes that are happening in the discipline, thus continuing to adhere to the professional practices many anthropologists today would find questionable" (1991, 676).[23] He objects, consistent with Foucault's perspective on social science knowledge, to any pretensions of scientific objectivity in anthropology because of "its embeddedness in power-knowledge systems" (676). And he argues for a type of critical anthropological practice "that distances itself from mainstream development institutions and conceptions, even when working within the development field" (677). Such a practice would be more concerned with social movements, political struggles, and the reconstruction of identities.[24] The conventionally defined objective of modernization and of political and economic development should be abandoned. "In the Third World, modernity is not an 'unfinished project' of the Enlightenment. Development is the last and failed attempt to complete the Enlightenment in Asia, Africa and Latin America" (1995, 221).[25]

What should development economists make of the postcolonial, poststructural, and postmodern critics' efforts to deconstruct social science knowledge and practice in the field of development? Depictions of modern development theory in terms of monolithic hegemony— "oppressive, disciplinary, normalizing, totalizing, essentialist, knowl-

---

23. For a representative sampling of academic research by development anthropologists in the mid-1980s see Bennett and Bowen (1988). It is hard for me to believe that Escobar was familiar with some of the more rigorous research by development anthropologists. See, for example, the very impressive research on peasant decision making by Christina Gladwin (1979, 1989).

24. Escobar's work has been critically reviewed by anthropologists (Little and Painter 1995; Autum 1966). Both are critical of Escobar for what they characterize as a narrow interpretation of the historical context within which anthropology has developed. Both share, however, Escobar's view that the discourse of development functions as a politically constructed tool for controlling and expropriating from the Third World (Autum 1996, 480).

25. The Enlightenment was the product of seventeenth- and eighteenth-century European thought that sought to bring reason to bear on religion, politics, and the arts. The "Enlightenment Project" sought to develop a universal morality, political democracy, an objective science, and an autonomous sphere of the arts (Snedeker 2000). The excesses of the French Revolution induced a Counter-Enlightenment that renewed emphasis on revealed religion and the authoritarian state (Berlin 2000). The critique of the Enlightenment project was renewed during the late twentieth century by a postmodern critique of rational thought and of the idea of progress. Michel Foucault (1972) and Jean-François Lyotard (1984) have insisted, for example, that the horrors of violence and political repression that have defined much of the twentieth century are the unintended consequences of the Enlightenment project of reason and domination over nature.

edge in pursuit of power"—are caricatures. Their depictions of the development efforts of the last half century as unmitigated disasters for the world's poor are surely overdrawn.[26] Development success has not been limited to the tiger economies of East Asia. Throughout what used to be called the Third World, most indicators of social and economic development have improved, and some, such as literacy and life expectancy, have improved dramatically even in the poorest countries.

In spite of my generally negative assessment, there are elements of the postmodern literature that I find potentially useful for development economists. I find the later work of Foucault on the material conditions of discourse important. Insistence that scientists and technologists learn to value local knowledge, while conventional, can hardly be overemphasized. But anthropologists also must share in the responsibility to the communities in which they work by insisting that local knowledge meet the tests of coherence and correspondence.

**Constructing Culture**

In the earlier sections of this chapter I have been concerned with the impact of cultural endowments on economic development. In a remarkable book, *Economics as Culture,* Stephen Gudeman (1986) has explored how changes in resource endowments, technology, and institutions have induced changes in culture, using the construction by Panamanian peasants of a succession of "local models" of economic organization. During the period studied by Gudeman, peasants living in the remote region of Los Boquerones experienced a transition from a subsistence production system, based primarily on beans and maize, to wage employment for a government-managed sugar mill. As the production system changed, the emic model employed by the peasants to interpret the organization of economic activity also changed.

Household Economy

The traditional system of agriculture practiced in the region involved a swidden (slash-and-burn) system of crop production. The household,

---

26. I am amazed by the continuing assessment of the "seed fertilizer" or green revolution in crop production as a failed technology and a source of widespread immiserization in the critical literature. I can only interpret the ongoing hostility by many anthropologists as an example of cultural resistance to technical change. For a review of green revolution controversies see Hayami and Ruttan (1985, 56–59, 256–60, 403–5). For a more recent perspective see Conway (1997).

based on a division of labor between the sexes, constituted the primary unit of both production and consumption. The land was owned by absentee landlords who practiced an extensive cattle-grazing system. The peasants had established a symbiotic relationship with the landowners. They were allowed to live on the land in exchange for a nominal rent and for clearing the land of trees and brush. After clearing, the land was planted to rice and corn for several years. The cultivator then moved to another plot and repeated the same operation. The landlords' cattle grazed on the grass and shrubs that grew on the cleared land. After a decade or so, after bush and trees had largely displaced the grass, the peasant returned to clear the land again and initiate a new production cycle. The practice of swidden agriculture represented not so much an adaptation to ecological conditions as a response to the tenure arrangements imposed by the landowner. Both the peasants and the landlords were trapped in what economists have referred to as a "low-level equilibrium" trap.

The peasants had constructed a household economy model to interpret the economic relationships in which they found themselves. They saw the production of subsistence commodities as sustained by the inherent fertility of the land. Any allocation of the commodities produced on the land to maintain productive labor, to feed their own animals, to support leisure, or to exchange in the market for purchase of luxury items was viewed as an expense. Expense referred to anything that was used up in the process of consumption. Profit was viewed as arising from trade rather than production. The model described an economy characterized by production but not accumulation.

In the late 1950s new economic opportunities were opened up by improvement of the road network and by the location of two small sugarcane mills in the region. The mills made cash advances against the market value of the cane to gain access to land on which to produce cane. Sugarcane, which had previously been grown only for household use, became a cash crop. Land available to produce subsistence crops became scarcer. Population growth also pressed against the land resource base. Slash-and-burn agriculture began to be replaced by permanent cropping. By the mid-1960s the peasants had fenced nearly all the land in an attempt to establish permanent rights to the land that they cultivated. The enclosure movement, which had begun in response to population pressure against the land resource base, was reinforced by the opening-up of a commercial market for sugarcane—a commodity produced from the land.

These changes in turn induced the construction of a more complex model of an economy "in which two productive activities were carried out simultaneously but ranked and kept in separate domains" (Gudeman 1986, 151). Both subsistence production and market production were seen as sustaining household consumption. But access to the commercial market for sugarcane required the establishment of a claim on the land resource. Since much of the sugarcane was harvested by labor hired by the day, the new concept of a labor market, in which income was not directly tied to the land, had to be integrated into the peasants' subsistence-market dual-economy model. There was not, however, a clear conception of the specific contributions of land and labor to income or of investment in land improvement as a source of capital accumulation.

In the early 1970s the government imposed land reform and established a modern cooperative sugarcane mill in Los Boquerones. The land was acquired from the former owners by the land reform agency. An irrigation system was developed and the land parceled to facilitate the irrigation and transport of the cane. Land ownership rights by the beneficiaries of the land reform were converted to rights to participate in the profits of the mill. The mill was managed as a state enterprise. Production of subsistence crops in the area developed for sugarcane was prohibited. The peasants became completely dependent on wages for purchasing subsistence commodities for household consumption in the market.

One result of these technical and institutional changes was to induce the construction by the peasants of a new model of the economy. The peasants saw themselves related to the broader economy primarily through the market for consumer goods and the labor market. They also continued to think of themselves as landowners even though ownership was not attached to a specific parcel of land but rather to a share of the profit (if any) earned by the mill. The land is no longer a force (factor) of production to the individual peasant family. Land rent has become an abstract concept by which the former peasants interpret their legitimate claim on employment by the sugar cooperative. Household consumption remains, however the primary object of economic activity.

In Gudeman's series of models, land was first thought of in terms of the claims that outsiders had on the peasants through ownership of the land. In the third model the villagers had claims, through the land, on the government-managed sugar cooperative (1986, 25). In the first model consumption depended on the power of the land to produce subsistence commodities, whereas in the third model wage labor provided the cash return upon which almost all consumption depended. The local household economy had been superseded by a market economy. The changing models were externally induced. They were mediated through external changes in control of the land. In each case the peasants constructed mental models of the economy not in terms of impersonal market forces but in terms of their relations to an external source of power (125).

The significance of the Los Boquerones case described by Gudeman is the capacity of the peasant community to respond to the externally induced changes in economic organization by constructing new economic models (to remodel) that enabled them to understand the technical and institutional changes that they confronted. At the beginning the peasants would have been willing to defend their rights to cultivation against a landowner who tried to dispossess them from the land. By the end they would have been willing to conduct a strike against the government agency that manages the sugar mill if the agency had tried to dispossess them of their land-based employment rights.[27]

The Los Boquerones case also has broader significance for the argument between the materialist and the interpretive models of cultural change. These peasants demonstrated a capacity to construct new mental models—to remodel their culture, in this case their understanding of the effects of technical and institutional change, in a way that enabled them to achieve a reasonably accurate interpretation of the changes in their material environment. This ability is fundamental to the development process. Conversely, failure to achieve such congruence can lead to marginalization and decline.[28]

## Base and Market

In subsequent chapters I address the issue of what grows in the process of social, political, and economic development. In anthropology, evo-

---

27. In a more recent book Gudeman (2001) has presented examples of how the Schumpeterian cycle of invention, innovation, and extramarginal returns works its way out among small-scale indigenous potters, metal fabricators, and brickmakers in Guatemala. Gudeman's peasant entrepreneurs were not, however, the heroic figures depicted by Schumpeter (chap. 5, this vol.) but were embedded in community relations that sustained their entrepreneurship. See also the discussion of the role of peasant entrepreneurs in the modernization of agricultural marketing in Indonesia by Hayami and Kawagoe (1993).

28. The Los Boquerones peasants' ability to reconstruct their mental models to achieve congruence with change in economic organization stands in sharp contrast to Paraguay's Guayaki Indians' culture, which was so "tightly structured" that they were not only unable to adapt to European penetration but had historically been unable to adapt to penetration of their territory by the more aggressive Tupi-Guarani (Clastres 1972; Geertz 2000, 110). See Embree (1950) for the concept of tightly and loosely structured societies.

lutionary perspectives have largely been abandoned.[29] The subfield of development anthropology has had difficulty in finding a secure niche. The concept of *base* advanced in a new work by Stephen Gudeman (2001) has, however, opened the door to an anthropological response to the question, What grows (or decays) in the process of development?

Gudeman has articulated an anthropological model in which the economy, in its broadest sense, consists of two interrelated realms—community and market. A community economy constructs and shares a base (or commons) that includes material, institutional, and cultural elements. Gudeman draws an analogy between the role of the base in a social system and capital in a market system. As a community confronts the forces of modernization or attempts to respond to the opportunities opened up by economic change, the base may be strengthened or it may erode.

During the process of development the base grows through the renovation of traditional institutions or the incorporation of new institutions for the generation and transmission of knowledge, for civic engagement and governance, and to manage resources and facilitate material production and trade. But the base owes much of its strength to the fact that it was not initially constructed to serve market functions.[30] Gudeman warns that an unintended consequence of deliberate instrumental exploitation of the base to advance economic development may be the erosion (or debasing) of community integrity and capacity.

**Perspective**

I now return to the issue that motivated my interest in anthropology. What is the role of cultural endowments in the process of economic development? My conclusion in an earlier article, on which this chapter is partially based, was agnostic (Ruttan 1988). I concluded that development economists could expect few dividends from pursuing the literature in anthropology in an effort to understand the role of cultural endowments in economic development until anthropologists developed more rigorous methods for incorporating cultural endowments and change in cultural endowments in their own research.

---

29. See, however, Richerson, Boyd, and Paciotti (2002).

30. The importance of what Gudeman refers to as a base has received some attention in economics. For example, Kenneth Arrow has argued that "commercial morality" represents a precondition for efficient markets (1990, 139). See also Hayek's discussion of the institutions that have emerged out of "spontaneous order," that is, due to human action but not to human design (1967).

I also argued that the United States and other developed countries had realized substantial benefits from ethnographic research conducted by anthropologists. Franz Boas used the results of comparative ethnographic research to challenge racist views of human behavior. Anthropologists "have been the first to insist . . . that the world does not divide into the pious and the superstitious; that there are sculptures in jungles and paintings in deserts; that political order is possible without centralized power and principled justice without codified rules; that the norms of reason were not fixed in Greece and the evolution of morality not consummated in England. Most important, we were the first to insist that we see the lives of others through the lenses of our own grinding and that they look back on ours through ones of their own" (Geertz 2000, 65). Cultural anthropology has also provided empirically based support for a liberal reform agenda in broad areas of social policy in the United States. It is doubtful, however, that interpretations of ethnographic studies can again play a similar role. Appeals to the primitive or exotic now encounter greater skepticism. Differences between U.S. and Japanese economic performance are, for example, much more complex than earlier appeals to cultural differences suggested (Benedict 1946).

I remain skeptical about the value to development economists of anthropologists' contributions to understanding of cultural differences or of sources of cultural change. The materialist research agenda has been valuable in confirming the impact of differences, and sometimes of change, in resource endowments and technology on institutional and cultural change. But materialist anthropology, whether drawing on ecological, Marxist, or neoclassical traditions, has little to offer in helping to understand the effects of cultural endowments, or of changes in cultural endowments, on resource endowments or on technical or institutional change. Materialist anthropologists have avoided, almost as thoroughly as economists, attempts to understand the sources of change in cultural endowments and the impact of cultural change on economic development (Kuran 1995b, 328).

My own perspective, as suggested in chapter 1 (fig. 1.1), is that the relationship among changes in cultural endowments, resource endowments, technology, and institutions is dialectical rather than linear. In anthropology a similar view has been associated with the work of Leslie White (1949). Sahlins, in criticism of White, commented that "the technological determinism of culture in White's evolutionary theory lives side by side with the cultural determinants of technology in his symbolic theory" (1976, 45). This should, in my view, be considered a merit rather than a fault. The dialectical relationship suggested in figure 1.1 is also troubling to many economists. It implies great difficulty in resolving the "identification" problem in empirical research on the sources of change among resource endowments, technology, institutions, and culture.

Interpretive anthropology, despite its tendency to slip into idealism and romanticism, places cultural endowments at the center of its research agenda. Over the longer run it is possible that advances in knowledge generated by research conducted within the framework of interpretive anthropology, particularly if they could be cast in a broader context of political and economic change than Geertz was willing to do in his own work, could become more helpful to those of us working in the field of development economics than research conducted within the materialist paradigm

I find the postmodern critical and social constructionist perspectives even more problematic than interpretive anthropology. The emphasis on the role of power in the construction of social science theory, as in the later work of Foucault, and the criticism by Escobar of development anthropologists for neglecting power relationships represent important perspectives. The efforts of postmodern anthropologists to shift the margin between anthropological and humanistic approaches, as in the area of cultural studies, represent an unproductive detour (Clifford 1997). In contrast I find that the effort by Gudeman to construct an anthropology of economy (and of economists) widens the opportunity for fruitful dialogue. His insistence on the capacity of peasant communities to construct coherent material models to interpret technical and institutional change and his exploration of the relationships between community and market economy provide a foundation for productive discussion between anthropologists and development economists.

The incorporation or disappearance of primitive cultures has been a source of continuing concern in debates about the future of the discipline. "Primitives . . . are a bit of a wasting asset" (Geertz 2000, 92). The response of Lévi-Strauss, the leader of the French structuralist school, was to call for an intensification of traditional ethnographic work so that knowledge of cultural diversity could be preserved, at least on paper and tape, before the last primitive cultures disappeared (1966a, 1996b). It was in that spirit that he sent one of his most promising students, Pierre Clastres, off to a remote corner of Paraguay to study the few survivors of the once populous Guayaki. Geertz, whose interests have focused on traditional rather than primitive societies, favors a more contemplative approach—"to clarify what on earth is going on among peoples at various times and draw some conclusions about constraints, causes, hopes, and possibilities" (2000, 138).

At the opposite extreme is the passionate romanticism of Arturo Escobar. Starting from a completely valid, and important, emphasis on the role of local knowledge, he moves on to insist that the object of grassroots movements and indigenous political mobilization should be the rejection of the entire paradigm of modernization or the project of development. It is only as an afterthought to a polemic against development that he suggests the possibility of a "complex process of cultural hybridization encompassing manifold and multiple modernities and traditions" (Escobar 1995, 218). But he does not suggest how anthropologists can bring their knowledge of cultural change to bear on the emergence of "multiple modernities" that might enable the traditional cultures they study to prosper in the world that they confront.[31]

My greatest disappointment with anthropology—"the science of man"—has been its failure to make the knowledge that it has acquired accessible for the development of the societies that have been the object of attention. The failure of anthropologists to engage more directly on issues of development has been a continuing puzzle (Redfield and Warner 1940; Rhoades 1984; Hackenberg and Hackenberg 1999). Much anthropological research is carried out in communities in which most families are engaged in food production. Apparently the interpretation of farming practices and resource management arrangements has been less challenging than the order of kinship, descent groups, and ritual patterns (Netting 1974). An important exception has been the extended research by a series of anthropologists working in interdisciplinary teams at the International Potato Center (CIP) since the mid-1970s. The result has been a series of important studies of the role of introduced potato varieties in Andean farming systems and of post-harvest potato storage technology and marketing systems. This research led directly to the design and adoption by Peruvian potato producers, and by potato producers in highland areas in Asia and Africa, of improved household storage systems for potatoes used for household consumption, market

---

31. The emerging field of "political ecology" represents a potentially promising attempt by some anthropologists to repair the neglect of the macropolitical context in the study of cultural change. At present, however, an action-oriented "green romanticism" diverts attention from attempts to develop a deeper understanding of the sources of economic and political change that must be taken into consideration in policy and institutional design (Escobar 1999; Vayda and Walters 1999). See, however, Arce and Long (2000).

disposal, and planting (Rhoades 1984, 20–30).[32] Over the last several decades anthropologists working in national and international agencies have forced their colleagues to pay much closer attention to local knowledge and practice in the design and implementation of relief and development projects. But these efforts, however valuable, have rarely been rewarded by high professional status (Cernea 1991a, 1996).

If anthropologists are to resolve the continuing crisis that confronts the discipline they must embrace rather than resist the "development project." My own perspective is not too different from that articulated by Lévi-Strauss in the mid-1960s. "It is out of deep respect for cultures other than our own that the doctrine of cultural relativism evolved; and it now appears that this doctrine is deemed unacceptable by the very people on whose behalf it was upheld . . . who prefer to look at themselves as temporarily backward rather than permanently different" (Lévi-Strauss 1966, 125). The crisis will not be resolved by the "deep hanging out" of the interpretive anthropologists; by the "endangered species" approach of the critical postmodern ideologies; nor, least of all, by anthropologically informed critical studies or travel writing.

What traditional cultures, peasant societies, and ethnic enclaves need from anthropology is the knowledge that will enable them to engage the broader national and international worlds in which they exist. This is also what would be most useful to development economists and to national and international development assistance agencies. If anthropology is to respond to these demands it will have to acquire a capacity to respond positively rather than reluctantly to the possibilities of institutional design.

---

32. I have been criticized, correctly, by a reviewer of an article on which this chapter draws, of neglecting the impact of anthropological research on practice. In addition to the work of Rhoades and his associates I could have selected examples from the work of other agricultural anthropologists. I have in mind, for example, the work in 1950s Peru by Allan Holmberg (1955); the more recent work by Lansing (1991) of the role of water temples in irrigation management in Bali; the studies by Netting (1993) on smallholder sustainable agriculture in Switzerland; and work on peasant decision making by Gladwin (1979, 1989). I have also been impressed by the rich anthropological literature that has addressed indigenous knowledge systems (Warren, Slikkerver, and Brokensha 1995) and the policy issues involved in the provision of relief and the protection of human rights for refugees from natural disasters and civil disorder (Moran 1996; DeWaal 2002; Messer 2002).

CHAPTER 3

# The Sociology of Development and Underdevelopment

Economists and sociologists have traditionally tended to avoid confronting each other on their home ground.[1] Since the Robbins-Parsons exchange in the early 1930s about the appropriate subject matter of economics and sociology, each field has been viewed by its practitioners as occupying almost completely autonomous roles. This has enabled each discipline to treat the other as largely separable, at least for short-run analytical purposes. But development economists and sociologists work in a world in which neither social norms or individual rationality can be ignored. Piore suggests that the relationship between the two disciplines should be conceptualized in the form of a transformation function involving continuous trade-offs rather than sockets where the knowledge from the other discipline can be either ignored or plugged in as convenient (1996, 743).[2]

This isolation has changed substantially over the last several decades. Economists have attempted to colonize territory formerly regarded as the domain of sociology, such as discrimination and family behavior. Sociologists have investigated the organization of markets, firms, and work (Baron and Hannon 1994; Smelser and Swedeberg 1994b, 17–20). Sociologists who have worked within dependency and world systems theory, and postmodernist critical theory, have challenged economists' understanding of economic organization and of the process of economic development. In this chapter I attempt to respond to the question, What can, or should, economists working

---

1. For an earlier draft of this chapter see Ruttan (1992). I am indebted to Alessandro Bonanno, Lawrence Busch, Tamar Khitarshvilli, Enzo Mingione, and Cornelia Butler Flora for comments on earlier drafts.

2. A similar point had been made earlier by Wrong (1961) and Granovetter (1985)—that economics employs an undersocialized interpretation and sociology an oversocialized theory of human behavior.

within the field of development economics learn from contemporary research in sociology?

**Why Sociology?**

Why should economists concerned about the development of poor countries be interested in sociologists' contributions to development theory and knowledge? In sociology, in contrast to anthropology, development has been a central concern. The founding fathers of sociology, from Marx (1818–83) and Spencer (1820–1903) through Durkheim (1858–1917) and Weber (1864–1920), were unreservedly committed to development—conceptualized in terms of the transition to capitalism.[3]

One reason is that a synthesis of economic and social development theory could provide greater depth to attempts to understand the development process. A second is to draw on sociological knowledge for development policy or development planning. Knowledge of the interplay between social structure and the response to policy initiatives could improve the effectiveness of policy and program design. A third reason is concern about the social impacts of economic growth. The impact on material culture could be so destructive of social structure that it generates a political backlash capable of disrupting the capacity to pursue policies leading to sustained development.[4]

Ever since economists began to concern themselves with issues of economic development, the mutual interaction between economic development and change in social structure has been recognized. But economists have seldom introduced sociological knowledge into their analysis of the development process (Swedberg 1990a). Kindle-

---

3. A common feature of the work of classical sociology was commitment to a single unified theory to account for the process of development (Nisbet 1970). Marx, for example, held that "the country that is more developed industrially only shows to the less developed the image of its own future" (1867, 8–9).

4. Aside from the problem of translation between the two loosely related languages of economics and sociology there is the problem that the discipline of sociology is incredibly fragmented. The twenty-four sections and fifty-four specialty areas within the American Sociological Association fail to reveal the full diversity of the field (Cappell and Guterback 1992; Ennis 1992). Much of the programming at the sessions of the annual meeting is controlled by the formally recognized sections. Some traditional fields such as social work, child welfare, and even rural sociology have completely withdrawn, or been expelled, and have formed separate departments, associations, and journals (Horowitz 1993). Survey research, which once served as a unifying methodology, similar to fieldwork in anthropology, no longer occupies a central role in a number of sociological subdisciplines or specialties.

berger's 1952 comment on early World Bank country analysis reports remains apt.

> These are essays in comparative statics. The missions bring to the underdeveloped country a notion of what a developed country is like. They observe the underdeveloped country. They subtract the former from the latter. The difference is a program. Most of the members of the missions come from developed countries with highly articulated institutions for achieving social, economic and political ends. Ethnocentricity leads inevitably to the conclusion that the way to achieve comparable levels of capital formation, productivity, and consumption is to duplicate these institutions. (391–92)

One of the more ambitious attempts by an economist to draw on sociology to interpret the process of economic development was by Bert Hoselitz (chap. 2, this vol.). Hoselitz, founder of the journal *Economic Development and Cultural Change,* drew particularly on the set of "pattern variables" in the structuralist-functionalist model outlined by Talcott Parsons. He hypothesized that an "advanced" economy could be expected to

> exhibit predominantly universalistic norms in determining the selection process for the attainment of economically relevant roles; the roles themselves are functionally highly specific; that the predominant norms by which the selection process for those roles is regulated are based on the principle of achievement, and that the holders of positions in the power elite, and even in other elites, are expected to maintain collectivity oriented relations to social objects of economic significance. In an underdeveloped society, on the contrary, particularism, function diffusion, and the principle of ascription predominate as regulators of social-structural relations. (1960, 41–42)

Hoselitz attempted to apply the pattern model to interpret the literature on the role of elites, particularly the entrepreneur, as deviant personalities who play a critical role as actors in the transition of a society from a traditional to a modern structure.

My own interest in exploring the sociological literature arises in large part out of concern over the limited success of development economists in specifying the sources of institutional innovation—of the actions needed to set in motion the process of institutional innovation. In earlier research, with Hayami and Binswanger, I argued that advances in social science knowledge had the effect of shifting the supply curve for institutional innovation to the right, thus reducing the cost of institutional change (Binswanger and Ruttan 1978; Ruttan and Hayami 1984, 203–23; also chap. 1, this vol.). My personal answer to the question "Why sociology?" is similar to that articulated very clearly by James S. Coleman, the leading social theorist of the last generation. Coleman was explicitly critical of his colleagues in sociology for their failure to address the sources of change in social norms and of institutional innovation (1990b, 4–5). He went on to argue:

> A major question that a theory of institutions should answer is how and under what conditions a formal institutional structure comes into being, buttressed by formal laws or rules rather than by an informal structure supported by norms. This is part of a broader agenda for sociology, that of developing theory for the constructed social organization that is coming to replace the primordial or spontaneous social organization that was the foundation of societies of the past. The institutional structuring that Parsons had in mind were these whose control was based on norms, not formal rules or laws. The social organization upon which these structures developed was spontaneous, not formal. Yet societies are undergoing a major change from the form of organization that generates norms and customs from which institutional structures grow to a form of organization more fully based on purpose or design. (1990a, 337)

An effective response to the research agenda outlined by Coleman is precisely what development economists should find most useful from sociology.[5]

## What Happened to Modernization Theory?

When economists began after World War II to extend the analyses of economic growth and development in the Third World, they carried

---

5. In a review of Coleman's *Foundations,* Robert Frank notes: "Economists have largely ignored the existence of such (social) norms; and when they have addressed them specifically, it has usually been to assert that rational agents would never follow them. Sociologists, by contrast, often seem to believe that social norms are the only important determinants of human behavior" (1992, 149).

with them the social (or economic) accounting system that had been developed in the 1920s and 1930s by pioneers such as Simon Kuznets and Richard Stone. By the late 1930s the new metric had been extended by Colin Clark's massive scholarship to include a large number of developed and developing countries and colonial territories (1940). This research enabled economists to map, however crudely, levels of comparative development and begin to trace, in quantitative terms, changes in rates of economic growth.

When sociologists ventured into the same territory, they did not attempt to directly address the question, What grows in the process of social development? They did bring with them to the study of development a set of empirical generalizations from classical nineteenth-century sociology that characterized distinctions between "traditional" and "modern" societies.[6] Hegel had proposed four characteristics of modernity: (1) individualism: each person is entitled to his own subjective freedom; (2) the right to criticism: nothing need to be taken for granted; (3) autonomy of action: the individual is responsible for his own actions; and (4) philosophy of reflection: the subject can know himself without having to rely on explanations grounded in religion (Oberroi 1995, 102; Habermas 1990).

Modern societies were everything that traditional societies were not. They possessed "a high level of differentiation, a high degree of organic division of labor, specialization, urbanization, literacy, and exposure to mass media, and [were] imbued with a continuous drive toward progress. . . . Above all, traditional society was conceived as bound by the cultural horizons set by its tradition, and modern society as culturally dynamic and oriented to change and innovation" (Eisenstadt 1973, 10).[7]

---

6. Coleman has advanced an induced institutional innovation interpretation of the development of sociology. "Sociology as a discipline came into being and grew during the period when the constructed social environment began to grow and displace the natural social environment" (1990b, 610). Sociologists would identify Coleman's assertion as a functionalist explanation.

7. The "classical" source of modernization theory is the work of Max Weber. "Weber's basic problem was how to explain the specificity and uniqueness of European modernity. Why it was that only in the West—and not in other civilizations—that the specific 'radical' tendency to a rationalization of the world developed and the major manifestations . . . could be found in all spheres of social life—in the emergence of capitalist civilization; the bureaucratization of different forms of social life; the secularization of the world view; the development of modern science and of the so-called scientific world view" (Eisenstadt 1987, 2).

## Structural-Functionalism

Sociologists also brought with them to the study of development a "structural-functionalist" or "systems" theory of social organization and action that had been elaborated by Talcott Parsons during the 1930s. In the structural-functionalist perspective:

> Societies are more or less self-sufficient, adaptive social systems, characterized by varying degrees of differentiation, and with roles and institutions . . . as their principal units. The balance or equilibrium of the various parts of the whole is maintained for as long as certain functional prerequisites are satisfied and, generally speaking, an institution is "explained" once the functions it fulfills are satisfied. Finally, the entire system, or any part of it is kept together through the operation of a central value system broadly embodying social consensus. (Harrison 1988, 6)[8]

As conceived by Parsons, society was divided into a series of distinct and relatively autonomous subsystems, each related to the other through a limited number of particular linkages or connections, which bind the subsystems into a larger whole (Alexander 1990). This conception of society was intended to legitimate the organization of the several social sciences, which appeared to reflect the natural divisions of the social world. The other social sciences would study the variables that economics took as exogenous, such as technology and tastes (Piore 1996, 742–43).

Those sociologists closest to the Parsonian tradition stressed the transformation of structure—the modernization of social systems. Those who were more strongly influenced by psychology stressed personal transformation—the modernization of the individual. It was also possible to distinguish two schools: those who were mainly concerned with aspects of modernization most closely related to economic development and those whose focus was primarily on aspects most closely related to political development.

---

8. Harrison notes that Parsons's views on social structure were strongly influenced by the writings of Bronislaw Malinowski based on his field research among the Trobriand Islanders during World War I. Malinowski related the "basic needs of individual to the derived needs that have to be met for the continued survival of entire cultures and societies. . . . Initially, there are the individual needs for food, drink, sleep, and sex. These are related to the needs of all members of society for safety, bodily comfort, and health. At a cultural level, there are derived needs for reproduction through kinship and health through the practice of hygiene" (Harrison 1988, 6).

Modernization theory was rapidly adopted by political scientists working in the area of political development. It became a central conceptual framework for the program of research carried out under the auspices of Social Science Research Council Committee on Comparative Politics (chap. 4, this vol.). Economists, as usual, resisted any transfer of knowledge from sociology; they even avoided, by and large, the use of the term *modernization.*

One of the earliest and most influential studies of modernization, drawing on both the traditional/modern dichotomy and the Parsonian structural-functionalist theory, was carried out by Daniel Lerner in the mid-1950s. In *The Passing of Traditional Society* (1958), Lerner examined the process of modernization in a number of Middle East countries and arrived at a particular world perspective.

> Modernization is a global process. . . . Traditional society is on the wane, and Islam is "defenseless" against the "rationalist and positivist" spirit. In particular, the role of the mass media is crucial, and is associated with a cluster of other indices of development: Urbanization, accompanied by an increase in literacy, leads to an increase in exposure to the mass media. At the same time, the increasingly literate and urbanized population participates in a wider economic system. Modernity comes about through changes in institutions but also in persons. (16)

For Lerner a crucial aspect of modernization is the development of personality characterized by rationality and empathy that "enables newly mobile persons to operate efficiently in a changing world" (49–50).

By the mid-1960s the wealth of empirical detail generated by modernization research in sociology, and in political science, was leading to a critical reassessment of the empirical generalizations and to a reformulation of the structural-functionalist model. The criticisms have been summarized by Eisenstadt: (1) Great variation with regard to the degree to which their traditions impacted or facilitated the transition to modernity, even if traditional societies were topologically different from modern ones. (2) A distinction between tradition and traditionalism, with traditionalism defined as the more extremist, negative reaction to forces of modernity and tradition as the general reservoir of behavior and symbols of a society. (3) Recognition that in modern or modernizing societies there was a persistence of strong traditions of behavior rooted in the past. (4) Documentation of the ability of many traditional groups, such as castes or tribal units, to effectively reorganize themselves in modern settings. (5) A growing recognition that in many new states, whose independence movements had been shaped by modern Western models, older traditional modes or models of politics tended to assert themselves (Eisenstadt 1973, 102).

These criticisms of the traditional-modern generalizations lead to the recognition of what Eisenstadt regards as two critical aspects of institutional development associated with modernization. "First, was the recognition of the possibility that partial 'modernization' might reinforce traditional systems by infusion of new forms of organization. . . . Second was the growing recognition of . . . the systematic viability of . . . transitional systems . . . by emphasizing that these societies may develop in directions that do not necessarily lead to any given end stage envisaged in the initial model of modernization. . . . these analyses have undermined some of the basic assumptions of theories of convergence" (1973, 102). Eisenstadt argued that these criticisms of the validity of the traditional-modern dichotomy, combined with the increased dissatisfaction with the social systems assumption of the structural-functionalist approach, weakened the commitment of sociologists to what appeared to be the excessively deterministic implications of modernization theory.

But Eisenstadt was not clear on where these criticisms leave modernization theory. In an earlier paper he treated the transitional society as an intermediate evolution from a traditional to a modern society. In the transitional stage "the main social functions of major institutional spheres of society became disassociated from one another, allocated to specialized collectives and roles, and organized in relatively specific and autonomous symbolic and organizational frameworks within the confines of the same institutionalized system" (1973, 102). If society is to avoid disintegration or "regression," a continuous process of reintegration of the social system must occur—and may give rise to "new types of social, political, or cultural structure, each of which has different potentialities for further change, for breakdown or for development" (1969, 376). In *Tradition, Change, and Modernity* (1973) Eisenstadt promised a definition of modernity less subject to the criticisms listed previously. It is possible that he has done so. But if so, it is not apparent in my reading of his work.

Evolutionary Theory

Parsons's response to the deepening of knowledge about the social systems of new societies was to introduce an evolutionary orientation into the structural-functionalist model (1964, 339–57).[9] In this model even the simplest social systems include four requisites—culture, communication, social organization, and technology. Societies that break out of the "primitive" stage of social evolution are characterized by evolution along four sets of evolutionary universals: (1) social stratification and cultural legitimization; (2) bureaucratic organization and money and the market complex; (3) generalized universalistic norms; and (4) democratic association.

*Stratification* provides a form of status differentiation that permits hierarchical differentiation that is independent of kinship. In the initial stages of development, it opens new opportunities, other than ascription, for the assumption of specialized responsibility. But as a society evolves toward full modernity, "stratification often becomes a predominantly conservative force" (Parsons 1964, 345). *Legitimization* is closely related to stratification. It is necessary that societies provide a rationale for differentiated roles such as the separation of political and religious leadership. "As evolutionary universals, stratification and legitimization are associated with the developmental problems of breaking through the ascriptive nexus of kinship, on the one hand, and of 'traditionalized' culture, on the other. In turn they provide the basis for differentiation of a system that has previously, in the relevant respects, been undifferentiated" (Parsons 1964, 346).

The second pair of evolutionary universals are administrative bureaucracy and money and markets. The crucial feature of *bureaucracy* is the institutionalization of the authority of the office—"the differentiation of the role of incumbent from a person's other role-involvements, above all from his kinship roles" (Parsons 1964, 347). *Money and market exchange* releases the mobilization of resources from excessive reliance on the two alternative systems available to society: (a) the direct or forcible requisitioning of resources and (b) the activation of nonpolitical solidarities and commitments (such as those of community, caste, or ethnic identity). Money and markets permit the "emancipation of resources from ascriptive bonds" (Parsons 1964, 349–50).

The last two evolutionary universals are generalized universalistic norms and democratic association. An example of *generalized universalistic norms* is a formal legal system. It is "applicable to the society as

---

9. Some students take the position that Parsons had abandoned the structural-functionalist paradigm by the early 1960s (Boudon and Bourricaud 1989, 183). I prefer, at this stage, to view the "evolutionary universals" as the attempt to construct a dynamic structural-functionalist model.

a whole rather than to a few functional or segmented sectors, highly generalized in terms of principles and standards, and relatively independent of both the religious agencies that legitimize the normative order of the society and vested interest groups in the operative sector, particularly in government" (Parsons 1964, 351).[10] The basic argument for considering *democratic association* a universal "is that the larger and more complex a society becomes, the more important is effective political organization, not only in its administrative capacity, but also, and not least, in its support of a universalistic legal order. Political effectiveness includes both the scale and operative flexibility of the organization of power." Nevertheless, power, "as a generalized societal medium, depends overwhelmingly on a consensual element" (355).[11]

The addition of an evolutionary dynamic to the pattern variables of the Parsonian structural-functionalist model was clearly a major advance. The older traditional-modern dichotomy was a black-box comparative static model in which diffusion of technology, institutions, and culture provided the mechanisms to force the transition from traditional to modern. The specified evolutionary pattern variables—social stratification, cultural legitimization, bureaucratic organization, money and the market, universalistic norms, and democratic association—represent a separate but closely related set of variables with which it is possible to trace social development. The social systems perspective occupied a role somewhat similar to that of equilibrium in economics. Disequilibrium—or lack of articulation—among the several evolutionary universals would set in motion evolutionary changes that would lead in the direction of equilibrium and closer articulation (Moore 1964b, 888; also chap. 1, this vol.).[12]

From today's perspective it appears that by the mid-1960s a theoretical base had been established in sociology for the pursuit of a highly

---

10. Parsons is even more explicit: "English common law, with its adoption and further development in the overseas English-speaking world, not only constituted the most advanced case of universalistic normative order, but was probably decisive for the modern world. . . . I think it is legitimate to regard the English type of legal system as a fundamental prerequisite of the first occurrence of the Industrial Revolution" (1964, 353).

11. Etzioni expresses this point more aptly: "Ultimately, there is no way for a societal structure to discover the members' needs and adapt to them without the participation of the members in shaping and reshaping the structure" (1968, 626).

12. The point has been emphasized by Wilbert E. Moore: "Much of modern sociology has been built upon the conception of society as a system characterized by functional interdependency of major elements and relationships, and characterized by an orderly and persistent balance, a kind of equilibrium. . . . Dysfunctional consequences of particular patterns of action were recognized and identified as potential sites of change" (1964b, 888).

productive development research agenda. But as early as the late 1950s modernization theory was being subjected to intense political and intellectual criticism. By the end of the 1960s both the theme of modernization and the evolutionary version of the structural-functionalist model had largely been abandoned as guides to research by sociologists concerned with third world development. Why did development sociology turn away from what appeared to be such a promising research agenda?

One explanation is *intellectual*. It was argued that the structural-functionalist model lacked sufficient scientific rigor (Merton 1968, 73–138). Structural-functionalist explanations "reverse the time sequence of conventional causal analysis" by treating the effects as the cause of action (Ingham 1996, 251). In retrospect it seems apparent that further advances in structural-functionalist analysis would have required a level of formalization comparable to general equilibrium theory in economics. But sociology, as a discipline, was not prepared to move in the direction of greater abstraction and formalization.[13]

The second, arguably more important explanation is *ideological*. The structural-functionalist theories appeared to reinforce commitment to the existing social order (Goulder 1970; Giddens 1977; Peet and Hartwick 1999, 85–88). Evolutionary metatheory in the Parsonian tradition provided little insight into the design of the institutions needed to respond to the social and economic disintegration of the 1930s. Nor was it able to address the social problems of the welfare state in the 1950s and 1960s—racial conflict, deviant behavior, delinquency, crime, and the social consequences of poverty (Goulder 1970, 159–62).[14] Many sociologists, prompted in part by the Vietnam conflict and Cold War tensions, abandoned their attempts to understand social change in favor of advancing radical political and economic change.[15]

For both reasons, the systems or equilibrium implications of the structural-functionalist model had become increasingly unacceptable

---

13. For a very useful analytical interpretation and defense of structural-functional analysis see Levy (1968) and Cancian (1968). Cancian discusses early attempts to develop a more formal functional analysis. See also Faia (1993, 1–55).

14. Goulder comments, "The infrastructure of Parsonianism remained pre-Keynesian, insofar as it concerns the relations among institutions or actors, in the model of a spontaneously equilibrated *laissez-faire* economy rather than a state managed welfare economy" (1970, 162).

15. Faia has pointed out that the critics of the alleged role of functionalist theory for its conservative bias—"its alleged utility as an ideology of rule" in justifying colonial administration or the social legitimacy of the middle class—have employed a "functionalist explanation of functionalism" (1993, 172).

to many development sociologists by the mid-1960s (Coleman 1986, 1310–11). Development thought became dominated, as in anthropology, by a plethora of antipositivist, subjectivist, interpretive, and constructionist perspectives (see box 2.1). In the rush to abandon Parsonian analysis, sociologists also largely abandoned attempts to construct a macrosociology.

**Alternative Sociologies**

The critics who rejected the ethnocentrism of the Parsonian model are, as Harrison (1988, 40) suggests, faced with a difficult problem. If they believe that development is in any way "progressive," but reject the evolutionary perspective, what do they put in its place?

The search for an alternative was the product of profound disillusionment among many social scientists with the impact of Western economic cultural and military penetration into non-Western societies. In the United States this disillusionment was associated with efforts to resist radical revolution and reform in Latin America and Southeast Asia (Horowitz 1982, 79). Students of modernization who had viewed their research as a contribution to U.S. development assistance efforts were discredited. Irving Louis Horowitz insisted that consensus theory provided an ideological cover for support of conservative or authoritarian regimes (1972, 487). He argued that "the most important task for sociology today is to fashion methods adequate for studying social order in a world of conflicting interests, standards, and values" (490).

Dependency Theory

One response to these concerns was to embrace a new radical macrosociology that owed more to economists and historians working within a neo-Marxist paradigm than to the work of sociologists themselves. The speed with which these new perspectives, variously labeled "dependency" and "underdevelopment" theory, were embraced by many sociologists was surprising, even to many radical critics of modernization theory (Horowitz 1972, 509). To an economist, it is surprising how a school of economics, radical political economy, largely ignored or viewed as "bad economics" by mainstream economics, so rapidly established a bridgehead and then set an agenda for theory and policy research in sociology (and in political science). In his text entitled *The Sociology of Modernization and Development,* David Harrison (1988) devotes more pages to underdevelopment theory than to modernization theory—and relatively few of the references are to works by sociologists.

The underdevelopment approach to the sociology of development represents a synthesis of three separate traditions. One is the Latin American structuralist school, represented by the work of Argentine economist Raul Prebish and the UN Economic Commission for Latin America (Hayami and Ruttan 1985). The structuralist school employed the conventional tools of economics, particularly supply, demand, and trade elasticities, to argue that during the early postwar period the new income streams generated by productivity growth in commodity production in Latin America were being transferred to the developed countries of North America and Europe in the form of lower prices while the productivity gains in manufacturing in the developed countries were, as a result of monopoly organization, retained and shared among workers and owners, rather than being passed on to customers in Latin America. Their policy prescription was import substitution.

The second tradition was a variant of the Marxist theory of imperialism advanced in the mid-1950s by the Stanford University economist Paul Baran (1957). Lenin had stressed that imperialism was the instrument by which capitalism would be transmitted to the Third World—and would ultimately weaken the domination of the advanced capitalist nations (Harrison 1988, 68). Baran stood Lenin on his head. Baran saw it as both in the interests and within the power of monopoly capitalism to permanently extract surpluses from the raw material–supplying countries of the Third World. "For Baran the only way Third World countries could escape from the economic impasse was to withdraw from the world capitalist system completely and introduce socialist economic planning" (Harrison 1988, 71).

The third tradition was the world systems perspective advanced by the social historians Samir Amin (1976), Immanuel Wallerstein (1979), and Arghiri Emmanuel (1972). The world systems perspective, particularly in the work of Wallerstein, insisted that the developing world had been intimately linked to the world capitalist system since at least the sixteenth century. The core capitalist economies had become developed by exploiting the periphery.

But it was the vigorous attack by Andre Gunder Frank, drawing his intellectual inspiration from Baran and his empirical evidence from the work of the ECLA economists, that was most influential in popularizing the underdevelopment and world systems perspective among a younger generation of anthropologists and sociologists (Booth 1975,

68). The central theme of Frank's work was that it was world capitalism that created and maintained the conditions of underdevelopment in the Third World. Capitalism has simultaneously generated both economic development in the center and underdevelopment on the periphery (Frank 1967, 1969).

Why did Frank's work, particularly the two books published in the late 1960s, generate so much attention from sociologists? It was not his originality that carried his work across disciplinary boundaries—his inspiration, Baran's earlier work, had been largely neglected. Rather, it was Frank's role as the great popularizer: "It was his voice—strident, passionate, dogmatic, contemptuous and insistent—to which students of the late 1960s and 1970s responded" (Harrison 1988, 81). But this cannot be a complete answer. The insistence on the negative effects of technical and cultural diffusion that class relations extended across national boundaries was particularly appealing to an already radicalized younger generation of American and European scholars (Benton 1978, 217–36; Collins 1986, 133–55). They were prepared to believe not only that the developed countries enjoyed an "unequal exchange" but that capitalist development in the First World was responsible for the "underdevelopment" of the Third World. The fact that underdevelopment and world systems theory was based to a more significant degree than modernization theory on the work of third world scholars also added to its appeal.

By the mid-1980s commitment to the "development of underdevelopment" perspective was beginning to erode. This was in part due to vigorous criticism by scholars committed to classical Marxism. Marxist scholars were particularly critical of the implication that the capitalist world system had existed well before the industrial revolution and of the neglect of third world class structure conflict as a source of change.[16] They were also critical of the neglect of internal class structure in favor of the tensions resulting from increased intensity of external linkages.

More important, however, was the widening discrepancy between some of the more extravagant implications of the theory and the record of economic and political development in the 1970s and 1980s, particularly in Latin America, which had earlier served as the incubator for underdevelopment theory. The assertion that increased external linkage resulted in retrogression on the periphery could not be sustained.

---

16. Bender notes that Frank "has been treated by more sophisticated Marxist theorists like some sort of country bumpkin who has marched into the living room without removing his muddy galoshes" (1986, 3–33).

Industrialization occurred most rapidly in those third world nations, particularly in East Asia, that had established relatively strong and open linkages to the world economy (Krueger 1983; Portes 1997). In addition, the posture of the United States toward military regimes in Latin America and elsewhere shifted from support to restraint.

But the underdevelopment–world systems perspective did result, at least temporarily, in an enlargement of the research agenda for the sociology of development (Horowitz 1982, 89–111). For the modernization theory of the 1950s and 1960s, linkage between developed and developing countries was primarily a one-way street. Modernization was a consequence of the diffusion of technology, institutions, and culture from the developed to the developing world. By the 1970s it was clear that this model was inadequate. Underdevelopment theory itself was an intellectual import into the developed world from the undeveloped. And, remarkably, it retained intellectual currency in the developed world after it had lost much of its intellectual and political appeal in its centers of origin. There is no longer any serious disagreement that development sociologists must take a much broader range of international influences into account in their analysis of domestic economic development whether they are working in the more developed or less developed countries of the world.[17]

The broader agenda must go well beyond the impact of the penetration of international capital to include the effects of such influences as international migration of refugees, workers, and intellectuals; the rising protests against modernization such as the revival of fundamentalist orientations in the world's major religions; the shifting emphasis in the struggle among major nations between emphasis on commercial and ideological advantage; the continuing force of apartheid—of racial and ethnic discrimination—in postmodern as well as in modernizing societies; and the transnational transfer of social and biological pathologies such as the drug trade and AIDS (chap. 8, this vol.).

## The Sociology of Knowledge

For a brief moment in the late 1980s and early 1990s, it appeared that critical and other postmodern perspectives would penetrate sociologi-

---

17. For an alternative perspective see So (1990, 135–65). So argues that dependency theory has retained its currency by incorporating concepts such as associated-dependent development, the bureaucratic authoritarian state, and the triple alliance among state, local capital, and international capital. My own sense is that to the extent that dependency theory has retained its currency it has been absorbed into concerns about globalization (Arrighi 1999).

cal thought about development as thoroughly as in anthropology (chap. 2, this vol.). In its more radical form postmodernist theory viewed all development thought, including modernization, Marxist, world systems, and dependency theories, as exploitative and coercive. More sophisticated and less ideological postmodernists saw social structures, norms, and institutions emerging out of negotiations or discourse among principles and agents (box 2.1, chap. 2). The diffusion of postmodernist cultural studies into sociology was resisted by mainstream sociologists in large part because of its reliance on persuasive discourse rather than a more formal methodology, but also because cultural sociologists resented the effort of cultural studies to extend their territorial claims (Hays 2000). One of the most active, and arguably the most successful, postmodern research programs in sociology has been in the sociology of knowledge and the related area of social studies of science and technology.[18]

The traditional view of the process of scientific discovery was that scientific advance was guided primarily by internal criteria. Scientific knowledge was generated by curiosity and validated by norms of coherence, correspondence, and esthetics internal to each scientific discipline (Popper 1935; Boudon and Bourricaud 1982, 213–18). In the mid-1940s Robert Merton advanced a sociological interpretation of the scientific research process. His early work focused on the institutionalization of the capacity to advance science in Western societies. In later work he interpreted the structure and functioning of modern research communities in sociological terms. Sociological communities were governed, in his view, by internally generated and reinforcing social norms.[19]

The norms of scientific research emphasized by Popper and Merton did not mean that science was unresponsive to the environment external to science. But external criteria were primarily associated with the

---

18. During the 1940s and 1950s sociological research on the adoption and diffusion of technology was regarded as a major contribution by sociologists, particularly rural and health sociologists, to advancing the modernization agenda. During the 1970s, however, the adoption-diffusion research agenda was subject to increasing criticism from a postmodern perspective. I discuss adoption-diffusion research in chapter 6.

19. The four norms emphasized by Merton were: (1) *universalism*—the particular attributes of the scientist (race, religion, class, or nationality; (2) *communalism*—the discovery of scientific knowledge is a product of a common effort, and recognition and esteem rather than property rights are the sole reward of the scientist; (3) *disinterestedness*—scientists are ultimately accountable to their peers, rather than the market, for the evaluation of their results; (4) *skepticism*—all scientific knowledge is subject to challenge and revision (Merton 1968, 604–15).

choice of scientific research agendas. In his historical study of science in seventeenth-century England, for example, Merton found that the range of problems investigated were influenced by social, economic, and political interests. The effort to design a satisfactory way of finding longitude, particularly at sea, was driven by all three considerations. But the method that ultimately prevailed was determined by which was the most accurate (Merton 1968, 661–81; Sobel 1995).

The Popper-Merton paradigm articulated and reinforced a perspective that has remained pervasive in the science community and in science policy. It was articulated most forcefully in the 1945 Bush Report that became the charter for U.S. post–World War II science policy (Bush 1945).[20] By the late 1960s, however, the emerging community of sociologists working in the field of science and technology studies was moving away from the Popper-Merton view of science. The publication in 1962 of *The Structure of Scientific Revolutions* by the historian of science Thomas S. Kuhn provided a new interpretation of the process of scientific progress that had the effect of widening the agenda for social studies of science. Kuhn argued that the falsification tests assumed by Popper and Merton were not consistent features of "normal science." It is only when normal science, the solving of puzzles within the framework of a scientific paradigm, is unable to accommodate new empirical findings that normal science gives way to revolutionary science and a search for a new paradigm. Then scientists engage in intensive testing and challenge of scientific findings.

The significance of Kuhn's work for the sociology of knowledge is his insistence that scientific knowledge, like other forms of knowledge, is "socially constructed." Kuhn emphasized that knowledge and competence in a mature field of science are transmitted through a dogmatic and highly structured training, which inculcates an intense commitment to existing modes of perception, beliefs, procedures, and problem solutions (1962, 5). Kuhn's work was greeted with substantial antagonism by many philosophers of science and with considerable skepticism by many members of the scientific community, but its significance was quickly recognized and embraced by sociologists.[21]

---

20. I have discussed the Bush Report, its influence on U.S. science policy, and its criticisms by economists, in Ruttan (2001b, 79–81, 535–42).

21. Barnes has argued that Kuhn's work represents one of the "few fundamental contributions to the sociology of knowledge" (1982b, x). "The idea of social construction is fundamental to sociological analysis. . . . Its application in modern science studies has drawn attention to the moment-to-moment activities of scientists as they go about producing and reproducing scientific culture. This is the significance of social construction and not its alleged relativistic implications" (Restivo 1995, 107).

By the early 1980s sociologists, anthropologists, and historians had extended the social construction perspective to research on technology development (Bijker, Hughes, and Pinch 1987; Latour and Woolgar 1979). The effect was to further challenge the linear model of the relationship between advances in science and technology in favor of an interactive model (fig. 3.1). The interactive model has also been extended beyond the relationship between science and technology to other areas of knowledge to explore other interactive relationships between science and society—between science and the political system, for example (Barnes 1982a, 1982b).

There has been a vigorous, and sometimes shrill, reaction by many natural scientists to the postmodern social studies of science literature.[22] Much of the reaction by natural scientists has been directed at what has been termed the Strong Program in social studies of science that has had as its objective the delegitimization of the "privileged claim" to authority by natural scientists and engineers.[23] There has, however, been very little negative (or indeed any) response to the social constructivist perspective by engineers and other technologists.

Postmodernism prescribes skepticism toward all forms of general metatheory, whether based on religion, social science, or natural science: "I define postmodern as an incredibility toward metanarrative" (Lyotard 1990, 330). Postmodernists defend the more local and particularistic character of knowledge (Watson-Verran and Turnbull 1995). It is this view that is often regarded as a challenge to the privileged status of scientific knowledge. But rejection of the Strong Program does not mean that it is necessary to reject the insight that scientific knowledge is socially constructed. Nor does appreciation of the empirical truth of much local knowledge represent a threat to scientific inquiry. Rather it implies an obligation on the part of natural and social scientists and technologists to attempt to understand the logical coherence and empirical correspondence of local knowledge.

---

22. See, for example, Weinberg (1992); Wolpert (1992); Gross and Levitt (1994); Gross, Levitt, and Lewis (1996). See also the notorious case of Sokal's Hoax. Alan Sokal, a physicist, offended by what he considered uninformed attempts to deconstruct scientific knowledge, submitted a paper, "Transgressing the Boundaries: Transformative Hermeneutics of Quantum Gravity," to *Social Text*. It was peer reviewed and published (Sokal 1996b). Following publication Sokal revealed that the article was intended to expose the scholarly claims of postmodernist social studies of science. "My article is a mélange of truths, half-truths, quarter-truths, falsehoods, non sequiturs, and syntactically correct sentences that have no meaning whatsoever" (Sokal 1996a, 93).

23. The Strong Program is generally traced to the seminal work of Karen Knorr-Cetina (1981) and Bruno Latour and Steve Woolgar (1986). For a collection of major works see Jasanoff et al. (1995).

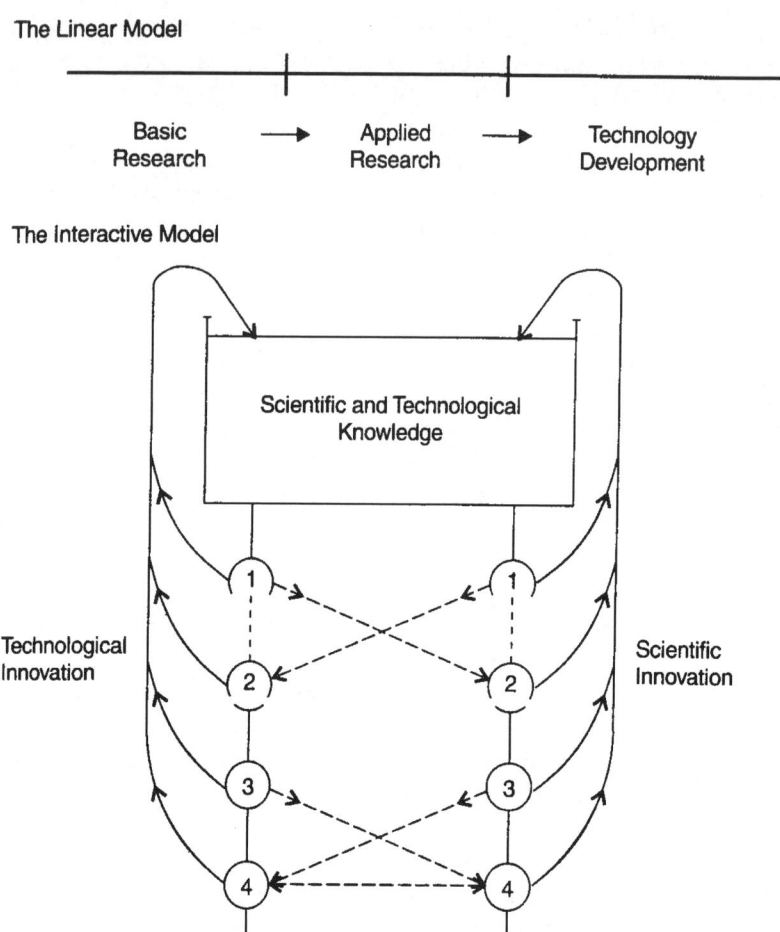

Fig. 3.1. Linear and interactive models of the relations between advances in scientific and technical knowledge. (Vernon W. Ruttan, *Technology, Growth and Development: An Induced Innovation Perspective* [Oxford: Oxford University Press]; adapted from Donald E. Stokes, *Pasteur's Quadrant: Basic Science and Technological Innovation* [Washington, DC: Brookings Institution Press, 1997], 10; Vernon W. Ruttan, *Agricultural Research Policy* [Minneapolis: University of Minnesota Press, 1982] 57.)

Postmodernism has not penetrated sociology as deeply as anthropology. It has, however, substantially dampened interest among sociologists in advancing development thought and practice. The claim that in the "present political and intellectual climate, dominated on the right by neoliberalism and on the left by postmodernism . . . development, understood as organized, collective interventions into social, cultural and economic processes on behalf of defined political goals [has] been silenced almost to the point of disappearing from memory" (Peet and Hartwick 1999, 197) is clearly a gross exaggeration. But Peet and Hartwick are correct that the normative approach to development, running from Durkheim through Parsons, no longer represents a major research agenda in American sociology.[24]

**Rational Choice, Social Norms, and Development**

I was once tempted to conclude that development economists had little to gain from attempting to incorporate this social metatheory into development economics (Ruttan 1992). The Parsonian structural-functionalist research agenda had been largely abandoned, wrongly in my judgment, by sociologists themselves. Dependency theory, which proved so attractive to a number of younger sociologists in the 1960s and 1970s, was borrowed uncritically from economists and historians. It has largely been abandoned in its centers of origin—in Latin America—and is actively pursued primarily in the intellectual backwaters of sociology and political science departments in the United States (Peet and Hartwick 1999, 118–22). Postmodernist perspectives have been so thoroughly antidevelopment in perspective that they have little to offer a world that continues to be committed to modernization and development.

Efforts to develop new metatheory to replace the discredited Parsonian structural-functionalism have, however, been singularly unsuccessful. Several social theorists, termed by Axel Van Den Berg "Grand Synthetic Theorists" (particularly Jürgen Habermas, Pierre Bourdieu, Anthony Giddens, and Jeffrey Alexander), have attempted to bridge the gap that emerged since the demise of structural-functionalism between the nomothetic macrosociological tradition and the several

---

24. The decline in development as a research agenda in sociology is reflected in the content of the recent volume commissioned by the American Sociological Association to take stock of the legacy and future of sociology in North America (Abu-Lughod 1999). The closest any of the chapters come to addressing development in poor countries is in a chapter on globalization.

ideographic or interpretivist microsociologies. Working from quite different perspectives—Habermas from critical theory, Bourdieu from a materialist perspective, Giddens from conflict theory, and Alexander from a neo-Parsonian perspective—they have attempted to bridge "the chasm between interpretivist micro and explanatory micro approaches" (Van Den Berg 1998, 230). In the end each has concluded that it would be nice if the gap that James Coleman has termed the macro-micro-micro-macro problem (fig. 1.2, chap. 1, this vol.) could be closed!

New Foundations

James Coleman's *Foundations of Social Theory* (1990) represented the most ambitious attempt since Talcott Parsons's *Structure of Social Action* to reconstruct social theory. Coleman attempted to reconstruct social theory on a "rational choice" foundation. The focus of Coleman's effort, throughout his immensely productive career, had been to bring theoretically grounded social theory to bear on issues of public policy. His research had ranged from early research on the diffusion of medical technology to his massive studies of education and economic opportunity (Coleman et al. 1966; Coleman and Hoffer 1989). The policy implications of his work have frequently generated intense controversy.[25]

> **Box 3.1. Rational Choice**
>
> Rational choice is the standard behavioral theory employed in neoclassical economics. It has become widely adopted in political science, sociology, and anthropology (chaps. 2, 3, and 4, this vol.). It has also been controversial, especially among anthropologists and sociologists.
>
> The basic assumption of rational choice theory is instrumental rationality. This means that participants either have correct models that they employ to interpret the world around them or

---

25. In 1968 as Coleman was reporting the results of his research on the educational impact of bussing on school desegregation and white flight from inner cities at the annual meeting of the American Sociological Society a banner bearing a swastika was unfurled behind the podium to signify Coleman's nefarious admission of "what everybody except the ideological extremists within the profession knew to be the case: that racial differences in elementary schooling remained a social fact and that inequality in educational opportunity was not dissolved by mandated bussing policy" (Horowitz 1993, 104).

receive information feedback that they employ to (rapidly) revise and correct their initially incorrect theories. The implication is that individuals and organizations attempt to respond to new information in a coherent manner (North 1990, 356).

The major criticism of rational choice theory is that cognitive processes involved in individual decision making are much more complex than is implied by rational choice theory. Experimental studies regularly encounter inconsistent and seemingly irrational behavior. So why do economists and many other social scientists continue to employ rational choice models in their attempts to understand human behavior or to design institutional reforms? One answer is that analytically and empirically viable theories of irrational behavior are not available.

What are the alternatives to rational choice theory? Elster (1986, 22–27) has suggested three alternative theories. *Structural theory* suggests that human action is narrowly constrained by social structure; this leaves very little scope for rational choice. The concept of *social norms* suggests that many actions can best be explained by habit, custom, duty, or tradition. Behavior often does not respond to environmental change even over the long run. The theory of *bounded rationality* holds that people tend to seek not the best alternative but rather one that is "good enough" or satisfactory. Economists tend to view these alternatives as useful descriptions but lacking in analytical power. They imply that humans are largely unresponsive to change in their social, political, or economic environment.

This does not mean that economists are entirely happy with rational choice as an explanation for human action. Rational choice theory works best in interpreting behavior in institutional environments that are characterized by relative stability in expectations about the behavior of others. These include organized commodity markets and regulated financial markets. They also include political party behavior in democratic societies. In other situations where there are few environmental constraints, such as consumer behavior, theories built on rational choice foundations do more poorly at interpreting human behavior (Satz and Freejohn 1994).

In preparing this box I have drawn on Jon Elster, ed., *Rational Choice* (New York: New York University Press, 1986); Douglass C. North, "A Transaction Cost Theory of Politics," *Journal of*

> *Theoretical Politics* 2 (1990): 355–67; Debra Satz and John Ferejohn, "Rational Choice and Social Theory," *Journal of Philosophy* 91 (1994): 71–87.

Coleman's *Foundations* was also regarded as controversial by most sociologists. His advocacy of the rational choice paradigm in sociological research was criticized as leading to sociology becoming subsumed within economics (Horowitz 1993, 103–17; Smelser 1990; White 1990). In contrast, the book has generally been reviewed favorably by economists. Robert Frank argues, for example, that Coleman is the first sociologist since Parsons to have given economists a systematic and methodologically compatible way of incorporating issues such as social norms, social capital, and feedback loops into traditional economic models (Frank 1992, 169; see also Piore 1996). Similarly, the adoption of rational choice assumptions created an opportunity for sociologists to enter into a serious exchange about theory and method with economists. But this is a door that few sociologists have chosen to take (Frank 1992; Baron and Hannen 1994; Ingham 1996).

Coleman has much to say about issues that are of interest to both development sociologists and development economists. Examples include his application of the theory of rational individual behavior to issues of social exchange, the emergence of social norms and formal institutions, and the importance of social capital (Frank 1992).

## Social Capital and Social Development

In an article published in the late 1980s Coleman returned to an issue posed by Talcott Parsons in his later work: What grows in the process of social development? Parsons had answered this question in terms of four evolutionary universals. Coleman's answer was social capital (Coleman 1988; 1990b, 300–321).[26] Coleman argued that social capital, along with physical and human capital, should be considered a fundamental source of economic growth. "Physical capital is wholly tangible, being embodied in observable material form; human capital is less tangible, being embodied in the skills and knowledge acquired by an individual; social capital is less tangible yet, for it exists in the *relations*

---

26. Alejandro Portes traces the origins of the concept of social capital back to Durkheim and Marx. He insists that the term *social capital* "does not embody any idea really new to sociology" (1998, 27). Coleman refers to the earlier work of Loury (1977, 1981) on the sources of racial income inequity and of Granovetter (1985) on the embeddedness of markets in complex social relationships. For useful reviews see Woolcock (1998); Sobel (2002).

among persons" (1988, S100). Social capital grows when the relations among individuals (or organizations) change in ways that facilitate more effective social action. Social capital, unlike physical capital and like human capital, does not wear out with use but degrades with disuse. Conversely it may accumulate with use (Ostrom 1997, 162).

Social capital, as employed by Coleman, encompasses elements of what I have discussed earlier under the rubrics of both institutions and culture (fig. 1.1, chap. 1, this vol.) and what Gudeman had discussed under the rubric of base (chap. 2, this vol.). Social capital has traditionally accrued as a by-product of activity engaged in for other purposes. Only a small part of the benefits from efforts to design and introduce new institutional arrangements accrue to those whose efforts are involved in their invention. The benefits may be diffused widely among the persons or organizations that are part of the particular social structure. An example is the mother who takes an active role in the development of a school parents' association. Most forms of social capital are public goods; social capital differs from human capital in that it is a communal rather than a private resource. It is not the private property of the persons who benefit from it (Coleman 1990b, 315). And it cannot readily be transferred from one person to another (Arrow 1999).

The public goods aspect of social capital means that the incentives that induce its formation must operate in a fundamentally different way than the incentives for physical and human capital formation. Social capital emerges as a consequence of human action but is often not the result of deliberate design. It becomes an important resource for individuals and organizations. But because many of the benefits of the behavior that generate social capital are realized by other persons, the economic incentives to contribute to the accumulation of social capital are weak. As a result, both the accumulation and the degradation of social capital occur as a by-product of other activities. The public goods characteristic of social capital means that, like open access natural resources, it is much more subject to erosion than either physical or human capital and, once degraded, is more difficult to reaccumulate or renovate.

Social capital has been characterized as "arguably the most influential concept to emerge from economic sociology in the last decade" (Woolcock 1998, 184). The concept has diffused with exceptional speed to achieve widespread academic and popular currency well beyond its sociological origins. A prominent example is the landmark book by political scientist Robert Putnam, *Making Democracy Work: Civic*

*Traditions in Modern Italy* (1993), which links contemporary differences in the performance of regional governments in Italy to the strength of civic traditions of trust, cooperation, and associational activity (choral societies, sports clubs, political associations, and others) extending back to the early Middle Ages.[27] Economists have found that social capital, particularly trust and civic cooperation, is a significant source of intercountry differences in economic growth rates (Knack and Keefer 1997). Later I present a case study of the construction of social capital and its role in the transformation of a traditional young people's social organization into a development-oriented institution in northern Burkina Faso (chap. 7). Development assistance agencies have embraced social capital as a missing link in development thought. They have sponsored studies designed to explore whether social capital can only be accumulated as an unintended consequence of activity undertaken with other objectives or if assistance resources can be brought to bear to deliberately enhance the accumulation of social capital and to exploit it to achieve development objectives (Narayan and Pritchett 1999; Dasgupta and Serageldin 2000; Narayan 2000; Isham, Ramaswamy, and Kelly, 2003).[28]

But while many social scientists working in the field of development have found the concept of social capital intuitively appealing, there is also a substantial critical literature. Coleman's work has been criticized on methodological grounds, for defining social capital in terms of its functions (Portes 1998, 5). A more serious concern has been the excessive attention given to the positive effects of social capital while neglecting its negative effects—"exclusion of outsiders, excess claims on group members, restrictions on individual freedoms, and downward leveling norms" (15).[29] The logic of aggregation of social capital, from its embodiment in the relationships among individuals to its embodiment as a structural variable of larger aggregates such as communities and nations, has remained unclear (Portes 1998, 21; Sobel 2002, 151–52). Dasgupta has insisted that "the idea of social capital sits

---

27. In a more recent book Putnam (2000) traces the effects of economic and social change on the growth, erosion, and renewal of social capital in the United States.

28. The rapid diffusion of the concept of social capital has been accompanied by a "proliferation of capitals." Thorsby has proposed "cultural capital" as an asset that embodies, stores, and provides cultural value (2001, 46). While social capital depends on the existence of social networks and relationships of trust between citizens, cultural capital also includes tangible assets (paintings, artifacts) as well as the intangible assets commonly identified with culture—ideas, beliefs, practices, values (Thorsby 2001, 44–60). For a closely related concept see the discussion by Robert Fogel (2000, 202–15).

29. For a more polemical assessment and criticism see Fine (2001).

awkwardly in contemporary economic thinking. . . . It is difficult to measure because we do not quite know what should be measured" (2000, 326).

Several suggestions have been made for how the several elements of social capital might be aggregated to form a measure of social capital. Narayan and Pritchett (1999) have, for example, suggested measuring the social capital of a community as a weighted sum of the sizes of its various social networks. Negative weights could be assigned to networks that have malign impacts—networks of drug dealers or terrorists, for example (Fukuyama 1995). In the United States, the National Commission on Philanthropy and Civic Renewal (1998) developed an Index of Civic Engagement based on five dimensions: the giving climate, community engagement, charitable involvement, the spirit of voluntarism, and active citizenship. It seems doubtful that a single measure of social capital will prove feasible (Woolcock and Narayan 2000; Narayan 2000; Inkeles 2000).

The relationship between social capital and economic and political development rests on a rather narrow empirical foundation.[30] It is unlikely, in the near future, that it will be possible to incorporate a measure of social capital, along with physical capital and human capital, into a production function and measure the contribution of social capital to output or productivity growth. My sense is that the best that can be done at the present time is to attempt to understand the relationships between the individual components of social capital and the broader measures of economic development (Temple and Johnson 1998; Aron 2000).

We are even further from an understanding of how to design and introduce incentive-compatible institutions capable of contributing to the growth of social capital. In spite of these qualifications it seems to me that the concept of social capital has at least opened the door to answering the question, What grows in the process of social development? It also raises the question of what forms of social capital are subject to decay in the process of development. It may be even more pertinent to ask whether the exploitation and erosion of some forms of social capital are inherent in the process of constructing the more encompassing forms of social capital that become the base, in the sense used by Gudeman (chap. 2, this vol.), for modern market economies.

---

30. It is only a slight exaggeration to say that almost the whole edifice of proof rests on Robert Putnam's *Making Democracy Work* (Inkeles 2000, 265). See, however, chapter 7 in this volume.

## Markets as Social Capital

It seems clear from Coleman's reference to the role of trust in the operation of the New York wholesale diamond market that he intended to include markets as a form of social capital (1988, S99). The emphasis on the public goods aspect of social capital in much of the social capital literature has, however, deflected attention from the role of markets, the central institution of capitalism, in the growth of social capital. Yet a fundamental issue for the leaders of both the new nations that emerged from colonial rule in the 1950s and 1960s and the nations that began a transition from socialism in the 1980s and the 1990s has been the construction of efficient market systems (Platteau 2000, 241–80).

In the early 1950s Peter Bauer, in his now classic *West African Trade* (1954), castigated the inefficiencies associated with the suppression of commodity and financial markets by the British colonial administration. The leaders of most new states, reacting against colonial practice and drawing inspiration from both Marxist theory and Soviet example, were deeply skeptical about the role of impersonal markets in the allocation of resources in production and the distribution of income among classes and individuals. In the early 1980s Robert Bates found that many postcolonial African governments continued to employ the repressive market practices that they had inherited from colonial administration (1981).

In the early 1990s American advisers recommended to Russian reformers that the economy be thoroughly and rapidly privatized (Boycko, Schleifer, and Vishny 1995). The recommendations, even when adopted, have rarely been effectively implemented. In retrospect it is apparent that the cold turkey market reforms were advocated and carried out with only a limited understanding of the institutions of capitalist development (Williamson 2000). The designers of the reforms were not sensitive to the extent to which modern capitalist institutions are embedded and rest within a system of cultural endowments such as trust, reciprocity, tacit knowledge, and risk-sharing arrangements.

In the early 1980s the New Economic Sociologists began to address what they considered a deficient economic understanding of markets and market development: "There does not exist a neoclassical theory of the market" (White 1981, 83). This meant that, by treating markets as impersonal disembodied exchange, economists working within the neoclassical tradition have failed to understand that markets and market behavior, even in modern economies, are embedded in complex social relations and institutional arrangements: a democracy polity, enforced and exchangeable property rights, and constitutional protection of a private sphere of individual activity (White 1981; White and Eccles 1987; Granovetter 1985; Smelser and Swedeberg 1994a, 1994b; Caldwell 1997; Guillén, Collins, England, and Meyer 2002). The new economic sociology was conceived as a direct challenge to the economic understanding of production and market processes.[31]

The concept of social capital sheds new light on the concerns discussed in the opening section regarding the tension between the institutions of traditional society and the forces of modernization. The social capital—the thick networks of social relations that function to allocate and distribute resources in small-scale societies with weakly developed state and market institutions—may become dysfunctional during the process of development. It may become an obstacle to the development of the forms of social capital needed to sustain more impersonal systems for the administration of justice and effective market behavior.[32]

As powerful as social networks or informal associations may be in sustaining civic and economic functions during the early stages of economic development, they will, as social and economic systems mature, increasingly be replaced by formal institutions of governance and impersonal market mechanisms. Although the shift toward more formal institutions of governance and market organization may weaken or deplete older forms of social capital, this weakening may be necessary for the emergence of new forms and types of social capital that become embedded in governance, civic, and economic institutions (Hirschman 1982; Moore 1994; Stiglitz 2000b; Serageldin and Grootaert 2000; Chen 2001). But the transition from old mental models to more impersonal models may be accompanied by great individual and social stress. Examples include the case of the transformation in property rights described by Gudeman (1986) in his study of Panamanian peasants (chap. 2, this vol.) and by Ensminger (1992) in a study

---

31. This criticism applies most aptly to economists working in the field of neoclassical growth theory (chap. 5, this vol.). The analytical foundations for interpreting the determination of prices and rents in terms of the impersonal forces of factor and product supply and demand emerged in classical growth theory during the four decades between the publication of Adam Smith's *Wealth of Nations* (1776) and David Ricardo's *Principles of Political Economy and Taxation* (1817). See Tribe (1978, 110–46).

32. The voluntary networks, associations, and informal norms are a source of social capital in small societies and of excluded groups in larger societies. However, they often reinforce social stratification, prevent mobility, and are a source of corruption and power on the part of dominant members (Narayan 2000).

of the transformation of property rights and labor relations by Orma pastoralists in Kenya.

There remains great ambiguity, however, as to whether social capital can be purposefully constructed or must arise largely out of endogenous evolutionary processes—a product of "spontaneous order" (Hayek 1978b). In his presidential address to the American Sociological Society, James S. Coleman (1993), correctly in my opinion, insisted that the social structures needed to meet the needs of complex modern societies must be constructed—spontaneous order is no longer adequate! But the New Economic Sociologists have yet to respond to Coleman's challenge to extend their analysis to include institutional design (Portes 1997, 240–46). Nor have the practitioners of the new political economy or the new growth economies (chaps. 4 and 5, this vol.). I return to the issue of institutional design, or renovation, in chapter 7.

Doing Development Sociology

Most mainstream sociologists are not engaged in attempting to advance what is sometimes referred to as metatheory, whether of the Parsons structural-functionalist or the Coleman rational choice variety. Nor do they spend much time considering the nuances of the several postmodern perspectives or the growth of social capital. Sociological research in the subfield of development sociology has often differed in choice of problem from research by applied sociologists working in developed countries. But it has differed little in terms of concept and method. Applied sociologists are sensitive, however, to the neglect of their research by social theorists (Rossi 1980; Berk 1981).

Much of the sociological research in developing countries prior to the early 1980s, particularly by scholars from North America, was conducted within an area studies framework, often within the context of the training of students from developing countries (Prewitt 1982). More recently there has been a growing attention to the area of project analysis and implementation. It has become commonplace that development project performance has failed to meet expectations with unacceptable frequency, often because they were "sociologically ill-informed and ill-conceived" (Cernea 1991b, 12–15). By the mid-1980s implementation failures had largely discredited the integrated rural development and other poverty-oriented program thrusts that had dominated development assistance policy, at least at the rhetorical level, from the early 1970s (Hayami and Ruttan 1985, 403–8).

One result has been an increasing, if somewhat reluctant, sensitivity on the part of national and multilateral development assistance agencies to the importance of human and social capital formation for both project design and implementation. It has not been easy, however, for either the development assistance agencies or sociologists to find ways to effectively bring sociological knowledge to bear within the project analysis, design, and implementation cycles employed by the assistance agencies as they attempt to translate development assistance policy into action.

In part, this difficulty reflects the traditional style of research within the discipline of sociology—much of sociological research on development projects was initially conducted more in the spirit of social criticism than with the objective of contributing to design or implementation (Selznik 1949; Cernea 1991b, 1–41). In part, this is because sociologists have been and continue to be brought into the process only at the end of the project cycle process, at the evaluation stage. More recently, however, as sociologists have begun to colonize national and international development assistance bureaucracies, at least some sociologists have learned how to complement their critical capacities with constructive contributions to project design and implementation. The result has been the emergence of a small body of literature on the practice of sociology within development assistance organizations.

**Perspective**

In this closing section I return to the questions posed at the beginning of this chapter. What should development economists learn from sociology? And what do sociologists have to contribute to development practice?

The major theoretical debate in sociology in the 1950s and early 1960s centered on the dominant role of Parsonian structural-functionalist theory. The logical and substantive difficulties raised by critics led most sociologists to abandon the functional (and evolutionary) interpretations of change in the structure of social relations (Ingham 1996, 252). I have insisted that the demand for social science knowledge is derived from the demand for institutional change (chap. 1). This carries an explicit assumption that a major purpose of institutional reform and design is to achieve more effective institutional performance. This design perspective implies that institutional structures can be purposefully "socially constructed" with the objective of achieving more effective performance. The constructed market for tradable pollution permits discussed in chapter 1 is an example. It is important that development economists be able to draw on sociological knowledge conducted within the functionalist tradition.

From the perspective of development economics, the alternative sociologies discussed in this chapter—dependency and the sociology of knowledge—represent stimulating, and at times exciting, diversions. It is possible that a new and more analytical dependency theory will be able to complement research by political scientists and economists on the unequal power relationships associated with globalization. But these contributions remain in the future. The postmodern research agenda on the sociology of knowledge has even less to offer development economists. The capacity to achieve access and to advance scientific and technical knowledge is fundamental for the development of poor countries. Acceptance by developing country intellectuals and political leaders of the perspective on the role of science and technology advanced by the postmodern Strong Program could become a serious obstacle to the design and adoption of the policies and institutions necessary to achieve such capacity. But I have no trouble in accepting the assertion that more empirically grounded ethnographic research on the social construction of indigenous technology could represent a potentially valuable contribution to the design and management of national technology systems in poor countries.

I find the research agenda pursued by James Coleman, particularly the rational choice approach advanced in his *Foundations of Social Theory* (1990b), an important step in facilitating communication among sociologists and economists. The suggestion by Piore that the relationship between knowledge in the two disciplines be conceptualized in the form of a transformation function represents an important analytical interpretation of the scope for more effective articulation of advances in knowledge in the two fields. The concept of social capital has resulted in the rapid penetration of sociological concepts into both political science and economics—sometimes to the discomfort of sociologists. The interpretation of economic markets as embedded in social relationships, even in advanced developed economies, opens up the possibility of important sociological contributions in the design of market reform in developing and transitional economies.

There can be no question, however, that the empirical research in rural sociology, industrial sociology, labor markets, and related areas in developing countries has been, and continues to be, of great value to development economists and practitioners. It is the value of this often mundane research that has enabled some sociologists and anthropologists to successfully colonize the staffs of international and national development assistance agencies since the early 1980s.

Let me add one qualification to the preceding assessments. The task of sociology (and of social science more generally) has been characterized as involving three interrelated elements: (1) preanalytical—the framing of the central issue to be investigated; (2) theoretical analysis—the construction of theoretical models and the derivation of theoretical predictions; and (3) empirical analysis—the empirical testing of the predictions derived from theory (Jasso 2000). In this book I insist on the importance of an additional element—(4) the design and reform of policies, mechanisms, and institutions.

Why haven't sociologists, as a profession, been more aggressive in bringing social science knowledge to bear on issues of institutional design, policy, and reform? Let me suggest three answers. One is aversion to rational choice theory—a commitment to an "oversocialized conception of man" (Wrong 1961). A closely related reason has been the tendency to conceptualize social action almost entirely in terms of "spontaneous order" induced by differences or changes in the environment. This view has in turn been associated with an exaggerated concern with the unintended consequences of social action. Concern that the consequences of action cannot be anticipated has resulted in reluctance to extend sociological analysis to include the design or reform of social institutions or to engage in social action.

CHAPTER 4

# What Happened to Political Development?

The subject matters of economic development and political development intersect over a broad front.[1] Economic policy is made by incumbent politicians in the context of political institutions. The analysis of the economic effects of alternative policies is the stock-in-trade of the economist. The choice of the alternative policies that are subjected to economic analysis is influenced by the agendas of political parties and interests. The subject matter of political science includes the political decision process by which policies are adopted and implemented. It also includes the social consequences and the public response to policy.

There is a deep fault line that divides scholarship in the two fields. Each field tends to treat the knowledge it draws on from the other as implicit rather than explicit. Political scientists and economists loosely grouped within the public or collective choice school of political economy have advanced our understanding of the process by which economic resources are translated into political resources and political resources are translated into economic resources (Downs 1957; Olson 1965; Krueger 1974; Bates 1983). But it is only recently that the implicit theorizing by economists about political development and of political scientists about economic development is beginning to be replaced by more explicit attempts to develop more integrated approaches to political and economic development.

This chapter represents an attempt to assess what development economists could learn from theory and research in the field of political development to advance knowledge and policy in the field of economic development. I proceed by first reviewing the contributions of several development economists who have attempted to give explicit attention to the political preconditions or conditions for economic

---

1. This chapter is a revised version of Ruttan (1991). I am indebted to Irma Adelman, John R. Freeman, Robert T. Holt, Maureen Kilkenny, Cynthia Taft Morris, and James Walker for comments on earlier drafts of this chapter.

development. I next review the evolution of thought in the field of political development, then address the question of what grows in the process of political development. In a final section I turn to the issue of the design of political institutions.

## Political Development in Development Economics

Economic development theory and analysis have been concerned primarily with the surface patterns and the proximate sources of economic growth. Patterns have been described in terms of the transformation of structure[2] and the succession of stages.[3] Sources of growth have been analyzed primarily in terms of the response of output to investment in physical and human capital (chap. 5, this vol.).

When development economics emerged as a subdiscipline in the 1940s and early 1950s there was a pervasive view among economists that the late industrializing countries required strong, authoritarian state institutions in order to mobilize the resources required for growth.[4] Democracy was a "luxury" that could not be afforded by poor states. This view drew on and was reinforced by the apparent success of centralized planning in Stalin's Russia. By the early 1970s there was increasing skepticism among development economists about the merits of forced draft mobilization and the efficacy of central planning. A view emerged that success in economic development could be more readily achieved, or at least sustained, in an environment characterized by a liberal economic and political order.

In general, these views emerged more out of experience and casual observation than serious scholarship. A few economic historians and development economists did, however, attempt to explain the relationship between political and economic development in somewhat greater depth. In this section I refer to the work of Alexander Gerschenkron,

---

2. One of the great traditions in economic development is the quantitative analysis of the transformation of economic structure. The classical treatment is Colin Clark (1957). Other major works in this tradition are Simon S. Kuznets (1966b, 1971); Hollis Chenery and Moshe Syrquin (1975); Walt W. Rostow (1978); and Angus Maddison (1982).

3. Stage theories occupied a prominent place in the work of the German historical school. Marx employed a system of five stages. During the 1950s and 1960s the stage system proposed by Walt W. Rostow became both a permanent feature of the language of development and the subject of substantial professional debate (Rostow 1956, 1960).

4. Kuznets notes, for example, that "clearly some minimum political stability is necessary if members of the economic society are to plan ahead and be assured of a relatively stable relation between their contribution to economic activity and their rewards. One could hardly expect much economic growth under conditions of turmoil, riots and unpredictable changes in regimes" (1966b, 451).

Karl de Schweinitz, Jagdish Bhagwati, W. W. Rostow, and Irma Adelman and Cynthia Taft Morris.

The theme that late industrializing countries benefit from the evolution of strong state institutions with the capacity to intervene directly in economic activity is a pervasive theme in Alexander Gerschenkron's studies of European economic history (1962, 1968). A major organizing principle in Gerschenkron's work is the continuing tension between change and continuity in history. Industrialization occurs in rapid "spurts" along the lines suggested in the "takeoff" or "big push" views of economic development. The more backward the economy, the more likely that industrialization would occur discontinuously—as a sudden great spurt.

In the case of the early industrializing countries, Gerschenkron argued that it was sufficient for the state to pursue policies aimed at creating a suitable environment, through an appropriate legal framework and the supplying of physical infrastructure, for the growth of industrial enterprise. But in the more backward economies of Russia and of eastern and southern Europe, successful industrialization would require more than simply introducing the institutional framework that suffices for the purposes of industrialization in an advanced country. The state must have the power to pursue "forced draft" industrialization, to extract surpluses from a reluctant peasantry, and to direct capital into industrial development.

Gerschenkron displayed considerable caution in drawing the implications of his analysis for development policy. Other scholars who share Gerschenkron's historical perspective have been less reticent. In *Industrialization and Democracy* Karl de Schweinitz argued that economic growth and democracy are complementary in the advanced Western economies. But in the early stages of economic development, democratic institutions often represent a constraint on economic development. The Euro-American route to democracy is closed to the presently less developed countries. The impulse for industrialization must come from the center of political power and spread outward into society, rather than coming from society itself, as was the case in the West during the nineteenth century. If the underdeveloped countries are to grow economically, they must limit democratic participation in political affairs: "Justice must take a back seat to growth objectives" (de Schweinitz 1964, 277).

Jagdish Bhagwati was even more explicit. In his introductory text, *The Economics of Underdeveloped Countries,* Bhagwati insisted that "socialist countries, such as the Soviet Union and Mainland China, have an immense advantage: their totalitarian structure shields the government from the . . . reactionary judgments of the electorate. The Soviet government's firm control on expansion in consumption over the last few decades could hardly ever be attempted by a democratic government. Another advantage of the socialist countries is their passionate conviction and dedication to the objective of economic growth, which contrasts visibly with the halting and hesitant beliefs and actions of most democracies" (1966, 203)

I cite Gerschenkron, de Schweinitz, and Bhagwati here not to criticize their work from the vantage of the early twenty-first century but to emphasize the pervasiveness of the view that authoritarian regimes, whether capitalist or socialist, were more effective at mobilizing resources for development than democracies. The belief that authoritarian regimes are conducive to economic growth was pervasive not only among students of economic and political development but also among the political elites and enterprise managers in developing countries, as well as among the officers and technocrats in the international financial institutions and assistance agencies (Freeman 1985; Ruttan 1996, 253–332; Kapur, Lewis, and Webb 1997, 208–11). Research by Rostow and by Adelman and Morris advanced a more complex set of relationships between political and economic development.

During the 1950s, W. W. Rostow elaborated and popularized a theory of stages of economic growth (1956, 1960). The stages were denominated (1) traditional society, (2) preconditions for growth, (3) takeoff, (4) drive to maturity, and (5) high mass consumption. The stage characterization drew primarily on the history of Western economies for empirical support. Rostow extended the implications of the stage perspective for development policy in contemporary developing countries. This was followed in the early 1970s by an effort to relate his stage perspective to the process of political development (1971). In this effort Rostow drew on the experience of a broad range of developed and advanced developing countries (England, United States, France, Germany, Soviet Union, Japan, China, Turkey, and Mexico). Rostow's approach was to explore the tension between traditional political culture and the new problems or tasks that the process of economic growth imposes on the polity as a society moves successively through the development stages.

Cycles in economic and political development emerge as the characteristic pattern in the traditional society. Custom and the elders rule in the clan and tribal societies, but conflict and conquest lead to kingdoms and empires. The expansion of resources during the upswing leads to improvement in administration and security. But the growth of bureaucracy and the resources required to maintain security lead to the erosion of political freedom and impose economic burdens on the peasantry that cannot be maintained in the presence of static agricultural technology. During the period when societies are establishing the preconditions for growth, the most dominant characteristic is political and economic insecurity. There is unstable competition among the aristocracy, the new entrepreneurs, the bureaucracy, and the military, with the military often emerging in leadership roles because of their mastery of technology and organization.

During the takeoff stage, growth of agricultural and industrial production begins to generate new income streams. Competition between classes (labor and capital) and estates (military and civil bureaucracies) over control over the new income streams becomes a source of political stress and crisis. But, Rostow argues, by the time the drive to maturity is well under way, constitutional issues of justice and equity and economic policies toward growth and welfare become easier to manage, and a liberal democratic order emerges as the most effective political system to sustain takeoff.

The age of high mass consumption places new stress on political institutions. The demand for improvements in the quality of life replaces the older demand for commodities. This extends to the public sector where demand for the quality of public services confronts the use of public resources to pursue security or other international political objectives. Experience with this stage is so limited, in Rostow's view, that it is difficult to foresee adequately either its political or economic implications.

During the 1970s and 1980s Irma Adelman and Cynthia Taft Morris pursued an exceedingly ambitious research agenda designed to explore the role of "initial institutions" on the pace and structure of economic development (Adelman and Morris 1967, 1973; Morris and Adelman 1988; Morris 1992). A variety of statistical methods were employed to identify configurations of economic and political change and to group closely related variables for different country types and groups. An attempt is made to capture the role of political institutions by variables measuring (1) the domestic economic role of government, (2) the socioeconomic character of political leadership, (3) the strength of national representative institutions, (4) political stability, and (5) foreign economic dependence and colonial status.

The Adelman-Morris results for the 1850–1914 period are consistent with the perspective of the modernization school in sociology and political science (chap. 3, this vol.). "At critical junctures . . . political institutions mattered greatly. With rare exceptions, economic growth and its benefits did not diffuse far where domestic landed elites aligning with foreign export interests dominated the political process. In all countries undergoing substantial industrialization, domestic commercial and industrial classes had or gained significant power in national leadership. In more politically diverse country groups—for example, land-abundant dependent countries—economic growth spread far only when landed elites no longer dominated domestic economic policies" (Morris and Adelman 1988, 211). The emergence of a legal system that strengthened property and market institutions was important for market expansion, industrial development, and agricultural development.

In an earlier book by Adelman and Morris, which focuses on more recent economic history, the impact of socioeconomic and political development was less clear. In their short-run analysis there was no apparent association between socioeconomic and political development. But in their longer-run analysis there was a positive association between socioeconomic development and political institutions. Beyond some critical level of economic and social development, socioeconomic change induces more effective participatory political institutions. The most striking pattern that emerges from their empirical analysis is the progressive differentiation and separation of the social, economic, and political spheres. Furthermore, this differentiation begins to emerge relatively early in the development process.

It is difficult to discover broad agreement among economic historians and development economists who have given explicit attention to political development. It does appear, however, that there would be fairly broad assent to the proposition that authoritarian regimes in which command over economic and political resources was relatively undifferentiated characterized the societies from which the presently developed market economies emerged. Furthermore, in these societies the emergence of capitalism preceded the emergence of democracy.

Since the early 1980s events have conspired to force development economists to expand their interest in institutions from attempts to understand the institutional sources of economic development to issues of institutional design. The disappointing results of attempts to implement the policy reforms loosely described by the term *Washington Consensus,* particularly in new African states and in the former centrally planned economies following the end of the Cold War, have focused attention on issues of institutional design.[5]

In an extensive review of reform efforts beginning in the 1960s, Clague found: (1) In almost every case involving an attempt to introduce major policy reform there was at least a temporary consolidation of authority. (2) A key feature of the reforms that were successfully implemented was emergence of a central policy agency dominated by economic technocrats. (3) The capacity of the system to sustain a market-friendly policy regime involved the consolidation of democratic political reform (1999, 15–16). Chile has often been taken as the model case for such reforms.

The implications that were initially drawn from such experience were to recommend getting on with economic policy reforms by whatever political means were at hand and to hope that the favorable economic performance resulting from their successful implementation would induce the democratic reforms necessary to assure the continued political viability of the economic reforms. Economists who drew their lessons from the implementation of reforms in Latin America were less successful when they attempted to apply the Latin American lessons to the new states in Africa and to the former centrally planned economies. In retrospect it appears that there was a lack of sensitivity to deficiencies in the capacity of market institutions and the institutions of democratic governance—institutions that were presumed in the Washington Consensus (Harrison and Huntington 2000). I return to the issue of institutional design in a later section of this chapter.

**Political Science and Political Development**

The 1960s was a period of intense intellectual ferment in the field of political science. Insights based on advances in the understanding of individual and group behavior, drawing on psychology, sociology, and economics, were incorporated into the theoretical domain of politics. The concept of political system was elaborated and distinguished from the changes in the environment in which political activity takes place. New quantitative methods from statistics and economics were adopted

---

5. The Washington Consensus refers to the following issues: fiscal deficits, public expenditure priorities (less subsidies; more education, health, and physical infrastructure), tax reform (broader base and lower rates), interest rates and exchange rates (close to market-determined levels), relatively free trade, free foreign direct investment, privatization of state enterprises, deregulation, and secure property rights (Clague 1999, 2).

to explore the relationships between the political system and its environment.[6] The emergence of new states turned the attention of political scientists to applying these advances in theory and methods to the problem of political development, often cast in terms of "nation building" (Almond 1966; Huntington and Dominiquez 1973).

One indication of the interest and ambitions of political scientists in political development was the major research effort organized by the Social Science Research Council Committee on Comparative Politics (SSRC/CCP) to "generate a doctrine of political development that would prove as powerful an analytical tool as economic theory had provided in its assault on problems of national poverty" (Montgomery 1969). In the United States, interest in political development was given further impetus in Title IX of the Foreign Assistance Act of 1966. The U.S. Agency for International Development (USAID) was instructed: "In carrying out programs authorized in this chapter emphasis shall be placed on assuring maximum participation in the task of economic development on the part of people in developing countries through the encouragement of democratic, private and local government institutions" (Braibanti 1969, 15).[7]

By the mid-1970s, however, interest by political scientists in issues of political development was rapidly eroding. Scholarship in the field of political development found itself facing a series of methodological, empirical, and ideological challenges. A review by Robert T. Holt and John E. Turner vigorously challenged the *methodological* foundations of the research carried out under the auspices of the SSRC/CCP. They characterized the methodology employed in the research carried out under the auspices of the committee as "persuasive discourse." The process by which the authors move "from the raw material to the theory is never made explicit. Rules of inference are not spelled out. Intuitive processes are apparently considered to be more important" (1975,

---

6. I do not, in this chapter, give specific attention to the implications of political theory for political and economic development. In political science, political theory has traditionally been identified with political philosophy and more recently with the history of political philosophy—"the political theorist often is not a theorist but a historian" (Gunnell 1993, 50).

7. Title IX was introduced by Congressman Donald M. Fraser (D-Minnesota) and cosponsored by Congressman Clement G. Zablocki (D-Wisconsin). Braibanti, in a burst of enthusiasm, characterized Title IX as "the most important element of doctrine in U.S. foreign assistance policy" (1969, 13). It is somewhat surprising that the political development research agenda of the 1960s gave so little attention to the lessons from attempts after World War II to redirect or "force" political development in Japan and Germany along democratic lines (Montgomery 1957).

987).[8] Holt and Turner insisted that progress toward the objectives that the committee set for itself, a collective effort to construct a theory of political development, could only be advanced by use of a more formal analytic deductive approach to theory construction and empirical analysis. While these criticisms were never explicitly acknowledged, it is clear that by the early 1980s there was increasing skepticism within political science about the usefulness of both the concept of political development and the methodological approach employed in the SSRC/CCP project as the focus for a research agenda (Eckstein 1982; Wiarda 1985).

The *empirical* challenge centered around the continued relevance of the Anglo-American model of political development. Prior to the mid-1950s, the literature was dominated by what Robert A. Packingham has termed the "legal-formal approach." Political development was identified with the attributes of English and American liberal constitutional democracy. The level of political development of a country could be measured by its linear distance from the Anglo-American model (Packingham 1973). Dissatisfaction with the Anglo-American model was stated as a challenge by Gabriel A. Almond in his 1966 presidential address to the American Political Science Association: "Enlightenment theory began with the leviathan state and postulated as the legitimate problem of political theory that of bringing the leviathan under control through institutional and legal checks and balances, and through popular processes. Modern political theory has to ask how the leviathan itself comes into existence, in order to cope with the intellectual problems of understanding the political prospects and processes of the new nations (877).

A second generation of scholars attempted to move away from what was regarded as the excessive parochialism of the linear model. Political development was defined in terms of multidimensional categories and analyzed as dependent variables responsive to changes in economic, sociological, and psychological variables.[9] Political scientists working within the framework of the modernization paradigm,

---

8. For a more detailed discussion of the methodological perspective that Holt and Turner brought to bear in their review, see Holt and Turner (1970) and Holt and Richardson (1970).

9. Lucian W. Pye has listed the different approaches to the definition of political development: (1) the political prerequisite of economic development; (2) the politics of industrial societies; (3) political modernization; (4) the operation of a nation-state; (5) administrative and legal development; (6) mass mobilization and participation; (7) the building of democracy; (8) stability and orderly change; and (9) mobilization and power (1966, 33–45).

adapted from sociology, viewed the emergence of market exchange and economic development as an important requisite to political democracy. Changes in social structure and the erosion of traditional social institutions were emphasized. Changes in the political culture, at both the level of beliefs and attitudes and the level of personality, were held to play independent roles in the process of political development, still largely associated with the transition to democracy (Packingham 1973, 195–240).[10]

These views were, in turn, criticized by a third generation of scholars who took their cue from the seminal article by Samuel P. Huntington, "Political Development and Political Decay" (1965). Huntington defined "political development" as the institutionalization of political organizations and procedures. Furthermore, he identified political development with the strength or capacity of government institutions—"whatever strengthens governmental institutions" (393). The level of institutionalization was measured by the adaptability, complexity, autonomy, and coherence of government institutions. He went on to argue that if these criteria can be identified and measured, political systems can be compared in terms of their institutionalization (412).

Harry Eckstein suggested an even more ambitious agenda. He assigned to political development, as a subfield of development theory, the task of developing a theory of political stages that identifies distinct stages that link primal to modern society (Eckstein 1982). The political domain must be distinguished from other dimensions of modernization: "What grows in political development is politics . . . the political domain of society" (470).[11] That domain includes legitimate power, conflict management, and the regulation of social conduct.

Eckstein returned to the history of the English polity to sketch out a prototype set of stages in national political development. These include: (1) primal polity—characterized by symbolic leadership; (2) substantive primacy—involving legal and extractive power; (3) "prophylactic" policy—including maintenance of order and management of dissent; (4) the polity of interests— involving the pursuit and grant-

---

10. The modernization "paradigm" that became the dominant organizing concept of both the scholars associated with the SSRC/CCP project and the wider community of scholars working in the field of comparative politics drew very heavily on the work of the structural-functionalist school in sociology, particularly the work of Talcott Parsons and his associates. The work of Daniel Lerner (1958) was particularly influential. For a review of the sociological literature on modernization see chapter 3, this volume.

11. For alternative approaches to stage theories in political development see Rokken (1974) and Holt and Turner (1966, 39–50).

ing of privilege; (5) the politics of incorporation—the virtual total politicization of social life; and (6) political society—in which the realm of privacy is minimized and political density is maximized, and the role of power is pervasive. Modern democracies, in the historical perspective of the Eckstein stages, "simply are the gentler twins of totalitarian rule, mitigated by open competition, free communications, and a sense of rights and liberties" (1982, 476).

The *ideological* challenge grew out of the profound disillusionment on the part of many younger social scientists, anthropologists and sociologists as well as political scientists, with the impact of Western cultural, economic, political, and military penetration into non-Western societies (chaps. 2 and 3, this vol.). It was reinforced by the political fallout resulting from the intellectual commitment of a number of leading scholars in the field of political development to influence U.S. policy in support of political development in countries such as Vietnam, Chile, Brazil, and Egypt. The consequences of U.S. interventions in Vietnam contributed to the discrediting both of the subdiscipline and of the scholars who had contributed to the development of the field.[12]

The intellectual climate created by these experiences contributed to the rapid diffusion in both sociology and political science of what came to be known as the dependency (or underdevelopment or world systems) theory as an alternative to the modernization paradigm as a lens through which to interpret political, social, and economic change in Third World countries (chap. 3, this vol.). The dependency perspective initially drew on the work of the Latin American structuralist school of economists for its empirical support. It turned to neo-Marxist theories of imperialism, particularly to the work of Paul Baran, for its theoretical foundations. But it was the colorful rhetoric employed by Andre Gunder Frank in his attacks on the modernization school that was

---

12. The perspective that was widely prevalent in the 1960s has been aptly summarized by Huntington and Dominiquez: "Modernization poses . . . three major challenges to political systems. In the first phase, the need exists to break down traditional institutions and practices and to inaugurate modernizing reforms designed to rationalize and secularize the systems of authority, to develop an efficient bureaucracy and military force, to equalize the relations of citizens to government, and to extend the effective reach of the state. Achieving these modernizing reforms usually requires the centralization of power. . . . One can think of political modernization as involving successively the concentration, expansion, and dispersal of power" (1973, 53). For a stronger statement in the context of political advice to the government of Egypt see Eichelberger (1969). For a vigorous defense of political development scholarship against the charge of ethnocentrism and ideological bias, see Almond (1987).

most influential in diffusing the underdevelopment perspective among a new generation of students of comparative politics.[13]

The central theme in Frank's polemic was that world capitalism, and particularly trade between the countries of the center and the periphery, rather than being an engine of growth, was responsible for "underdeveloping" the Third World. The same historical processes responsible for the development and expansion of capitalism at the center were responsible for underdevelopment and for political and economic dependency in the periphery.

By the mid-1980s commitment to the dependency perspective was rapidly eroding, particularly in its centers of origin in Latin America. This was in part due to the criticism of scholars committed to classical Marxism. More important, however, was the widening discrepancy between some of the more extravagant implications of the theory and the record of economic and political development in the 1970s and 1980s. The assertion that external economic linkages resulted in economic and political regression in the periphery was not sustained by the historical record.[14]

Neither the modernization theory nor the dependency reformulation is satisfactory. Huntington's identification of political development with the strengthening of governmental institutions is unduly restrictive. And Eckstein's linear sequence of stages hardly passes the test that he imposes on earlier theories of political development: "What is absent . . . is a theory of the fundamental forces . . . that brought us to our political condition and continues to push us through political turns" (1982, 468). He is not able to escape the fundamental criticism of economic staging, as a convenient rather than an analytical way of slicing historical time. Dependency theory had put itself into a situation from which it was attempting to escape by jettisoning much of the baggage it acquired during the 1970s (chap. 3, this vol.). It had, however, diverted a generation of younger political scientists from the more rig-

---

13. See especially Baran (1952, 1957) and Frank (1966, 1967). For further discussion of the intellectual sources of the dependency perspective see Hayami and Ruttan (1985, 34–39). For a review of the attack by the dependency and related underdevelopment and world systems schools on the modernization paradigm, see Harrison (1988, 62–148).

14. One response to the lack of congruence between the theory and economic history was to argue that dependency theory was intended to connote a general "frame" rather than a precise "data container" and, hence, cannot be subject to the normal empirical tests (Duvall 1978). For attempts to test dependency theory against historical experience see the several articles in Deyo (1987). It is remarkable that in the Deyo index there is not a single reference to the work of Frank.

orous theory construction and testing that Holt and Turner had called for in the mid-1970s.

## The Political Basis of Economic Development

To what extent is economic development conditioned by or even dependent on political development? Must political and economic development processes be highly articulated? I have noted earlier in this chapter the assumption by development economists in the early post–World War II period that authoritarian governments were more effective at mobilizing economic and political resources for development. In contrast there was a strong theme in early post–World War II political development literature to the effect that democratic governance had a positive impact on economic development (Huntington and Dominiquez 1973). Kwame Nkrumah, the first president of Ghana, put the same point more dramatically: "Seek ye first the political kingdom and all things will be added unto it."[15]

What have political scientists and economists have been able to learn about the implications of political development for economic development?[16] I focus on three bodies of literature: historical studies, studies of political culture, and quantitative studies.

### Historical Studies

An early and exceptionally ambitious attempt to explore the political basis for economic development was made in the mid-1960s by Robert T. Holt and John E. Turner. In *The Political Basis of Economic Development* (1966) they accepted the Rostow stage model of economic development as a working hypothesis and attempted to examine the political developments in the preconditions stage that precedes the takeoff stage.

> In order to manipulate in a defined manner the variables that concern us, we have concentrated attention upon France, 1600–1789; China, 1644–1911; Japan, 1603–1868; and England, 1558–1780. Shortly after the final dates listed, Japan and England entered upon a period of rapid industrialization, whereas

---

15. Quoted in Huntington and Dominiquez (1973, 10).
16. I do not address the issue of the relationship between democracy and war. The issue is often cast in terms of whether war is less likely between democracies than between democracies and dictatorships. For a useful review see Mansfield and Snyder (1995).

France and China lagged behind in varying degrees. If there are significant political requirements for economic growth, and if there were no obvious technological factors at work we should expect to find certain political similarities between France and China . . . that distinguish them from Japan and England. (1966, 5)

It is difficult to capture the richness of the Holt-Turner analysis. Yet a major conclusion that emerges from their work is that the premature aggregation or concentration of political resources represents a constraint on economic development. The governments of France and China were much more active in infrastructure development than those of Japan and England. They had greater fiscal capacity, provided a more secure legal framework for economic activity, and were more active in the organization and regulation of economic activity. But the contributions to development that derived from the aggregation of political and economic power became a constraint on local or regional political development and on private economic activity.

## Cultural Endowments

The assumption that cultural endowments play a fundamental role in political development is so pervasive that it often remains implicit rather than explicit in the political development literature. This is in sharp contrast to the treatment of cultural endowments in economics, where cultural considerations have traditionally been cast into the "underworld" of development thought and practice (chap. 2, this vol.). It is still difficult to find a leading scholar in the field of development economics that would be willing to make a commitment in print to the proposition that "in terms of explaining different patterns of political and economic development . . . a central variable is culture—the subjective attitudes, beliefs, and values prevalent among the dominant groups in the society" (Huntington 1987, 22)

In American political thought, it is assumed by almost all schools, from the adherents of the older legal-formalist tradition to the more recent behavioral and public choice schools, that the ideas embodied in the U.S. Constitution were a product of the political culture of the Enlightenment. This is illustrated by the 1976 exchange about constitutional choice between Vincent Ostrom and William Riker (Ostrom 1976a, 1976b; Riker 1976). Ostrom argued that the design of the U.S. Constitution, with a federal structure, separation of powers, and democratic election of the legislature and the executive as central elements, has been responsible for the maintenance of a system in which broad areas of political and economic activity remain outside the direct purview of public authority. Riker argued, in contrast, that it is not that we are free because of constitutional design, but rather that the constitutional design was a reflection of the political culture of a free people. He argues, more broadly, that constitutional design is derivative of political culture rather than being its source. But there is greater agreement between Ostrom and Riker than either concedes. Ostrom did not argue that U.S. constitutional design was cut from whole cloth. The political culture of the Enlightenment was the product of several centuries of thought and practice in the evolution of systems of governance that could function effectively without the unlimited exercise of sovereign authority.[17]

The view that political culture forms a coherent pattern that informs political thought and governs (but does not determine) political behavior in any society represents the central core of the research program pursued by Lucian W. Pye, beginning with his studies of politics and personality in Burma and his active role in the SSRC/CCP.[18] Political development is conceptualized primarily in terms of nation building. In spite of his emphasis on the role of cultural endowments, Pye drew more directly on the work of Talcott Parsons and the structural-functionalist tradition in sociology than on research by anthropologists.[19]

During the 1960s and 1970s, interest in cultural endowments as a source of political and economic development waned (chap. 2, this vol.). As the Cold War wound down, however, ideological concerns eroded, and differences in cultural endowments again attracted students of political development. Pye (1985) argues that the most

---

17. On the Enlightenment, see footnote 25, chapter 2. The impact of Enlightenment political thought was tempered by American political practice. De Tocqueville noted that "the doctrine of the sovereignty of the people came out of the townships and took possession of the state . . . the township was organized before the county, the county before the state, and the state before the nation" (1835, 59).

18. See Pye 1962; Pye and Verba 1965; Pye and Pye 1985. There was an apparent decline of interest in the political culture research agenda for about a decade following the critical review by Holt and Turner of the research program of the Committee on Comparative Politics. In his criticism of the critics, Pye notes that most scholars of comparative politics have not been attracted by "convoluted ways of elucidating the obvious by mathematical formulas" (Pye and Pye 1985, 10). It is hard to avoid the inference that the leaders of the SSRC/CCP school of political development and their critics were unable to join issues because of fundamental disagreement about how to advance knowledge in political science.

19. I find it somewhat surprising, for example, that he does not even refer to the research on the political systems of highland Burma by anthropologist Edmond Leach (1954).

significant aspect of East Asian culture, particularly the Chinese culture that draws on Confucian thought, is the tendency to place more value on the collective and to be less sensitive to the values of individualism than the West. This results in a style of political leadership that is, at least from a Western perspective, highly paternalistic. Ritual plays a particularly strong role in assuring the legitimacy of political power in the ethical, moral, and sociopolitical sense characteristic of the Chinese cultures of East Asia, in the ritualized power of South Asia and the patron-client style of personal power in Southeast Asia. Pye concedes that cultures do converge during the process of modernization. Yet economic growth will be associated with polities in which the structure of political power will reflect attitudes toward power and authority very different than in the West.[20] I return to this issue in chapter 8.

It was, however, Samuel Huntington's insistence that cultural clashes among the world's great civilizations—Western, Islamic, Chinese, Hindu, Orthodox Judaism, Japanese, and African—would become the fault line for post–Cold War global tensions that ignited an intense popular and professional debate about the role of cultural endowments in political development (1993, 1996).[21] Huntington rejects the modernization assumption of a broad secular trend toward a higher level of global civilization. He grants that the process of modernization has been associated with an enhanced level of material well-being throughout the world. But he is unwilling to concede that modernization is leading to the common values that would enable a global civilization. In what amounts to an extended sermon Huntington insists that "a multicultural world is unavoidable because global empire is impossible" (1996, 318). Furthermore: "In the emerging era, clashes of civilizations are the greatest threat to world peace, and an international order based on civilizations is the surest safeguard against world war" (321). He also insists that the emergence of multiculturalism at home is a threat to the political viability of Western civilization.

---

20. In the late 1980s the government of Singapore launched a campaign to strengthen the commitment of the Singapore Chinese to Confucian values and philosophy in an effort to provide an ideological foundation for the authoritarian style of democracy that the Lee government established in Singapore (Huntington 1991b).

21. Huntington locates his work squarely in the interpretive tradition rather than in the empirical-analytical tradition of political science. "This book is not intended to be a work of social science. It is instead meant to be an interpretation of the evolution of global politics after the Cold War" (1996a, 13). It also represents a vigorous refutation of the triumphalist "end of history" theme advanced by Fukuyama (1989, 1996).

The Huntington thesis has been the subject of vigorous popular and professional criticism (Ajami 1993; Fukuyama 1996; Shweder 2000a). In contrast to Huntington's pessimism about the universality of Western values, Francis Fukuyama, a former U.S. State Department policy analyst and planner, insists that the universalization of Western liberal democracy is the triumphant form of human governance. In spite of disagreement about the path of political development, both Huntington and Fukuyama share a perspective on the primacy of cultural endowments in political development. Both avoid serious discussion of the interaction between the political and economic dimensions of development.

## Quantitative Studies

During the late 1950s and 1960s, political scientists made a large number of efforts, using cross-country statistical analysis, to test assumptions about the impact of political institutions on economic growth.[22] Most of these early studies, summarized by Huntington and Dominiquez (1973), involved little more than attempts to establish statistical association between indicators of political and economic development. Efforts were also made to estimate more complex causal models that specify the mechanisms by which different types of political regimes influence economic development. An early example was the attempt by Robert M. Marsh to test the "authoritarian model" of the relationship between political and economic development, using data for 1955–70 (1979). He interprets his analysis as providing "some support" for the authoritarian model: "Political competition/democracy does have a significant effect on later rates of economic development; its influence is to retard the development rate rather than to facilitate it" (244). One is left with the impression that Marsh is somewhat reluctant to accept the results of his own statistical results. He emphasizes that his static model provides less support for the authoritarian model than the tests of his dynamic model.

It is difficult to be certain, from this distance, whether the results of the empirical studies conducted in the late 1950s and 1960s were a source of the shift in interest among political scientists toward prob-

---

22. Empirical cross-country studies became a major growth industry in the 1960s and 1970s in both political science and sociology. See, for example, the collections edited by Holt and Turner (1970) and by Inkeles and Sasaki (1996). In contrast, cross-country empirical studies did not blossom in economics until after the emergence of the "new growth economics" in the mid-1980s (see chap. 5, this vol.).

lems of political order and stability or whether the studies were, themselves, a reflection of the shift. Nevertheless, they tended to be quite consistent with the conclusions reached by economists who addressed the same issues during that period. One inference that can be drawn from these early quantitative studies is that a poor country that fails to establish a reasonable degree of political stability imposes a severe burden on the forces conducive to economic growth. A number of countries that appeared in the 1960s to be characterized by relatively stable authoritarian regimes had, by the late 1980s, experienced cycles of both (1) political stability and instability and (2) rapid and slow economic growth. In authoritarian regimes of both the Right and the Left, the successful transition to more open political systems characterized by multiple centers of power has been exceedingly difficult. Autonomous centers of power, in either the public or the private sector, were viewed as potential threats to political regimes that were attempting to husband the limited political resources available to them at the center.

**The Economic Foundations of Political Development**

A more productive line of empirical inquiry has been the attempt to explore the impact of economic development on political development. In a now classic article published in the late 1950s Lipset presented data that he interpreted as showing that economic development stimulated the growth of democratic institutions (1959). Subsequent studies by political scientists and economists, using more refined statistical techniques and with greater attention to causal mechanisms underlying the links between democracy and development, have confirmed the Lipset thesis—there is a stable political association between economic development and democracy (Rueschemeyer, Stephens, and Stephens 1992, 13–39; Barro 1997, 49–87).

What is the mechanism that accounts for the association between economic development and democracy? Two responses, one structural and the other cultural, have been advanced. Drawing on a series of cross-country historical studies, Rueschemeyer, Stephens, and Stephens (1992) argue that a change in the balance of class power is the mechanism by which economic development induces political development. Industrial development weakens the political power of traditional elite classes, particularly the landed class, and strengthens the power of the working class. It brings members of the working class together in factories and cities and improves the means of communication and transportation that facilitate worker organization. But they also argue that the organization of the working classes, while a necessary condition for the emergence of democracy, has never been sufficient. When important elements of the middle classes—tradespeople and professionals—conceive of it as in their interest to join forces with the working class, the prospect for a successful transition to democracy is enhanced.[23] "Capitalist industrialization promotes democracy because it weakens the agricultural elite, strengthens the bourgeoisie, strengthens the working class and bolsters civil society" (Schneider 1995).

In contrast, Ronald Inglehart, drawing on a series of cross-country quantitative studies, offers a cultural interpretation (1977, 1990, 1997).[24] "With rising levels of economic development cultural patterns tend to emerge that are increasingly supportive of democracy" (1997, 330). He traces the effect of economic growth on democracy to a broad pattern of cultural changes: "Economic development leads to cultural changes that make mass publics more likely to want democracy and more skilled in getting it" (330).[25]

Inglehart characterizes these cultural changes as a shift, at the individual level, toward postmaterialist values that become institutionalized in the form of political priorities—women's rights, gay and lesbian emancipation, more active political participation of ethnic minorities, and support for nuclear disarmament and environmental protection (1997, 240–56).[26] At the societal level Inglehart characterizes these changes as reflecting a diminishing marginal utility of economic growth. In the high-income postindustrial societies political conflict over life-style and worldviews have replaced the older conflicts over the ownership of the means of production and the flow of income streams

---

23. Rueschemeyer, Stephens, and Stephens provide a more nuanced class-based analysis than Barrington Moore (1966). Moore places major emphasis on the "bourgeois revolution" but largely ignores the role of the working class and the independent peasantry.

24. The data source for the Inglehart studies is the World Values Surveys of attitudes, values, and beliefs. The study was carried out in three waves of representative national surveys: in 1981–82, 1990–92, and 1995–98. For further information about the surveys see the World Values Survey web site (<http://wvs.isr.umich.edu>). The factor analysis method employed by Inglehart is similar to that used earlier by Adelman and Morris (1967, 1973). Inglehart does not refer to the Adelman and Morris work.

25. For other recent research that has supported a positive, and relatively strong, impact of economic growth on democracy see Pourgerami (1988); Helliwell (1992); Burkhart and Lewis-Beck (1994); and Barro (1997).

26. In his later work Inglehart (1997) refers to the cultural changes in the wealthy societies as postmodernist rather than postmaterialist. Since his use of the term *postmodernist* is not congruent with the way the term is employed in literary criticism and in anthropology (box 2.1) I continue to use the term *postmaterialist* in referring to Inglehart's work.

associated with economic growth. Inglehart concedes, however, that in the poor and middle-income countries of the world materialist values and economic tensions associated with class structure may retain their potency (256–65).

It is now hard to avoid concluding that there are few penalties, at least in terms of economic growth, associated with a transition toward democracy. Democracy can be pursued for its own sake—without sacrificing economic growth! Barro (1997, 49–87) cautions, however, that countries that are more democratic than predicted by his cross-country quantitative model have a tendency to regress toward non-democratic regimes. Przeworski counters, however, that waiting for the long-term structural or cultural changes that lead to democracy to work themselves out can hardly be recommended when a dictator is killing people (1997).

A clear inference that may be drawn from the studies reviewed here is how little we still understand about the recursive relationships between democracy and development. Our understanding of the role of those cultural endowments that are influenced by and in turn influence both development and democracy, however important empirically, is even less adequate. My own sense is that the structural interpretation—change in the structure of power—is most relevant for countries in the early stages of economic development. In contrast, cultural interpretations that emphasize changes in patterns of consumption and values associated with income growth are more important during the latter stages of development.

## Political Power and Political Development

It is hard to escape the conclusion that the scholars who have been engaged in advancing knowledge in the field of political development have been reluctant to confront the central question of political development. Unless political development has little meaning other than political change, it is necessary to answer, What is it that grows in the process of political development? In the case of economic development, the answer is fairly straightforward. What grows is socially productive economic capacity measured in terms of its physical, institutional, and human resources. If instead of the development of a society's economy we are concerned with the development of its polity, we must also attempt to identify what grows. I will argue that the most obvious candidate for what it is that grows in political development is power!

## Power Defined

But what is power? Power occupied a central role in the literature of political science in the 1950s.[27] It was viewed as an instrument or a resource to be used in advancing other objectives or values. It was also viewed as the central phenomenon to be explained by political science. But the concept of power has also been the subject of considerable professional controversy. Some scholars have suggested that the concept is so ambiguous that it should be abandoned. In contrast Lucian Pye has argued that power should serve as the central concept in a revitalization of scholarship in the field of political developments (Pye and Pye 1985).

The traditional definitions of power in the political science literature share a view of power as an instrument that enables agents to alter the behavior of other agents. "A has power over B to the extent that he can get B to do something that B would not otherwise do" (Dahl 1957, 202–3). This definition has been extended to include not only overt constraint but also indirect or latent constraints.[28] To paraphrase: A has power over B if A can affect the incentives facing B in such a way that it is rational for B to do something he would not otherwise have chosen to do. This has served as the conceptual foundation for efforts to develop empirical measures of power (Simon 1953).[29] It is this concept of power, power as a private good, that has informed the studies

---

27. Lasswell and Kaplan (1950, xiv). Lasswell and Kaplan distinguish between (1) political science—the study of the shaping and sharing of power; (2) political doctrine—the justification of existing or proposed political structure; (3) political philosophy—the logical analysis of political science and doctrine; and (4) political theory—including political science, doctrine, and philosophy (xi).

28. The suppression of conflict as an aspect of power is emphasized by Bachrach and Baratz (1962). The exercise of power to influence the "objective interest" rather than the perceived interest of an agent has been emphasized by Lukes (1974, 22–23). These several extensions have been reviewed and criticized by Isaac (1987a, 1987b).

29. For discussion of the role of power in economic analysis, see the articles in the book of readings by Rothschild (1971). See also Harsanyi (1962) and Zusman (1976). I have not found the literature on the role of power in economics particularly helpful. Mainstream economics has generally taken the power system as either appropriate or institutionally sanctioned. In the neoclassical model, homogeneous unit firms and households try to improve their lot within the constraints of existing technological and market conditions. Nor have attempts to recast the analysis of individual and group behavior in a broader rational choice perspective given explicit attention to the role of power. The dominant framework for policy analysis treats policymakers as intervening to maximize a "social welfare function." Attempts to build endogenous policy models have generally assumed that the costs and benefits to the participants' efforts to change the distribution of resources and the partitioning of income streams could be evaluated in terms of economic outcomes.

of the impact of political development on economic development and the impact of economic development on political development in earlier sections of this chapter.

This instrumentalist (or manipulative) approach to the definition of power has been criticized as excessively empirical.[30] Jeffrey C. Isaac argues, on methodological grounds, for a "thicker" structuralist or realist approach to the nature of power. Theories of power should depend not on the unique characteristics of individuals but on "their social identities as participants in enduring, socially structured relationships" (1987a, 24). Isaac argues that social or political power should be defined in terms of the capacities to act—the "power to" possessed by social agents (23).[31] He did not, however, provide us with a contemporary road map to guide investigation along the research agenda he has proposed.

In searching for an approach to understanding the role of power in political development, I find it useful to draw on an important, though generally neglected, paper by Talcott Parsons (1963; reprinted in Parsons 1969). The significance of Parsons's paper for the analysis of political development is that he regards power as a system resource that is capable of expansion or growth. He directly challenges the "limited good" or zero-sum definition of power. In Parsons's view, the political system or the polity of a society is composed of the ways in which the relevant components of the total system are organized to achieve effective action, that is, the "power to" achieve collective goals. Political development can be viewed as a positive-sum rather than a zero-sum game (Buchanan and Tullock 1962, 23–24).

Parsons also raises the question of whether the hierarchical organization of political systems necessarily implies that political resources must, by their very nature, be distributed more unequally than economic resources. He does not believe so. He suggests two constraints on the hierarchical ordering of power. One is the franchise. In "the leadership systems of the most 'advanced' national societies, the power

---

30. "The behaviorist foundations constrained [the instrumentalists] from conceiving power as anything more than a behavioral regularity and prevented them from seeing it as an enduring capacity" (Isaac 1987a, 19). "In the realist view, social science would be . . . concerned with the construction of models of the social world and its lawful structure. The primary object of theoretical analysis would not be behavioral regularities but the enduring social relations that structure them" (1987a, 18; 1987b).

31. Ball insists that power is not traceable to some (set of) antecedent events, but is better described in terms of capacity. Thus we can speak of "the President's power to veto legislation and the Senate's power to pass legislation over a Presidential veto. This does not readily translate into the President's power over the Senate or its power over him" (1975, 210).

element has been systematically equalized through the device of the franchise" (Parsons 1969, 118). Hannah Arendt, writing from an intense normative perspective, makes the same point more elegantly: "Under conditions of representative government, the people are supposed to rule those who govern them" (1986, 62). In a democratic system, the franchise is the one resource that is distributed equally (Wrong 1979, 197). And even in authoritarian systems, as Pye has noted in his study of Asian political systems, it is a great illusion of politics to assume "that power flows downward from the ruler through the elite to the masses, whereas in actual fact the process is precisely the reverse" (Pye and Pye 1985, 130).[32]

The second constraint stems from the interpenetration of economy and polity. The interpenetration of economy and polity plays a critical role in the expansion or growth of political resources or power. The structural requirements for the organization of a productive economy place limits on the ability of the political system to obtain control over commodities and services. The productivity of the economy is in turn dependent on an economic organization that is capable of mobilizing the productive effort or the competence of its constituents. Parsons argues that this requires equality of opportunity—the equalization of opportunities for citizens to participate meaningfully and effectively in the shaping of the polity of which they are a part (1969, 122–25).

We are now ready to return to the question that was posed at the beginning of this section. What is it that grows in the process of political development? It seems clear that my initial intuition has a solid basis in the literature of political science if not in the literature of political development. *What grows in political development is power!* Furthermore, its growth must be measured in terms of both its concentration and its distribution.

If this is correct, the distribution of political power must be given much more attention than is reflected in the political development literature.[33] By conceptualizing power in terms of both its amount and its

---

32. Pye interprets the history of Japanese modernization as a particularly strong illustration of the growth of power flowing upward from its base. That power in Japan is built upward from the motivations of subordinates and local networks of relationships "explains in no small measure why historically the Japanese have gone through dramatic changes in the form of their political and economic system without experiencing similar changes in their social dynamics" (Pye and Pye 1985, 177).

33. Montgomery argues that political scientists traditionally were more interested in the distribution than in the growth of political power. In commenting on the literature on political development he notes that "it is surprising for political scientists, whose traditional concern in the liberal West has been the distribution or counter-balancing of power, to concentrate on the accumulations of power" (1969, 290).

distribution, it is possible to make two important theoretical propositions about its growth: (1) power that is closely held, or highly concentrated, faces severe constraints on its growth; and (2) power that is loosely held, that is, equally distributed, also faces severe constraints on its growth. In both cases the growth of power, primarily along a single dimension, runs into diminishing returns. If one accepts these two propositions, then it is possible to maintain that political development has advanced (1) if the amount of power available to a society grows with no worsening of the distribution of power or (2) if the distribution of power has become more equal, with no decline in the amount of power available to a society.

Huntington was surely correct in insisting that the concept of political development should be reversible—that it should be broad enough to cover the possibility of political decay (1965). The preceding definition meets this test. If changes in political structure are not accompanied by sufficiently rapid institutionalization of political processes, greater equity in distribution of political power may be accompanied by loss of power at the center and reduction of aggregate political power. Conversely, the aggregation of political power at the center may be acquired at the cost of a less equal distribution of political power. In both cases the result may be political decay.

A further advantage of this definition is that it can incorporate the effects of geographic expansion (or contraction) of states and empires as well as changes over time. If geographic expansion, in addition to aggregating power, improves the distribution of power by creating new opportunities for the population incorporated into the larger unit, geographic expansion can be said to contribute to political development. The creation of the European Economic Community may be an example. In some cases, political development may be associated with a reduction in the geographic extent of a nation or an empire. It can be argued, for example, that the breakup of the French colonial empire contributed to the political development of France. But did it contribute to political development or decay in Algeria?

In the first decades following World War II it was generally believed that corporatist mixed economies were capable of superior economic and political performance (Freeman 1989). The recent experience of Eastern Europe and the successor states of the former Soviet Union suggests that when a society experiences rapid change in the distribution of political and economic resources it is also likely to experience substantial erosion of both political and economic resources. Francis Fukuyama, in an article that achieved considerable notoriety, characterized the political changes under way in Eastern Europe and in the

USSR in the late 1980s as "an unabashed victory of economic and political liberalism" (Fukuyama 1989). In my judgment it is premature to predict convergence toward any single trajectory of political development. Nations will continue to develop culturally distinct paths of political (and economic) development.

On Measurement

A continuing puzzle in the political development literature is why so little effort has been devoted to attempts to model and measure the political development process. The importance of measurement has been widely emphasized. Some of the conceptual problems of measuring power, conceived as a reciprocal but asymmetrical relationship between two parties, were outlined by Herbert A. Simon in the early 1950s (1953). The measurement of community power was the subject of considerable disagreement, centering on the work of Dahl and of Mills, in the late 1950s and early 1960s (Polsby 1960). Huntington stressed the importance of developing a definition of political development that would lend itself to quantification (1965, 412). Attempts have been made to measure particular dimensions of political development along one or more of the dimensions of development of the Parsonian structural-functionalist model (chap. 3, this vol.). One of the more successful attempts to quantify differences in the growth of the welfare state among the Western democracies has been the sectoral approach employed by Peter Flora and his associates (Flora and Alber 1981, 37–80). But the only broad-scale frontal assault on the problem of measurement has been the determinedly empirical efforts by Morris and Adelman (1988).

How important is the issue of measurement? Can power be treated in political science in a way that is analogous to the way utility is treated in economics? The inputs into the power function (analogous to a production function) for a particular society at a particular time are largely institutional.[34] Whether a nation's political culture and tastes are biased toward power concentration or power distribution will depend on the traditions and strength of a nation's civil and military bureaucracies and of its judicial system, as well as the degree of centralization or decentralization of its governance structure. But the

---

34. This point has also been made with respect to individual power by Harsanyi: "Power in a schedule sense can be regarded as a production function describing how a given individual can 'transform' different amounts of his resources into social power of various dimensions" (1962, 73).

problem of measurement of the growth and distribution of power, or of reasonable proxies, should not face unreasonable difficulties. It should not, for example, be too difficult to design measures of the degree of decentralization of power conceived in terms of "power to" among the several levels of government. Nor should it be too difficult to measure the extent to which power is concentrated within a small political elite. And it should be possible to estimate elasticities of substitution between the concentration and distribution of power in different societies.

**Institutional Design**

The experience of the last half-century confirms that adoption of the formal structure of democratic governance is not sufficient to assure political development (Montgomery 1990). It is necessary to acquire a political culture and to establish the institutions that assure the viability of the political and economic order. This is a long-term process. It is measured in terms of generations. Today's institutional innovations become the source of the political culture of the next generation (fig. 1.1, chap. 1).

But where does one turn for the knowledge and the experience needed to guide the design of the institutions that will enhance capacity for political development? My own sense is that modest beginnings have been made by students who have been employing the tools of what is variously known as the "political economy" or "new institutional economics" in attempts to understand the "collective action" problem—most notably Elinor Ostrom and her colleagues at the Workshop in Political Theory and Policy Analysis at Indiana University (Ostrom 1990, 1998, 2000; Ostrom, Gardner, and Walker 1994; Ostrom and Walker 1997; Ostrom et al. 2002).[35]

The single most powerful motivation for this new research agenda was the 1957 publication of Anthony Downs's *An Economic Theory of Democracy* (based on his Stanford Ph.D. thesis). He constructed a "comprehensive theory of democratic decision making that assumes rational self-serving behavior on the part of the range of political actors, including voters as well as party leaders" (Miller 1997, 1175). Downs employed an economic model of political behavior to interpret

---

35. My optimism is tempered, however, by the limited consideration of institutional change in the political economy literature. For example, the highly regarded text by Torsten Persson and Guidig Tabellini (2001) that attempts to present an integrated approach to political economy takes institutions as given.

a large number of political phenomena—such as political party policy convergence, voter turnout, and public ignorance—that had long puzzled political scientists.

A second major contribution was a seminal book by Mancur Olson, *The Logic of Collective Action* (1965). Olson asserted that no self-interested person would contribute to the production of a public good: "Unless the number of individuals in a group is quite small, or unless there is coercion or some other special device to make individuals act in their common interest, rational self-interested individuals will not act to achieve their common or group interests" (2). This "zero contribution" thesis lent itself to investigations of issues such as the problem of the exploitation of open access or common pool resources using the then popular Prisoner's Dilemma game.

The rapid penetration of the political economy perspective into the traditional territory of political science by Downs, Olson, and other "new institutional economists" was initially welcomed (or at least was not actively opposed) by some political scientists who found the new analytical tools useful (Almond 1993). This initial reception was followed, however, by concerns that the conclusions drawn from the zero cooperation thesis were contradicted by even casual observations of individual and group behavior. "Many people do vote, do not cheat on their taxes, and contribute to voluntary associations.... Individuals in all walks of life and in all parts of the world voluntarily organize themselves to gain the benefits of trade, to provide mutual protection against risk, and to create and enforce rules that protect natural resources" (Ostrom 2000, 137–38).[36]

The initial implications drawn from these findings were profoundly conservative. If what Friedrich Hayek (1978b) termed "spontaneous order," known in more recent literature as "Coasian bargains" (Olson 2000, 45–67) or self-organization (Ostrom 1990), arises without conscious human design and can be maintained without formal enforcement machinery, the scope for rational design of institutions is extremely limited (Sugden 1989). The contribution of Ostrom and her colleagues is that they have brought together the results of a massive body of field observations, extensive laboratory evidence, and careful theoretical analysis to distill a set of principles that are directly relevant to institutional design.

---

36. Ostrom did not attempt to provide an alternative theory to interpret the prevalence of group behavior. Social psychologists have argued that there is an evolutionary basis for the human proclivity for group behavior (Gupta 2002). In contrast, research by experimental sociologists is consistent with an institutional interpretation of differences among societies in group behavior (Yamagishi, Cook, and Watabe 1998).

**Box 4.1. Institutional Design Principles**

Elinor Ostrom and her colleagues at the Workshop in Political Theory and Policy Analysis at Indiana University have articulated eight design principles from their research on self-organized resource regimes.

The *first* design principle [is that] the presence of clear boundary rules . . . enables participants to know who is in and who is out of a defined set of relationships and thus with whom to cooperate.

The *second* design principle is that the local rules-in-use restrict the amount, timing and technology of harvesting the resource; allocate the benefits proportional to required inputs; and are crafted to take local conditions into account.

The *third* design principle is that most of the individuals affected by a resource regime can participate in making and modifying the rules. Resource regimes that use this principle are both able to tailor better rules to local circumstances and to devise rules that are considered fair by participants.

The *fourth* design principle is that . . . resource regimes select their own monitors, who are accountable to the users or are users themselves and who keep an eye on resource conditions as well as on their use.

The *fifth* design principle [is] that the resource regimes use *graduated sanctions* that depend on the seriousness and context of the offense. By creating official positions for local monitors, a resource regime does not rely only on willing punishers to impose personal costs on those who break a rule.

The *sixth* design principle [is] the importance of access to rapid, low cost, local arenas to resolve conflict among users or between users and officials. . . . By devising simple, local mechanisms to get conflicts aired immediately . . . the number of conflicts that reduce trust can be reduced.

The *seventh* design principle [is that] the capability of local users to deliver an ever-more effective regime over time is affected by whether they have minimal recognition of the right to organize by a national or local government unit.

An *eighth* design principle [that] characterizes successful systems . . . when common pool resources are somewhat larger [is] the presence of governance activities organized in multiple

> layers of nested enterprises. . . . Among long-enduring self-organized regimes, smaller scale organizations tend to be nested in ever-larger organizations.
>
> This box is based on the institutional design principles articulated in Elinor Ostrom, "Collective Action and the Evolution of Social Norms," *Journal of Economic Perspectives* 14 (2000): 137–58; quotation from 149–53. The design principles were first articulated in Elinor Ostrom, *Governing the Commons: The Evolution of Institutions for Collective Action* (New York: Cambridge University Press, 1990). Additional testing of the design principles since 1990 has resulted in only modest revisions.

Changes or differences in contextual variables can be expected to influence the success of efforts to implement the several design principles. These could include differences in resource and cultural endowments, differences in the growth, level, and distribution of income, and technical and institutional change (fig. 1.1, chap. 1). Ostrom and others have noted the effects of factors such as the failure of national governments and development assistance agencies to take into account indigenous knowledge and institutions.[37]

Self-Organization

One implication of Ostrom's research is that there is substantial scope for institutional innovation arising out of the spontaneous actions of individual agents. I have described the spontaneous institutional innovations in land and labor relations in a Philippine village (chap. 1, this vol.). Another example, one of the more dramatic spontaneous institutional innovations of the late twentieth century, was the "household responsibility system" of reforms in agricultural production relationships in China.[38]

---

37. Among other studies that have given careful attention to the concerns enumerated in the Ostrom design principles, see Hunt (1989) and Ensminger (1997). Hunt finds that many efforts by the administrators of national irrigation systems in Asia to introduce water user associations have been unsuccessful because of failure to incorporate the full range of design principles. Similarly Ensminger finds that in Africa efforts to introduce formal property rights (land titling and registration) have consistently failed to take into account customary distributional norms in the allocation of land tenure rights.

38. In the next several paragraphs I draw heavily on the work of Justin Yifu Lin (1994) and Shengen Fan (1991). See also the highly personal account by Kate Xiao Zhou (1996).

Following an initial "land-to-the-tiller" reform in the late 1940s and early 1950s Chinese agriculture was collectivized beginning in 1953 under the First Five-Year Plan. Initially, collectivization was surprisingly successful in enhancing agricultural production. But during the 1960s and 1970s production barely kept pace with population growth. By the late 1970s the government recognized that it would be necessary to provide greater incentives to agricultural producers. But it had no intention of reforming the structure of agricultural production, based on collectivized production teams that had been put in place in the 1950s.

Beginning in the late 1970s a small number of production teams, first secretly and later with the support of local authorities, began to subdivide collectively managed land and assign production responsibility to individual households. The arrangements were first worked out by production team members in the province of Szechwan. The authorities reacted by trying to restrict the diffusion of the household responsibility system to areas with poor agricultural resources. The restrictions were increasingly ignored by local authorities, and the system diffused rapidly to other provinces. By late 1981, when it was given belated approval by the central government as an acceptable institution for organizing agricultural production, 45 percent of the production teams in China had adopted the household responsibility system. The impact on production of loosening the bounds of the communal structure was dramatic. Agricultural production in China rose by more than 7 percent per year between the late 1970s and the mid-1980s.

In retrospect there had been a growing disequilibrium between the potential level of productivity and the levels actually being achieved by the production teams between the mid-1950s and the late 1970s. During this period technical advances in crop breeding and in agricultural production practices had raised potential yields, but these higher yields had not been realized by the production teams. As the gap between the yield frontier and the yields realized by the production teams widened, the gains that could be realized from the reform of the communal production system widened. The gap finally became sufficiently wide to induce spontaneous institutional reform on the part of the Chinese peasantry.

The Chinese government has failed, however, to complement the spontaneous institutional reforms initiated by the production teams by designing the more macrolevel institutional reforms needed to sustain rapid growth in agricultural production. Land, labor, and commodity markets remain repressed. In a complex modernizing economy the gains that can be realized from spontaneous order Coasian bargains or self-organization are not enough![39]

Formal Institutions

Beginning in 1970 Italy introduced a new system of autonomous regional governments. Responsibility for former centrally administered programs in fields such as agriculture, environment, health, education, and urban and regional development were transferred from the central to the new regional governments (Putnam 1993). The performance of the regional governments, measured in terms of the effectiveness of service delivery and citizen satisfaction, was strongly influenced by differences in civic culture between the north and the south. In the north, social and political networks are organized horizontally. Communities are characterized by dense networks of civic organizations—ranging from choral societies to soccer clubs. In the south, public life is organized hierarchically. Political participation is based on personal dependency and opportunities for private gain. Engagement in civic social and cultural activity is limited.[40]

The new regional governments incorporated to a substantial degree the design principles articulated by Ostrom. The first two decades of the regional experiment witnessed a dramatic change in political culture in all regions. There was "a trend away from ideological conflict toward collaboration, from extremism toward moderation, from dogmatism toward tolerance, from interest articulation toward interest aggregation, from concern with radical social reform toward good government" (Putnam 1993, 36). But the changes were much more dramatic in the north than in the south. The pace of both political and economic development was much more rapid in the north than in the south.

Putnam draws two conclusions from his analysis. One is that in societies where the norms and networks of civic engagement are lacking the prospects for democratic collective action is severely constrained. Cultural endowments influence the rate and direction of institutional

---

39. This point has been emphasized by Ostrom: "Making the choice of operational-level rules endogenous does not mean that the choice of collective-choice or constitutional-choice rules become endogenous at the same time" (1990, 52).

40. Putnam traces the differences in civic culture between the north and the south of Italy to the early Middle Ages. He traces the origins of civic republicanism in northern Italy to the rise of self-governing city-states following the decay of Byzantine control in the twelfth century. At the same time a highly centralized state built on Norman feudalism and Byzantine bureaucracy was emerging in southern Italy.

development. The second is more optimistic—the formal institutions of governance can be designed to change political practice and enhance political development (1993, 181–85). State intervention may be necessary to induce the changes in cultural endowments and institutions necessary for development (Platteau 1994b, 803).[41]

A major constraint on the implementation of several of the Ostrom design principles has been the reluctance of central governments in many developing countries to allow competing sources of political power to emerge at the local or regional level. The emergence of multiple sources of power is often viewed by fragile national governments as threatening to the nation-building agenda—as inconsistent with the traditional focus of political development on "whatever strengthens government institutions" (Huntington 1965, 393). The principles are, however, consistent with the definition of political development advanced in the previous section: what grows in the process of political development is power measured in terms of both its concentration and distribution. They are also consistent with the growing body of literature on social capital and the role of civil society in political and economic development (chap. 3, this vol.). Ostrom has shown that it is possible to design institutions at the local level that generate growth in social capital. But what are the design principles that can be drawn on, for the institutions necessary to aggregate social capital to improve institutional performance at the regional or national level? In spite of the powerful insight that can be derived from Ostrom and Putnam, their work provides few guidelines for macrolevel institutional design. And in the early twenty-first century it no longer seems adequate to refer such inquiry to the authors of the U.S. Constitution and the Federalist Papers!

**Perspective**

What are the implications of this review of the literature on political development for development economists who would like to draw on the political development literature in their own work? The last two decades have seen exceedingly fruitful collaboration between political scientists and economists in the modeling of economic and political activity in the subfield of public choice. Familiarity with the rational choice paradigm has provided the two disciplines with an increasingly

---

41. For other studies that give careful attention to the design principles (box 4.1) see Hunt (1989) and Ensminger (1997).

common analytical language shared even among those who disagree about the value of the paradigm. As yet, however, fewer linguistic or analytical bridges have been built between the subfields of political and economic development.

It is also useful to ask, What assistance can be drawn by practitioners in the development assistance agencies for assistance for political development? In the case of assistance for economic development, both sector development and policy reform efforts have been able to draw on a powerful body of economic thought—neoclassical economic theory—that provides a useful set of analytical tools with which to address the issues of development practice and the design of policies for economic reform. The application of these tools, even when used with skill and sensitivity to context, has not represented a guarantee against failure, particularly in countries in which development has been delayed by deficiencies in the political institutions of democratic governance and the economic institutions of capitalism (Ruttan 1996, 472–91; Clague 1999). There is, as yet, in the political development literature no body of theory on which to draw as a guide for programs to advance political development. There are only series of empirical generalizations and the beginning of some design principles.

From the experience of the presently developed countries, one empirical generalization that appears relatively secure is the apparent association between authoritarian political organization and rapid economic growth at the beginning of the development process. Reasonably firm evidence to support this view is found in both the economic development and political development literature. It also seems apparent, although the empirical basis for the generalization is less secure, that highly centralized political systems become an increasingly serious obstacle to economic growth as countries evolve toward middle-income status. The economic and political crises experienced by Germany, Italy, and Japan in the interwar period and by the former USSR and the eastern European centrally planned economies since the late 1980s are consistent with this generalization.

A second empirical generalization that seems reasonably secure is that democracy rests more securely in high-income rather than low-income countries. This generalization does not, however, support an argument that democratization must be delayed in the interest of economic growth. Democracy appears to impose no necessary burden on economic development. The evidence from the last twenty years does, however, suggest that a rapid transition from extreme authoritarianism can hardly be achieved without going through a period of economic regression. This is because of the internal disruption that authoritarian political systems undergo as they attempt to make the transitions to a polity in which political resources are more equitably distributed. Spain may be cited as a country in which this transition occurred while maintaining both political stability and rapid economic growth.[42]

We also have an exceptionally well-validated set of principles for the design of local resource management institutions. In my judgment these principles also apply to the design of other local institutions such as agricultural extension services, rural infrastructure, schools, and cooperatives. What is currently lacking is the understanding of the macro-micro-micro-macro linkages necessary to articulate these microlevel principles with the design of the national level institutions of governance.

One inference that might be drawn from this observation is that economists should continue their search for the success of economic development efforts, measured in terms of the growth and distribution of income, without too much help from the field of political development. Similarly, political scientists should continue their search for the sources of political development measured in terms of the growth and distribution of power without expecting too much help from the discipline of economics.

An even more pessimistic inference might be that both disciplines could abandon their attempts to achieve an integrated understanding of the processes of political and economic development and refocus their attention, using the tools of rational and public choice, on the analysis and design of public policy. This may be all that can realistically be asked for. The daily lives of people in both the developing and developed countries, whether characterized by authoritarian or liberal polities or by market or centrally planned economies, are conditioned by the policies pursued by their governments. Whether a society is pursuing an import substitution or an export-oriented trade policy has immediate implications for economic growth and the distribution of the dividends from economic growth. Very substantial gains in economic performance can be realized by policy reforms that erode the repression of market and civic activities, thus leaving greater scope for spontaneous order, even in the absence of major institutional innovations.

---

42. The last decade has seen a burgeoning literature on the transition from communism. For a sampling, see Kornai (1990); Clague and Rausser (1992); Murrell and Wang (1993).

However, I am reluctant to come to either of these conclusions. While the cross-elasticity between the growth of economic resources and political resources may be lower than Parsons suggested, it is clearly very high. Politically active agents are continuously engaged in translating economic resources into political resources. Agents are also involved in the employment of political resources to achieve access to economic resources. I am forced to conclude that scholars working in the fields of economic and political development must develop research agendas that will facilitate greater collaboration if they are to succeed at understanding the growth of either the economic or the political resources available to a society.

CHAPTER 5

# Growth Economics and Development Economics

Both growth economics and development economics emerged as distinct fields of inquiry in the early post–World War II period. Modern growth economics emerged out of a concern with the preservation of full employment in developed capitalist economies. Development economics focused on growth initiation and acceleration in less developed traditional societies. Growth economics, the province of the practitioners of "high theory," was committedly macroeconomic in orientation. Development economics was more microeconomic in orientation and drew on knowledge from related research in anthropology, sociology, and political science and on the insight of practitioners (Krugman 1995, 1–29).[1]

There has been an uneasy relationship between these subdisciplines. Growth economists have tended to view the development economics literature as lacking in rigor and burdened with irrelevant organizational and behavioral detail. Development economists have often felt that the only message growth economists were sending them was to get interest rates (and other prices) right. After a hiatus of over two decades there has emerged, since the mid-1980s, renewed interest in the theory of economic growth. With the emergence of a new and richer growth economics literature a more fruitful dialogue between growth economics and development economics may now be possible. The purpose of this chapter is to address the question, What should development economists learn from the new growth economics?

---

1. For earlier drafts of this chapter see Ruttan (1998, 2001b). I have benefited from comments on earlier drafts by Christopher Bliss, John Chipman, Yujiro Hayami, Bruce Johnston, Timothy Kehoe, Gopinath Munisamy, Mark Nerlove, Howard Pack, Tursynbek Nurmagambetov, Stephen Parente, Terry Roe, James Roumasset, Robert M. Solow, and Michael Trueblood.

## Classical and Schumpeterian Growth

In this initial section I briefly review the classical and Schumpeterian theories of economic growth. I review the classical theory because of the uncompromising pessimism of its economic vision. I review the Schumpeterian theory because of its radical challenge to the neoclassical interpretation of capitalist development. In this brief introduction I cannot, of course, capture the full richness of either the classical or Schumpeterian theory. I include them because themes from each continue to inform and challenge contemporary debates about the role of resource endowments, technology, institutions, and entrepreneurship in the process of economic development.

### Classical Growth

The classical theory of economic growth was the product of a remarkable series of intellectual innovations that occurred between the publication in 1776 of Adam Smith's *Wealth of Nations* (1937) and the publication in 1817 of David Ricardo's *The Principles of Political Economy and Taxation* (1911).[2] There was general agreement among the classicals that the growth of the labor force and the accumulation of capital were the fundamental sources of economic growth. There was also agreement that the possibilities of productivity growth in agriculture, from the division of labor and invention, were more limited than the possibilities in manufacturing. In manufacturing, the progress of invention might more than offset the tendency for diminishing returns. But in agriculture, and in the other natural resource sectors, it was held that the progress of invention would be incapable of offsetting the effects of diminishing returns. Finally, there was agreement that at the institutionally determined "natural" wage rate the long-run supply of labor was perfectly elastic.

The dynamics of the classical model, most fully developed by Ricardo, can be illustrated by tracing the effect of an increase in production resulting from a new invention. An example might be the new Watt-Boulton steam engine used to pump water from mines. A similar sequence might occur as the result of the discovery of new land or new raw materials.

---

2. For the contributions of Malthus, Anderson, and Torrens in addition to Smith and Ricardo see Tribe (1978, 110–46). The critical innovations involved advances in the understanding of the role of impersonal market forces in determining prices, rents, and income distribution.

- The increase in production creates a surplus over and above the amount necessary to cover the subsistence wage. This disposable surplus represents a "wages fund" that capitalists can use to hire more labor.
- The increase in the wages fund results in competition among capitalists for the inelastic (in the short run) supply of labor. The effect is a rise in the wage rate and a decline in the rate of return to capital.
- The higher wage rate induces an increase in the rate of population growth. The rise in wage rates and the increase in population generate a rise in the demand for food.
- The rise in the demand for food is met by bringing progressively lower quality land into production—land on which the marginal product of an incremental dose of capital and labor is lower than on the land already in use.
- The price of food rises in order to cover the cost of production on the marginal land. The effect of rising food prices is to reduce the real wage rate. As the wage rate declines toward the subsistence level the rate of population growth declines.
- The production surplus that gave rise to the higher profits and the higher wage rates initially realized by capitalists and workers is absorbed by a combination of higher land rents and the subsistence wages of a larger labor force. When the surplus has been fully absorbed, a new stationary equilibrium is reached at which all the surplus above the laborers' minimum subsistence is captured by landlords. A new round of growth is dependent on a new invention or new discovery.

In the classical model, diminishing returns to increments of labor and capital applied to an inelastic supply of land represented the fundamental constraint on economic growth. During the nineteenth century the exploitation of new resources and advances in agricultural technology released the constraints on growth posited in the classical theory (Hayami and Ruttan 1985).

Schumpeterian Growth

It was the radical insight of Joseph Schumpeter, first articulated in his classic work, *The Theory of Economic Development* (1912), that placed economic growth back on the economics agenda after nearly a century of neglect (Fogel 1997, 18). Schumpeter rejected the equilibrium economics of neoclassical economic theory. He insisted that the essence of capitalist development lies "in the inevitable tendency of that system to *depart* from equilibrium" (Rosenberg 1994, 49).

In the Schumpeterian system the entrepreneur and innovation, and credit and profit, were the central elements. Schumpeter focused his attention on the technological leaders—on innovative individuals—because of the growth forces released by their entrepreneurial activities. Innovation entailed the development of new products and new processes, the exploitation of new materials and new markets, and the carrying out of new organization of industry (1934, 57–94). These changes are introduced by new firms and are associated with the rise to leadership of new men.[3] The problem addressed in conventional equilibrium economics is how capitalism manages existing structures, "whereas the relevant problem is how it creates and destroys them" (Schumpeter 1942, 84).[4] The capitalist entrepreneur exists in a "perennial gale of creative destruction."

In Schumpeter's early work the entrepreneurial role was limited to those who direct profit-oriented activity in the private sector. It did not include inventive activities—it was "no part of his function to 'find' or 'create' new possibilities" (Schumpeter 1934, 88; Ruttan 1959). In his later work, particularly in *Capitalism, Socialism and Democracy* (1942), the large enterprise replaced the entrepreneur as the dynamic source of economic growth. Its functions include the development of new methods of production and new commodities. An even more inclusive interpretation of Schumpeter's thought would find it useful to extend the concept of entrepreneurship to include the engineers and scientists who are the sources of strategic technical and scientific innovations and to the bureaucratic entrepreneurs who are the sources of many of the innovations in the institutional infrastructure of modern societies.[5]

---

3. Schumpeter's discussion of the role of innovation and the innovator is most fully developed in *Business Cycles* (chaps. 3 and 4); see also *Economic Development* (1934, chap. 2).

4. Since the early 1980s there has been, following several decades of neglect, a neo-Schumpeterian revival. One focus has been the dynamics of the "carrying out of new combinations" and the resultant "creative destruction" of old products, processes, and firms. See, for example, Heertje and Perlman (1990). A second focus has been on the institutional environment that facilitates innovations in the corporate enterprise (Van de Ven, Angel, and Pool 1989). See also the attempts by Nelson and Winter (1982, 2002), drawing on inspiration from Schumpeter, to develop a new evolutionary economics.

5. I have in mind, for example, scientist-engineers such as William Shockley who, along with John Bardeen and Walter Brattain, produced the first working transistor, and the staff of the Defense Department Advanced Research Project Agency who were responsible for the Internet innovation (Ruttan 2001b, 324–25, 340–42). I have earlier discussed the lack of a theory of invention in Schumpeter's work (Ruttan 1959). Rosenberg (2000) has argued that in his later work Schumpeter embraced an endogenous view of technical change.

Schumpeter's fundamental insight was that technical change, implemented by the irrational "animal spirits" of the entrepreneur, is the source of extramarginal returns—the large new income streams—that become a source of disequilibria in capitalist economies. These new income streams are captured by modern growth accountants in conventional measures of productivity growth.

In his later work, Schumpeter argued that the rationalist attitude fostered by capitalism "does not stop at the credentials of kings and popes but goes on to attack private property and the whole scheme of bourgeois values" (1942, 143). The rational bureaucratization of the process of capitalist development would result in the challenging of the social and political credentials of the entrepreneurial class itself (121–30). The ensuing struggle over the partitioning of the new income streams generated by capitalist development would become a source of change in the institutional structure of capitalist economies (121–42). The struggle to create and partition the new income streams generated by a technically progressive capitalist economy became, in the mature Schumpeterian system, a dynamic source of institutional change.[6]

### Modern Growth Theory

In the new theories of economic growth that emerged in the mid–twentieth century, initial attention focused on the rate of capital accumulation relative to the rate of growth of the labor force. There have been three waves of interest in growth theory in the last half-century. The first was stimulated by the work of Harrod (1939, 14–33; 1948) and Domar (1946, 137–47; 1947, 343–45). The second wave began in the mid-1950s with the development by Solow (1956, 65–94) and Swan (1956, 343–61) of a neoclassical model of economic growth. The third wave was initiated in the mid-1980s by Romer (1983, 1986) and Lucas (1988, based on his 1985 Marshall Lecture).[7]

### Keynesian Growth Theory

The question posed by Harrod and Domar, using somewhat different terminology, was, Under what circumstances is an economy capable of

---

6. For the role of Schumpeterian innovation by "petty capitalists" in transforming a traditional economy see Hayami and Kawagoe (1993); Gudeman (2001), 110–20; chapter 2, this volume.

7. For a review of the new growth economics literature from a development economics perspective see Bardhan (1993, 1995). For reviews from an economic history perspective see Crafts (1995) and Williamson (1995).

achieving steady state growth? This question was forced onto the economic agenda by the Great Depression of the 1930s and the expectation that the end of World War II would be followed by renewed instability. In the Harrod-Domar view, instability in economic growth was the result of failure to equate a "warranted" and a "natural" rate of growth. The warranted rate of growth is dependent on the savings rate and on a given capital requirement per unit of output. The natural rate is the maximum long-run sustainable rate of growth. It is determined by the rate of growth of the labor force and the rate of growth of output per worker (box 5.1). This central proposition of the Harrod-Domar model arises from the assumption that investment is both capacity creating and income generating.

> **Box 5.1. The Keynesian (Harrod-Domar) Model**
>
> The question posed by Harrod and Domar, using somewhat different terminology, was, Under what circumstances is an economy capable of achieving steady state growth? This question had forced itself onto the economic agenda by the Great Depression of the 1930s and the expectation that the end of World War II would be followed by renewed instability.
>
> In the Harrod-Domar view instability in economic growth is the result of failure to equate a "warranted" and a "natural" rate of growth.
>
> When the warranted rate is given by $s/v$ and the natural rate by $n + m$ the equilibrium expression is:
>
> $$s/v = n + m$$
>
> where:
>
> $s$ = the saving rate (a fixed fraction of net output)
> $v$ = the capital requirement per unit of output
> $n$ = the rate of growth of the labor force (and population)
> $m$ = the rate of labor saving technical change.
>
> Thus, if the savings rate were 10 percent of income and the capital output ratio 4, the warranted rate of growth would be 2.5 percent. If the labor force was growing at 1.0 percent and labor productivity at 1.5 percent per year, the warranted and natural rates would be equal.

An attraction of the Harrod-Domar model was that it attempted to study long-run growth with the tools of Keynesian economics that had recently become familiar to economists. Use of the model diffused rapidly to the planning agencies of many newly independent countries. It seemed to confirm the widely held belief among development economists and planners that the transition from slow to rapid growth required a sustained rise in the rate of savings and investment.[8] It provided a rationale for interventions designed to raise savings rates and encourage investment in heavy industry in order to remove the constraints on production resulting from capital equipment. And it provided the conceptual foundation for the two-gap (in savings and foreign exchange) model developed by Chenery and associates to estimate foreign aid requirements of developing countries (Chenery and Strout 1966). It was also interpreted as consistent with the view that achieving sustained growth would be more difficult for capitalist economies than for economies where the central planning apparatus would have more direct access to the instruments needed to force a rise in the savings rate and to allocate investment to its most productive uses.[9]

Neoclassical Growth Theory

The second wave in the development of modern growth theory began with the neoclassical model introduced by Robert M. Solow (1956) and Trevor W. Swan (1956). Solow was motivated by skepticism that a sustained rise in the savings rate was the key to the transition from a slow to a fast growth path and by a concern that the capital-output ratio should be replaced by a richer and more realistic representation of technology (1988, 303). Solow's departure from the Harrod-Domar model was to substitute a variable capital-output ratio for the fixed coefficient capital-output ratio of the Harrod-Domar model. He insisted that the primary effort in his 1956 article "is devoted to a model of long run growth which accepts all the Harrod-Domar assumptions except that of fixed proportions" (Solow 1956, 66).

The initial version of the Solow neoclassical model (box 5.2) has

---

8. This view was articulated by W. Arthur Lewis: "The central problem of the theory of economic development is to understand the process by which a community which was previously saving 4 or 5 percent of its national income or less converts itself into an economy where voluntary savings is about 12 to 15 percent of national income or more" (1954, 155).

9. These views were argued most forcefully by the Indian planner P. C. Mahalanobis (1953, 1955). See also Bhagwati (1966, 203). In the development planning literature it became common to refer to the Harrod-Domar-Mahalanobis model.

been succinctly described by Prescott: "The model has a constant returns to scale aggregate production function with substitution between two inputs, capital and labor. The model is completed by assuming that a constant fraction of output is invested" (Prescott 1988, 7). Technical change was represented by a time trend in the constant term of the production function. The model was employed in a 1957 article in which an aggregate two-factor production function was used in accounting for growth in the U.S. economy. To Solow's surprise, and to the surprise of the profession generally, four-fifths of the growth in U.S. output per worker over the 1909–49 period was accounted for by changes in the technology coefficient. The two articles triggered a whirlwind of theoretical and empirical research that lasted well into the 1970s.

In the initial Solow-Swan neoclassical model, steady state growth can hardly be avoided. A country that succeeds in permanently increasing its savings (investment) rate will, after growing faster for a while, have a higher *level* of output than if it had not done so. But it will not achieve a permanently higher *rate* of growth of output (Solow 1988, 308). What were the implications of the Solow neoclassical growth theory and related growth-accounting exercises for development economics? The initial results seemed to completely reverse the earlier Harrod-Domar implications. Technological change replaces growth of capital equipment as the primary source of growth. Technical change must counteract the effect of diminishing returns to capital if output per worker is to continue to grow. Subsequent growth-accounting exercises employing broader definitions of capital resulted in somewhat lower estimates of the contribution of technical change. But technical change continued to outweigh growth of physical capital by a substantial margin in studies conducted in the United States and other presently developed countries.

---

**Box 5.2. The Neoclassical (Solow-Swan) Growth Model**

Solow's departure from the Harrod-Domar model was to substitute a variable capital-output ratio for the fixed coefficient capital-output ratio.

"The model has a constant returns to scale aggregate production function with substitution between two inputs, capital and labor. The model is completed by assuming that a constant fraction of output is invested" (Prescott 1988, 7).

$$c_t + i_t = f(k_t, n_t)$$
$$k_{t+1} = k_t + i_t$$
$$i_t = \sigma\, f(K_t, n_t)$$

where

$c$ = consumption
$i$ = investment
$k$ = capital
$n$ = labor
$\sigma$ = the fraction of output invested.

If factors are assumed to be paid their marginal product, then given $k_t$ and $n_t$ the date $t$, national income, and product accounts can be computed for the model economy.

In Prescott's exposition and in Solow's original model there is no technical change. Some authors, however, seem to interpret the Solow labor variable as "effective labor," as in the Harrod-Domar model.

Research on sources of growth in poor or newly developing countries have typically found that a much smaller share of economic growth has been accounted for by productivity growth than in developed countries. This has often been interpreted as an indication that inappropriate technology that was transferred from high-wage economies, where it had been developed, to low-wage economies failed to generate as high productivity gains in low-wage as in high-wage economies.[10] In addition, research carried out within the neoclassical framework has not shed much light on the driving forces behind the proximate sources of growth—on the determinants of the growth of physical and human capital and technical change.

## Endogenous Growth Theory

Massive, and continuing, divergence in absolute and relative per capita income across countries is a dominant feature of modern economic history (Kuznets 1955, 1966a; Maddison 1979, 2001; Pritchett 1995; Prescott 1997). The "new" growth economics literature was initially

---

10. See, for example, the literature on appropriate technology (Schumacher 1973; Eckhaus 1977; 1987; Stewart 1987a, 1987b, 101–9).

motivated by (1) the apparent inconsistency between the implications of the neoclassical theory and lack of evidence of convergence toward steady state growth even among presently developed economies (Romer 1983, 3) and (2) the inability to successfully account for differences in income growth rates or income levels across countries (Romer 1994).[11] "By assigning so great a role to 'technology' as a source of growth, the theory is obliged to assign correspondingly minor roles to everything else, and so has very little ability to account for the wide diversity in growth rates that we observe" (Lucas 1988, 15). Romer argued that what is needed is "an equilibrium model of endogenous technical change in which long-run growth is driven primarily by the accumulation of knowledge by forward-looking, profit maximizing agents" (1986, 1003).[12]

A primary goal of the new growth economics is to build models that can "ensure that the long run growth rate of income depends not only on the parameters of the production and utility functions, but also on fiscal policies, foreign trade policies, and population policies" (Srinivasan 1995, 46).[13] The effect was to challenge the neoclassical assumption that policy can affect the level of economic activity but not the rate of economic growth.

In the initial endogenous growth models advanced by Romer (1983, 1986), long-run growth is driven primarily by the accumulation of knowledge. The production of new knowledge exhibits diminishing returns at the firm level. However, the creation of new knowledge by one firm is assumed to generate positive-external effects on the pro-

---

11. There have been a number of useful surveys of the new growth economics literature (Verspagen 1992; Van de Klundert and Smulders 1992; Hammond and Rodriguez-Clare 1993; Amable 1994).

12. Romer and Lucas were not the first to attempt to endogenize the process of technical change. Kaldor (1957) advanced a Keynesian model with an endogenous "technical progress function" (Palley 1996). Romer (1986, 1990) and Lucas (1988, 1993) both acknowledge inspiration from Arrow (1962) and Uzawa (1965). But neither Romer or Lucas refers to the Kaldor article. In the 1960s several attempts were made to rescue neoclassical growth economics from the limitations of exogenous technical change under the rubric of induced technical changes (Fellner 1961; Kennedy 1964; Ahmad 1966). For reviews see Nordhaus (1973); Thirtle and Ruttan (1987); and Ruttan (1997). I find it somewhat difficult to imagine that Romer and Lucas were so unfamiliar with this earlier literature on endogenous technical change that they were forced to reinvent it from scratch.

13. Srinivasan (1995, 37–70) points out that the neoclassical growth models could also generate sustained long-run growth in per capita income, even in the absence of technical progress, provided the marginal product is bounded away from zero by a sufficiently highly positive number. He also notes that this is not a particularly attractive assumption "since it implies that labor is not essential for production!" (46). It also assumes that nonrenewable resources are either not essential or have easily available substitutes.

duction technology of other firms. Furthermore, the production of consumption goods, which is a function of both the stock of knowledge and other inputs, exhibits increasing returns (box 5.3).[14]

> **Box 5.3. Endogenous Growth Theory (Romer-Lucas)**
>
> The "new" endogenous growth literature was motivated by a presumed lack of evidence of convergence toward steady state growth in the presently developed economies and by the inability to successfully account for differences in growth rates or income levels across countries.
>
> The initial models are frequently referred to as $AK$ models after the assumed production function $AK$ where $K$ can be thought of as a proxy for a composite capital good that includes physical and human components.
>
> $$Y = K^{1-\alpha}(AL_y)^\alpha$$
> $$A = \delta L_A$$
>
> where
>
> $Y$ is output,
> $A$ is productivity, knowledge, or ideas,
> $K$ is capital.
>
> $\delta$ parameterizes the efficiency of R&D. Labor is used in two activities, the production of output ($L_y$) and the search for innovations ($L_A$) so that $L_y + L_A = L$.
>
> The increasing returns to scale in this production function reflect the nonrivalrous nature of knowledge: given some level of knowledge $A$, doubling capital and labor inputs into production is sufficient to double output; doubling the stock of knowledge as well would lead to more than doubling of output.

---

14. The initial Romer model and other closely related models are frequently referred to as closed economy $AK$ models after the assumed production function ($Y = AK$). In expanded versions of the model, $K$ can be thought of as a proxy for a composite of capital goods that includes physical and human components (Barro and Sala-i-Martin 1995, 146). Amable and Solow have pointed out that this initial Romer model has not been able to avoid the razor-edge balance of the older Harrod-Domar model. If the elasticity of production coefficients of the accumulated factors is greater than 1, the growth is explosive (Amable 1994, 30; Solow 1995, 51; 1997, 7–14).

The models advanced by Romer abandon the neoclassical assumption of perfect competition and require either constant or increasing returns to capital. An important implication of the model is that the market equilibrium is suboptimal since the external effects of the accumulation of knowledge are not considered by the firm in making production decisions. Another implication is that factor shares, typically employed as the elasticity coefficients in the neoclassical production function, can no longer be used to measure the contribution of capital and labor. Romer suggests that the typical capital coefficient (.25) severely underestimates the contribution of capital, and the labor coefficient (.75) severely overestimates the contribution of labor. In his model the capital coefficients, adjusted to take into account the accumulation of knowledge (or of human capital), would have to be (implausibly) close to 1 in order to generate the extremely high growth rates of the East Asian newly industrializing countries (NICs) (Romer 1987, 163–202).

Lucas (1988), drawing on Uzawa (1965), proposed a second alternative to the neoclassical model. In this model human capital serves as the engine of economic growth. He employed a two-sector model in which human capital is produced by a single input, human capital, and final output is produced by both human and physical capital. Two alternative human capital models are analyzed. In the first, the schooling model, the growth of human capital depends on how a worker allocates time between current production and human capital accumulation. In the second, the learning-by-doing model, the growth of human capital is a positive function of the effort devoted to the production of new goods.

As in Romer, in both Lucas models there are, in addition to the internal effects on the worker's own productivity, "external effects" that are the source of scale economies and that enhance the productivity of other factors of production.[15] Since the production of human capital is intensive in its use of human capital, its accumulation involves a significant sacrifice of current utility. In the first model this sacrifice takes the form of a decrease in current consumption. In the

---

15. "The spillover effect of the average stock of human capital per worker in the Lucas model and of knowledge in the Romer model are externalities unperceived (and hence not internalized) by individual agents. However, for the economy as a whole they generate increasing scale economies even though the perceived production function for each agent exhibits constant returns to scale" (Srinivasan 1995, 43). In effect Romer and Lucas have completed, or have attempted to complete, the agenda initially advanced by Jorgenson and Griliches (1969). They have substituted a new black box—termed *scale effects*—for the old black box of technical change as a source of productivity change.

second it takes the form of a less desirable mix of current consumption goods than could be obtained with slower human capital growth (Lucas 1988, 18). Lucas argues that this deficiency could, in principle, be solved in the first case by subsidizing schooling and in the second case by subsidizing research and development.

In 1990 Romer advanced an alternative endogenous growth model in which he followed Lucas in emphasizing the importance of human capital in the development of new knowledge and technology. He departed from Lucas, and from his own earlier work, by treating technical change as embodied in new producer durables. The basic inputs in the model are capital, raw labor, human capital, and an index of the level of technology. The technology component is disembodied nonrival knowledge that can grow indefinitely. Human capital is the cumulative embodied product of formal education, on-the-job training, and learning-by-doing. The model economy has three sectors: (1) a research sector that uses human capital and the stock of knowledge to produce new knowledge in the form of designs for producer durables; (2) an intermediate goods sector that uses the designs from the research sector together with forgone output to produce the producer durables used in the production of final goods; and (3) a final goods sector that uses raw labor, human capital, and producer durables (but no raw material) to produce final output—which can be consumed or saved as new capital (Romer 1990).

In this model, growth in the stock of capital used in the production of final goods takes the form of growth in the number of intermediate inputs rather than in the quality of each input. Growth in the number of intermediate inputs implies monopolistic competition in the market for producer durables and assures external scale economies as a result of the growth in output of each consumer durable. The critical allocative decision is the share of human capital employed in research. As in his earlier model and in the Lucas models, the optimum rate of growth exceeds the market rate since the externalities from knowledge creation are not considered by the firm making production decisions.[16]

As his work continued to mature, Romer turned to the contribution of ideas as the primary source of economic growth (1993, 1996). "Neoclassical growth theory explains growth in terms of interactions between two basic types of factors: technology and conventional inputs. New growth theory . . . divides the world into two fundamen-

---

16. A somewhat similar model has been proposed by Aghion and Howitt (1992) in which innovation takes the form of improvements in the quality of intermediate goods that, in turn, improve the productivity of the intermediate goods in final good production.

tally different types of productive inputs that we can call 'ideas' and 'things.' Ideas are nonrival goods. . . . Things are rival goods" (1996, 5–6). For Romer, scale effects are important because ideas, as nonrival goods, are expensive to develop but inexpensive to use. Their value increases with the size of the market. This implies that large countries, with large internal markets, have a greater incentive to produce ideas than small countries. As a result, large countries can be expected to grow more rapidly than small countries—particularly when small countries burden themselves with the control and regulatory structures that characterize large countries.[17]

What are the implications of the Romer-Lucas-inspired endogenous growth literature for development economics? Griliches notes the importance of their work in emphasizing "that (i) technical change is the result of conscious economic investments and explicit decisions by many different economic units and (ii) unless there are significant externalities, spillovers, or other sources of increasing returns, it is unlikely that economic growth can proceed at a constant undiminished rate into the future" (1992, 294). Bardhan insists that the most substantive contribution "is to formalize endogenous technical progress in terms of a tractable imperfect-competition framework in which temporary monopoly power acts as a motivating force for private innovations" (1995, 2985).

Bardhan is suggesting, in effect, that the Romer-Lucas research agenda represents an attempt to formalize the Schumpeterian insight on the role of technical change captured by the term "creative destruction" (Amable 1994, 33). This implies that the marginal product of capital need not decline over time to the level of the discount rate. Thus the incentive to accumulate human and physical capital and to invest in research and development may persist indefinitely, and long-run growth in per capita income can be sustained.

Even more important than the results of their own research has been the stimulus that the Romer-Lucas work provided for a new burst of theoretical and empirical research in the field of both growth and

---

17. There is, however, a large literature that suggests that ideas are much more expensive to transfer than implied by the literature that treats knowledge as a pure public good (Teece 1977; Hayami and Ruttan 1985, 255–98). In a series of yet unpublished papers Keller has shown that knowledge diffusion is severely limited by distance. Furthermore, trade patterns are the dominant source of differences in bilateral knowledge spillover, while foreign direct investment and communication differences together account for only about 30 percent (Keller 2001, 2002).

development economics (Sandler 2001, 236–40). By the late 1990s, several graduate-level textbooks had incorporated and extended the work in growth economics conducted over the previous decade (Barro and Sala-i-Martin 1995; Aghion and Howitt 1998; Jones 1998). At the theoretical level there has been a proliferation of models that introduced a large set of candidate measures of a country's social, political, and economic status in order to account for both a general failure of convergence or the "miraculous" growth of a few countries such as Korea and Taiwan. The theoretical literature has been complemented by empirical efforts to measure the effects of different national policies, tax rates, and trade policies, for example, in accounting for different national growth rates (King and Rebelo 1990, S126–50; Levine and Renelt 1992; Barro 1997). I later turn to the implications of the new growth economics for development.

An important extension of the new growth economics, particularly for development economics, is the synthesis of the new growth economics and the new trade theory by Gene Grossman and Elhanan Helpman. Their contribution is the result of an exceedingly ambitious research agenda that began in the late 1980s (1990a, 1990b, 1991b) and culminated in the publication of *Innovation and Growth in the Global Economy* (1991a).[18] Grossman and Helpman have pushed their analysis further than the other scholars working in the new growth economic tradition to incorporate the international movement of goods, capital, and ideas. Drawing on Romer and Lucas, they have formally incorporated "the process of introduction of an ever expanding set of new goods and technologies" (Bardhan 1995, 299). The results obtained by Grossman and Helpman represent an important contribution to the formal analysis of trade and growth. But their analysis is characterized by a curious industrial fundamentalism.[19] It will be necessary to extend their analysis beyond the rather narrow confines of the industrial sector and embrace a richer economic structure and institutional environment before their research agenda will be very useful to development economists (Krueger 1997).[20]

---

18. I have discussed the Grossman-Helpman work in greater detail in an earlier article (Ruttan 1998).

19. For an extension of the Grossman-Helpman trade and growth model to incorporate technical change, trade, and growth in an economy that includes both an industrial and an agricultural sector, see Roe and Mohtadi (2001).

20. I have in mind, for example, the integration of trade and development theory suggested by Findlay and Jones (2001).

## Dialogue with Data

The assault by the proponents of the new endogenous growth theory on the neoclassical model was beginning, by the early 1990s, to generate a substantial backlash. The qualifications have been both theoretical and empirical.

Solow pointed out that Lucas (1988) had acknowledged that "a touch of diminishing returns to capital (human capital in this case) would change the character of the model drastically, making it incapable of generating permanent growth. [But he] did not notice that a touch of increasing returns to capital would do the same" (Solow 1995, 49). Solow went on to argue that Lucas's version of the endogenous growth model was not robust. "It can not survive without exactly constant returns to capital" (Solow 1995, 51; 1997, 7–14).

Much of the empirical work has not attempted to directly challenge the endogenous growth models but rather to rehabilitate the neoclassical model (Pack 1994; Kenlow and Rodriguez-Clare 1997). An important landmark in this effort is the cross-country analysis by Mankiw, Romer, and Weil (1992).[21] They first reject the Romer-Lucas characterization of the neoclassical model. When they incorporate saving, population growth rates, and human capital accumulation into the cross-country regressions, along with physical capital and raw labor, Mankiw, Romer, and Weil find that the "augmented" neoclassical model accounts for about 80 percent of the cross-country variation in

---

21. See also Barro (1997). There has been a virtual explosion of aggregate cross-country studies since the early 1980s. Levine and Renelt (1992) subjected the studies that had been completed by the early 1990s to a careful sensitivity analysis. They found that almost all of the results were fragile. However, they did find three robust results: (1) a negative relationship between the initial level of per capita income and subsequent economic growth, (2) a positive correlation between the share of investment in GDP and growth, and (3) positive correlation between the investment share and the ratio of trade to output. Much of the more recent empirical research has abandoned the discipline of the production function for an unstructured "search for variables," such as equipment quality, market distortions, government spending, tax policy, financial capital, trade policy, ethnicity, legal culture, religion, and even distance from the equator (Sala-i-Martin 1997). A reviewer of an earlier draft of this chapter has pointed out that the negative sign on the per capita income term should not be taken as a sure indicator of rehabilitation of the neoclassical model because it could be picking up the dynamic effects of catch-up or structural change. It should be noted that the macro cross-country studies have completely ignored the sector-level cross-country studies by development economists. For example, a series of cross-country studies for the agricultural sector using augmented neoclassical production functions (that included land, labor, capital, intermediate inputs, education, and research and development) were highly successful in accounting for differences in output and productivity levels among countries (Hayami and Ruttan 1985; Lau and Yotopoulas 1989).

income. They also find that the augmented model predicts what they term "conditional convergence" across countries.

There have also been a number of studies that tried to explore the sources of growth of the newly industrializing countries of East Asia—Hong Kong, Singapore, South Korea, and Taiwan. The results seemed to challenge the conventional wisdom that high productivity growth in the manufacturing sectors largely account for the rapid overall rate of economic growth in these countries (World Bank 1993; Kim and Lau 1994; Easterly 1995; Young 1995). Young concludes, from a traditional neoclassical growth accounting exercise, that the very high rates of growth in output of the East Asian NICs are primarily accounted for by (1) a rise in labor force participation rates, (2) a rise in investment to GDP ratios, (3) improvements in education, and (4) the intersector transfer of labor from agriculture. In retrospect it appears that rapid growth in total factor productivity contributed to exceptionally high rates of economic growth in these economies. Total factor productivity was particulary important as a source of growth in output per worker (Crafts 1995; Pack 2001; Easterly and Levine 2001).

A second line of inquiry questions the ability of familiar public policy instruments to bring about permanent changes in growth rates, at least in the presently developed countries. Islam (1995) employs a dynamic panel data model (combining cross-section and time-series data) with individual country effects. This enables him to incorporate differences in the aggregate production function both across groups of countries (convergence clubs) and for individual countries. Islam's results suggest that growth in each country converges to its own steady state growth rate, conditional on differences in technology and institutions. He notes that this conclusion is hardly optimistic. "There is probably little solace to be derived from the finding that countries in the world are converging at a faster rate when the points (and growth paths) to which they are converging remain very different" (1995, 1162).

A number of recent articles have focused on the growth experience of particular countries or regions. Jones (1995a) argued that in spite of permanent changes in a growth-increasing direction of a number of the variables identified as potential determinants of long-run growth in the new growth economics literature (openness to international trade, human capital investment, population growth, and others) they have had little or no impact on growth in the OECD economies during the post–World War II era. The rate of growth in per capita income in the United States, for example, has apparently (and implausibly) remained essentially unchanged during the entire 1880–1987 period (fig. 5.1).

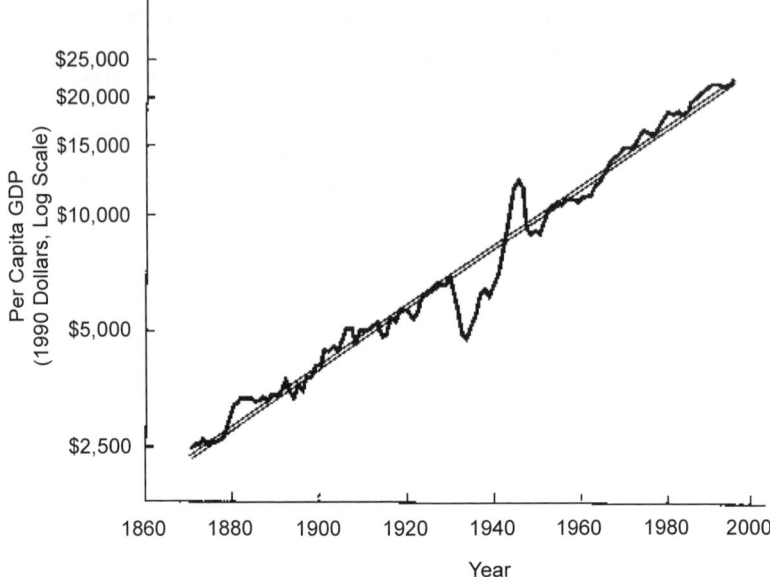

Fig. 5.1. Real per capita gross domestic product in the United States, 1870–1994. (Charles I. Jones, *Introduction to Economic Growth* [New York: W. W. Norton, 1988].)

Jones draws the startling conclusion that "either nothing in the U.S. experience since 1880 has had a large persistent effect on the growth rate, or whatever persistent effects have occurred have miraculously been offsetting" (1995a, 499; Pritchett 1997).[22] And Griliches insists even more strongly: "Knowledge externalities are obviously very important in the growth process but do not help us explain what has happened (in the United States) in the last two decades" (1994, 16).

A third line of inquiry has involved attempts to test the returns-to-scale hypothesis that is central to the new growth economics. In spite of its intuitive plausibility the evidence from cross-industry and cross-county studies provides only minimum support for the role of scale economies as an important source of growth (Burnside 1996; Backus, Kehoe, and Kehoe 1992; Jorgenson and Yip 2002). Backus, Kehoe,

---

22. Rostow (1956, 25–48; 1960) had earlier argued that the "preconditions" for economic growth in the United States were established between the 1780s and 1840s. He dates the U.S. takeoff into rapid industrial development between 1843 and 1860.

and Kehoe (1992) have attempted to test the sources of scale economies—learning by doing, investment in human capital, and specialized intermediate inputs—emphasized by Lucas and Romer. In their search for evidence of scale economies from these three sources they find modest support for scale economies in the manufacturing sector but fail to find evidence of economywide scale economies.[23]

Cross-country agricultural production function studies for 1960-80 have found scale economies across conventional inputs (land, labor, capital, and operating inputs) for developed countries but not for developing countries and across conventional inputs plus human capital for both developed and developing countries (Hayami and Ruttan 1985, 143-57; Kawagoe, Hayami, and Ruttan 1985). Similarly, when *AK* endogenous growth models have been augmented to include human capital and intangible capital as well as physical capital, and allowance is made for changes in government policies that affect the capital/output and labor/leisure choices, estimates based on the models capture both long-term trends and short-run deviations from trend (McGrattan 1998).

What should development economists learn from this more recent research carried out within a neoclassical framework? When the neoclassical model is extended to incorporate the variables of an augmented production function to explain differences in the level of per capita income or of partial or total factor productivity, reasonable results are obtained (Hayami and Ruttan 1970, 138-62; Kawagoe, Hayami, and Ruttan 1985, 113-32; Mankiw, Romer, and Weil 1992, 407-38; Mankiw 1995, 275-376; Pritchett 1997). These results are, however, less than a conclusive demonstration of the adequacy of the neoclassical model. There have been few attempts to employ econometric methods to directly test the results of the neoclassical and endogenous growth models against each other (Pack 1994; Brock and Durlauf 2001).

The recent research on growth empirics, stimulated by the work of Romer and Lucas, has identified a number of new, and sometimes unanticipated, growth patterns. These include the long-term persis-

---

23. Intraindustry trade is sometimes taken as evidence of scale economies. Chipman (1992a, 1992b) has argued, however, that evidence of intraindustry trade is frequently based on aggregation bias from an inappropriate industry classification system that obscures the fact that the great bulk of what is often classified as intraindustry trade consists of trade in distinct commodities. He then goes on to demonstrate that even when commodities are correctly classified intraindustry trade cannot be taken as evidence of scale economies. Burnside (1996) also stresses aggregation bias in cross-industry studies in the United States that purport to show economies of scale.

tence of growth trajectories, the relationship between democracy and growth, and linkages between physical geography and growth. These advances in our knowledge are important. But they are subject to two limitations. From the point of view of decision theory the variables employed are far too abstract to be used as policy instruments. At the same time they are largely a reflection of surface phenomena—the proximate sources of economic growth.

Answers to more fundamental questions, such as why some countries save and invest more than others, why some countries invest a larger share of gross national product (GNP) on education or on research and development, why some countries were able to put a package of high payoff inputs together more effectively than others, or why some countries have responded to shocks (the food crises of the 1960s or the oil shocks of the 1970s) more effectively than other countries, continue to remain beyond the reach of the models employed by both the neoclassical and the new growth economists (Nelson and Pack 1997; Temple 1999; Pritchett 2000; Easterly 2001b).

## Growth Economics as Development Economics

It is now time to turn to a fuller assessment of the implications of the new growth economics for development economics.[24] Robert Lucas attempted to answer this question in a 1991 lecture at the European meeting of the Econometric Society (Lucas 1993). He was very explicit about what he wanted to accomplish. He wanted to be able to encompass both the Korean "growth miracle" and the Philippine "growth failure." And he viewed growth miracles as productivity miracles. Just as Solow in the mid-1950s viewed his contribution as a modest modification of the Harrod-Domar model, Lucas viewed his contribution as adapting the neoclassical model to fit the observed behavior of both rich and poor economies. He saw the new growth economics as displacing not only the older neoclassical growth economics, but development economics as well. He insisted that the new growth economics includes "those aspects of economic growth we have some understand-

---

24. I do not, in this section, attempt to present an exhaustive exposition of the scope of development economics. A useful tour is presented in the three-volume *Handbook of Development Economics* (Chenery and Srinivasan 1988, 1989; Behrman and Srinivasan 1995). Apparently, as an afterthought, the editors commissioned a final article in volume 3B entitled "The Contribution of Endogenous Growth Theory to the Analysis of Development Problems: An Assessment" (Bardhan 1995). For an excellent introduction to development economics see Hayami (1997).

ing of, and development [economics includes] those we don't" (1988, 13).[25]

As an exercise in the integration of the several strands of the new growth economics into a coherent system, the Lucas article was an exciting tour de force. Lucas, Romer, and other practitioners of the new growth economics must be credited with trying to reach behind the proximate sources of growth and to treat as endogenous some of the more fundamental sources.[26] This was achieved by importing into growth economics several concepts that have been conventional in development economics. One was the concept of scale economies that occupied a prominent role in the early development literature.[27] A second was the insight into the role of human capital advanced initially by development economists.[28] A third was the concept of endogenous technical change that, under the rubric of induced technical change, had achieved substantial success in the hands of development economists and economic historians in interpreting the rate and direction of technical change (Hayami and Ruttan 1971; Thirtle and Ruttan 1987; Ruttan 1997). From the point of view of neoclassical general equilibrium economics, scale economies, human capital investment, and technical change were sources of externalities, not formally incorporated

---

25. Paul Krugman has been even more dismissive of development economics. "Once upon a time there was a field called development economics—a branch of economics explaining why some countries are so much poorer than others, and ... prescribing ways for poor countries to become rich ... That field no longer exists" (1995, 6–7).

26. Charles Manski notes that throughout much of the twentieth century mainstream economics has traded breadth for rigor. The narrowing of economics culminated with the formalization of the neoclassical theory of general competitive equilibrium (Arrow and Hahn 1971). Manski argues that since the early 1970s the discipline of economics has sought to broaden its scope while maintaining its rigor (2000, 115).

27. Economies of scale played an important role in what Krugman (1995, 1–29) has termed the "high development theory" of the 1940s and 1950s. Krugman identifies the period of high development theory as beginning in 1943 with the publication of the "big push" model by Rosenstein-Rodan (1943, 202–11), which drew on Harrod (1939) for inspiration, and ending with the popularization of ideas of forward and backward linkages by Hirschman (1958). Attempts to implement policies based on these ideas left many developing countries stuck in a low-level equilibrium, burdened with nonviable capital-intensive industries (Bardhan 1995). For an attempt to rehabilitate the big push approach within the framework of the new development economics, see Murphy, Shleifer, and Vishny (1989).

28. Emphasis on the role of human capital in development thought extends back to at least the 1950s. See, for example, the presidential address to the American Economic Association by Theodore W. Schultz (1961). In an early cross-country analysis, Krueger found that "differences in human resources between the United States and less developed countries accounted for more of the differences in per capita income than all other factors combined" (1968, 658). For additional examples, see Bardhan (1993). See also the articles on human resources and labor markets in Chenery and Srinivasan (1988).

into the theory of general competitive equilibrium. From the perspective of development economics, they were powerful sources of economic growth that were treated as exogenous in neoclassical theory.

Mancur Olson poses a challenge that, when viewed from the perspective of development economics, is a deficiency in both the old and the new growth economics. Olson noted that during the late twentieth century, "neither the old or the new growth theories predict the relationship actually observed: *The fastest growing countries are never the countries with the highest per capita incomes but always a subset of the lower income countries . . . that grow faster than any high income country*" (1996, 20). A second deficiency is the inability of growth theory to interpret the great instability in less developed country growth. Only a small subset of the developing countries that achieve high rates of growth in any one decade experience a high rate of growth the following decade (Pritchett 2000).

A remarkable new growth-accounting exercise by Edward Prescott (2002) makes it now possible to offer a more positive assessment. In his 2001 Ely Lecture to the American Economic Association, Prescott explores the sources of differences in GNP levels among a number of OECD countries. He finds that in the late 1990s GNP per worker in France was depressed relative to the United States by about 30 percent. The difference was largely accounted for by the higher marginal tax rates on labor income in France. He also found that in Japan GNP per worker is also depressed by about 30 percent relative to the United States. In the case of Japan, decline in the rate of growth in labor productivity since the early 1990s was the largest source of depression. Prescott's significance for this chapter's focus (what development economists can learn from growth economics) is that the methodology employed in his study opens up the possibility of achieving a deeper understanding of the cycles of growth spurts and depression that characterize the experience of even the more successful developing countries.

In earlier chapters I have attempted to address the issue of what grows in the process of social and political development. The practitioners of growth economics have rarely attempted to explore the relationships among social, political, and economic development. When they have incorporated variables related to social and political development in cross-country growth regressions the effort has typically reflected a "search for variables" approach rather than an effort informed by social and political theory.

An Unfinished Agenda

There are several other well-developed concepts that must be imported from development economics before the new growth economics can successfully lay claim to success as a "new development economics" or provide new insights for development practice. In this section I list some of the more fundamental concepts that, though well understood by development economists, have largely been neglected by practitioners of growth economics.[29]

*Structural Transformation*

The issue of structural transformation, the transition from a primarily agrarian to an industrial—commercial economy, has represented a core issue in development economics since the publication of Colin Clark's classic work in 1940 (Clark 1940; Jorgenson 1961; Ranis and Fei 1961). The assumptions of homothetic preferences and neutral technical change employed by most growth economists have led to the neglect of the problem of structural transformation (Matsuyama 1992; Bardhan 1995). Once these assumptions are abandoned, structural change emerges as a central feature of the process of development (Syrquin 1994; Echevarria 1997, 2000). An attempt to analyze economic development with a model in which there is no mechanism to generate structural transformation can hardly be regarded as serious. It resembles an attempt to perform Hamlet with no role for the Prince of Denmark.[30]

---

29. For a more complete list see Temple (1999).

30. It is hard to overemphasize the importance of structural transformation, particularly the transition from a predominantly agrarian economy to an industrial and then a service economy, in the development literature. The classic empirical studies are Clark (1940) and Kuznets (1966a). For the evolution of thought see Lewis (1954); Jorgenson (1961); Ranis and Fei (1961); Fei and Ranis (1964); Dixit (1973). See also the articles on structural transformation in Chenery and Srinivasan (1988). Because of the importance of structural transformation, development economists have generally preferred to work with two-sector models of the Lewis-Jorgenson-Ranis-Fei type rather than two-sector models in the Uzawa (1961, 1963) tradition. Failure to incorporate the roles of growth in agricultural production and of agricultural trade in the early stages of structural transformation represents a serious deficiency in any attempt to understand the development process (Echevarria 1995; Park and Johnston 1995; Tomich, Kilby, and Johnston 1995).

## The Demographic Transition

The demographic transition is one of the more familiar processes associated with economic development. It seems somewhat negligent that attempts to develop an endogenous theory of per capita income growth have failed to address the issue of growth and distribution of population and labor force, particularly given the attention that has been focused on East Asia, where population growth rates have declined dramatically, in the new growth economics literature. Development economists have made substantial progress in constructing endogenous models of family fertility decisions. Less progress has been made in our understanding of such factors as investment in health, nutrition, and education that influence infant and child mortality rates and the growth of human capital (Nerlove 1974, 1994; Binswanger et al. 1980; Birdsall, Kelley, and Sinding 2001).

## Natural Resource Constraints

Natural resource constraints have yet to be fully incorporated into modern growth economics (Musu and Lines 1995, 273–86; Echevarria 1997; Ruttan 2001b). At the very least it is important to incorporate land (and other natural resource endowments) and environmental constraints into growth models.[31] When resource and environmental effects are more adequately incorporated the comment by Solow in his classic 1956 article will become more apt: "The scarce-land case would lead to decreasing returns to scale in capital and labor and the model would become more Ricardian" (67). It is also important to separate those investments in technology development that represent maintenance research and development—the R&D necessary to offset declines in natural resource quality or loss of productivity in biological technology—from R&D investment in productivity-enhancing technical change.[32]

---

31. For an empirical exploration of the relationship between environmental indicators and economic growth see Grossman and Krueger (1995). For an attempt to incorporate environmental effects in a closed-economy endogenous growth model for the United States see Elbasha and Roe (1995).

32. Maintenance research can represent a relatively high share of R&D expenditures in biological technology. The resistance of new crop varieties to pests and pathogens is eroded by the evolution of new races. The effectiveness of new drugs to control animal and human disease is eroded by the coevolution of infectious disease organisms (Ruttan 1982, 59–60; 1999).

*Income Distribution*

The issue of the relationship between income distribution and economic growth has, until recently, been almost completely neglected in growth economics. Much of the early development literature focused on the Kuznets-shaped income distribution curve (Kuznets 1955, 1–28; Bacha 1977, 52–87; Galor and Tsiddon 1996). Kuznets suggested, and subsequent research seemed to confirm, that income inequality increases during the early stages of economic development but declines after income reaches a threshold level. By the late 1980s, however, income inequality was increasing in countries such as the United States and Japan that achieved the highest income levels. Numerous explanations have been advanced. The most prominant is that there is a "skill bias" in the demand for labor in countries in which growth is dominated by high technology information and communication industries (Conceição and Galbraith 2001; Acemoglu 2002).

The conditions under which economic growth leads to a widening or narrowing of income distribution and the conditions under which changes in income distribution enhance or threaten economic growth are the subject of a large literature in development economics (Kanbur 1997). The literature on sources of poverty and poverty alleviation has been enriched by the literature on entitlements stimulated by Sen (1981, 1983). Cross-country empirical studies have consistently found a negative correlation between income inequality and economic growth (Aghion, Caroli, and Garcia-Peñalosa 1999). But the causal relationship between economic growth and income inequality remains unresolved.[33]

### A New Development Economics?

A central analytical issue that has not been adequately addressed by the new economic growth literature is the issue of institutional change. "In the economic growth literature agents optimize but the institutional framework is given" (Tornell 1997). The challenge remains how to introduce induced or endogenous institutional change into an opti-

---

33. Economic historian Robert Fogel, recipient of the 1993 Nobel Prize in Economics, has raised the issue of the effect of unequal distribution of "spiritual resources," or what I would refer to in this context as unequal distribution of social capital. Spiritual resource deprivation—"a sense of purpose, a sense of opportunity, a strong family ethic, a strong work ethic, and high self esteem" (Fogel 2000, 178) acts as a barrier to participation in both civic and economic activity (202–14).

mizing growth model. In the mid-1980s Joseph Stiglitz initiated an attempt to extend the "New Institutional Economics"[34]—particularly from the literature on property rights, transaction costs, and information theory—to construct a "New Development Economics" (1986). This section gives a brief introduction to the property rights, transaction cost, and information theory concepts.

*Property Rights*

One of the basic premises of development thought during the first decade after World War II was that institutional constraints represented a major barrier to technical change and productivity growth in developing countries. Much of this concern focused on the issue of the reform of property rights, particularly land tenure arrangements, in developing country agriculture. The analytical foundation for this concern extends back to the classical economists who recognized that in a world of imperfect land, capital, and credit markets, sharecropping represented an improvement over wage labor because of its positive incentive effect. This classical perspective was reinforced and extended by Marshallian neoclassical analyses of tenure relations, which emphasized the productivity incentives of a owner-operated agricultural system (Hayami and Ruttan 1985, 389–98).

As early as 1950 D. Gale Johnson questioned the empirical validity of the Marshallian analyses (Johnson 1950). In the late 1960s Stephen Cheung, drawing on inspiration from Coase (1937, 1960), presented a formal proof that if private property rights are well-defined and the enforcement of contract terms is costless "the implied resource allocation under private property rights is the same whether the landlord cultivates the land himself, hires farm hands to do the tilling, leases his holdings on a fixed rent basis, or shares the actual yield with his tenant" (1969a, 4).[35] Cheung's research initiated a furious round of analytical and empirical research that has led to a reexamination by economic theorists, institutional economists, and development economists

---

34. The genealogy of the New Institutional Economics traces to Coase (1937). For an introduction to the New Institutional Economics see North (1981, 1990b) and Williamson (1979). For a collection of the seminal articles see Steven G. Medema (1995).

35. I had made essentially the same argument in an article published in the late 1960s (Ruttan 1969). I have, however, been quite critical of the implication that attempts to reform land tenure institutions are unwarranted, which drew on Cheung's analysis (Hayami and Ruttan 1985, 389–98). In subsequent work Cheung has explored the implications of transaction costs and risk aversion for the evolution and reform of land tenure arrangements (1969b).

of the role of property rights not only for the process of agricultural development but also across the entire range of contractual relations (Furubotn and Perjovich 1972; Stiglitz 1974). As this literature matured it became clear that transaction costs were an important source of the productivity differences associated with property rights and labor market arrangements (Otsuka, Chuma, and Hayami 1992; Hayami and Otsuka 1993).

*Transaction Costs*

By the mid-1980s it was possible to complement the property rights literature with a more mature understanding of the role of transaction costs in accounting for the wide differences in the efficiency with which resources were employed in production. Transaction costs include the time and expense necessary to obtain the information needed to make, negotiate, and enforce an exchange (Williamson 1985, 2). The transaction cost approach maintains that the economic institutions of capitalism emerged out of efforts to economize on transaction costs. It opened up powerful new insights in the field of industrial organization. Whether transactions occur primarily among firms, as in markets, or within firms, as in a multidivisional corporation, depends on which form is most effective in economizing on transaction costs (Williamson 1979, 1985).

The transaction cost approach also provides powerful insights into the organization of political "markets." In a seminal article published in 1990 Douglass North concluded that transaction costs represented an even heavier burden on the efficient functioning of political than economic markets. Dixit (1996) extended the transaction cost framework to analyze a broad spectrum of agency problems and contractual relationships in the political marketplace. Even in mature democratic systems, transaction costs impose a heavy burden on the design and implementation of efficient institutional arrangements (see, for example, the constructed market for emissions trading case, chap. 1, this vol.).

The implications of transaction cost economics for economic development arise from the new income streams generated by institutional innovations that reduce transaction costs. The constraints imposed by transaction costs on the capacity of a society to design and implement institutional innovations have, in the past, imposed a severe burden on economic growth (North 1990b; Eggertsson 1994, 1997, 2003). The design of institutional arrangements that reduce transaction costs is as surely a source of economic growth as is the design of new technologies that reduce the cost of material production.

## Information Theory

Since the late 1980s attempts have been made to complement the property rights and transaction cost literature with a new information-theoretic literature that goes beyond the new growth economics in attempting to clarify the sources of differences in productivity and economic growth rates over time and across countries. Stiglitz, drawing on research on information asymmetries in rural credit markets, has argued that the imperfect information perspective provides a unifying theoretical framework that incorporates the contribution of both the property rights and the transaction cost approaches (Stiglitz 1974, 1988, 2000a; Stiglitz and Weiss 1981; Hoff and Stiglitz 1993).[36]

Two core insights of the information-theoretic approach are key. (1) Information is scarce and the transaction costs involved in its acquisition are high. But information also has many of the properties of a public good. Although it is often possible to exclude others from accessing new knowledge (through intellectual property rights, for example), it may be inefficient to do so. (2) There are often important differences (or asymmetries) in access to information among principals and agents, between landlords and tenants, or between regulatory agencies and firms.[37] Stiglitz retains the rational actor assumption of neoclassical microeconomic theory—individuals and firms attempt to respond to new information in a coherent manner. But principals and agents often have a limited understanding of the environment in which they function. The prices at which markets clear may differ substantially from the prices that would prevail at competition equilibrium.

Integration of the property rights, transaction cost, and information-theoretic literature has provided new tools to interpret the different institutions that govern property rights, ranging from land tenure to intellectual property, and to enhance understanding of conflict situations where agents have asymmetric information about each other,

---

36. Analysis of the implication of asymmetric information on market behavior traces back to the now classic article "The Market for Lemons" by Nobel laureate George Akerlof (1970).

37. It is somewhat surprising that Stiglitz gave only limited attention to the debate initiated by Friedrich Hayek and Oscar Lange that ran from the 1930s to the 1950s about the information requirements of centrally planned and market economies. For a review see Caldwell (1997).

such as between terrorist groups and governments, in political science (Sandler 2001, 122–29). This has in turn opened up new insights into the sources of differences in productivity and income growth across industries and countries (Collier and Gunning 1999). It remains something of a paradox, however, that only endogenous institutional changes in property rights have been formally incorporated into growth models (Tornell 1997; Acemoglu, Johnson, and Robinson 2001). The potential of the new institutional economics and the new information-theoretic literature to contribute to incentive-compatible institutional design has only been incompletely realized.

### A More Comprehensive Growth Economics?

Why have growth economists been so slow to incorporate such fundamental issues and concepts, both from the old and the new development economics, into growth theory? Tractability, of both modeling and analysis, is one part of the answer. Insistence on working within the narrow constraints of steady state growth models has represented a fundamental obstacle to building on the rich body of literature advanced by development economists, institutional economists, and economic historians.[38] In retrospect it seems clear that a pervasive obsession with the conditions for convergence (or the traverse) to steady state growth accounts for much of the failure of both the old and the new growth economics to extend their reach to encompass some of the more fundamental sources of economic development.[39]

Furthermore, the distinction between *level effects* and *rate effects*, however important analytically, does not carry over well into development economics. The growth obtained by exploiting the transition

---

38. For a detailed iteration of both the rediscovery and the neglect of ideas initially advanced by development economists see Bardhan (1993). A similar perspective has been advanced by Romer (1993) in what appears to be a remarkable departure from his articles of the late 1980s. For an example, see Lucas's uncomfortable discussion of the difficulty of incorporating the rise in schooling levels in East Asia. He notes that "the percentage of school age children in school has little leverage in explaining differences in growth rates. The fast growing Asian economies are not, in general, better schooled than some of their slow growing neighbors" (1993, 257). He then goes on to suggest that although schooling levels are increasing in virtually all societies "it cannot be pursued within a steady state framework" (258).

39. More than twenty-five years ago Solow suggested that "the steady state is not a bad place for growth theory to start, but may be a dangerous place for it to end" (1970, 7; Hicks 1985, 10). Griliches (1994) made a similar point, somewhat more cautiously, in questioning an excessive commitment to equilibrium economics in his 1994 presidential address to the American Economic Association. See also Nelson (1998).

dynamics from one balanced growth pattern to another is as welcome to a developing country as a source of an improvement in welfare as growth along a balanced growth path (if such exists). For a low-income country, it is not particularly interesting to insist that the "pay-off from a higher saving rate is not a permanently higher rate of growth; it is a permanently higher output per man" (Solow 1970, 20). The distinction between a policy leading to a growth-rate effect rather than a level effect will not be obvious to even the best economists employed in national planning or finance agencies or in multilateral development banks (Solow 1997; Pack 2001; Yusuf 2001, 15–19).[40]

A second part of the answer is that the early growth economics literature was largely concerned with attempts to address the problem of economic stability that confronted modern industrial states characterized by liberal economic and political institutions. The models developed by the Keynesian and neoclassical growth economists were designed to address the problem of achieving short-run quantitative growth objectives. When the attention of economists shifted to the problems of achieving economic growth in the preindustrial economies of Latin America, Asia, and Africa it became apparent that the application of standard macroeconomic policy instruments often failed to generate either economic stability or economic growth. The design or

---

40. The timescale for transition effects in neoclassical models has generally been estimated to be quite long (Atkinson 1969). They may also be quite difficult to distinguish from rate effects. Rivera-Batiz and Romer (1991) present a model in which economic integration of countries "with identical endowments and technologies" can result in a permanent increase in growth rates primarily because it results in an increase in the extent of the market. For an excellent review see King and Robelo (1993).

A promising start toward a more fruitful articulation between growth economics and development economics has recently begun in the work of a group of "new neoclassical" growth economists associated with the Department of Economics at the University of Minnesota and Minneapolis Federal Reserve Bank (Parente and Prescott 1991, 1993, 1974, 2000; Backus, Kehoe, and Kehoe 1992; Prescott 1997; Schmitz 1993; Chari, Kehoe, and McGrattan 1996). Chari, Kehoe, and McGrattan depart from the traditional neoclassical model by abandoning the deterministic transition path between steady state growth paths in favor of stochastic transition probabilities. Their motivation is that they find little persistence in individual country growth rates. Parente and Prescott have advanced a research agenda designed to explain differences in per capita income levels rather than growth rates. They invoke institutional constraints, primarily policies designed to protect the interests of domestic suppliers of factor inputs, on the efficient international transfer and adaptation of new technologies to explain the large and persistent productivity and income gaps that cannot be explained by differences in physical and human capital (Parente and Prescott 1994, 2000; Prescott 2002). For an analysis that is quite similar in spirit to that advanced by the Minnesota school but argued more intuitively, see Olson (1996).

renovation of the institutional structure would have to be added to the economic policy agenda.

Following the end of the Cold War it became apparent that modern growth economics was equally unprepared to address issues of institutional reform in formerly centrally planned economies of Eastern Europe, the former Soviet Union, and East Asia (Sachs 1997; Clague 1999; Eggertsson 1994, 1997; Kolodko 2000; Hough 2001; Easterly 2001b). To proponents of the Washington Consensus, the dismantling of the existing institutions of central planning was a necessary first step for the rapid emergence of a market economy (Murphy, Shleifer, and Vishny 1989, 1992; Aslund 1995; Murrell 1995), and once prices were set free it was assumed that market institutions would emerge autonomously. In their advice they typically emphasized a cold turkey approach to reform—"getting on with the economic policy reforms by whatever political means are at hand in the hope that successful implementation of the reforms would induce the necessary political support needed for the reform of political and economic institutions" (Clague 1999, 16). But reforms based on the principles of the Washington Consensus in financially troubled less developed countries or formerly centrally planned transitional economies have rarely been successful in initiating or renewing economic growth.[41]

The success of the new growth economics in endogenizing investment in knowledge production and human capital acquisition has not been followed up by a successful effort to endogenize the process of institutional change or to incorporate institutional design theory.[42] At this point the challenge posed by Schumpeterian growth theory— identification of the institutional sources of the new income streams that have characterized progressive capitalist economies—still remains resistant to formal analysis.

**Perspective**

The new growth economics, like the neoclassical growth economics, has advanced our understanding of the process of economic growth in

---

41. For a highly personal jeremiad against the policies pursued by adherents to the Washington Consensus, particularly the International Monetary Fund, see the book *Globalization and Its Discontents,* by Joseph Stiglitz (2002), former chief economist at the World Bank.

42. For a review of attempts to treat institutional change as endogenous see chapter 1 of this volume. The concept of incentive comparable institutional design was introduced by Hurwicz (1972a, 297–333). For a more recent treatment see Groves, Radner, and Reiter (1987). For a review of the empirical literature on institutions and growth see Aron (2000).

industrial economies characterized by technical change and reasonable stability of expectations regarding factor and product markets, legal institutions, and civic culture. But it is not about the problems facing the poor economies of the world (Solow 1997, 71). Growth theory, even informed by the new growth theory, has not been able to meet the Lucas challenge. It provides little insight into the policy reforms necessary to translate the Philippine (or the Nigerian) growth failure into a growth miracle. And it provides even less insight into the institutional innovations that will be necessary for the former centrally planned economies to make a successful transition to market-oriented economies. The transition from highly centralized economic and political systems to the institutions of market capitalism involves more than changes in economic policies or institutional arrangements. It will involve constitutional changes that establish the fundamental parameters within which changes in economic policy and institutional arrangements will be permitted to occur (chap. 4, this vol.). It is hard to avoid concluding, in spite of the central role that knowledge spillovers play in the new growth economics, that inadequate attention has been given to the design of institutions that more effectively internalize these changes (Sandler 2001, 243).

My own sense is that the most significant advances in knowledge about economic development will continue to emerge from research conducted at the microlevel. I have emphasized the important role of investigations conducted by students of political development at the microlevel in clarifying political design principles (chap. 4). The real sources of economic growth are induced by changes in resource endowments, technology, institutions, and culture. These changes can best be understood by investigations conducted at the microlevel (Stiglitz and Weiss 1981). The effects of these microlevel changes on economic growth are captured at the aggregate level in measures such as total factor productivity growth that are the source of new income streams.

I am not arguing, however, that development economists and growth economists should continue to follow their natural inclination to ignore each other's work. There needs to be a continuing dialogue between development economists working in the fields of household economics, agricultural economics, labor economics, and industrial organization and the practitioners of growth economics (Stiglitz 2002a). There is too much interesting and important data being generated by the development process that is begging to be understood and interpreted to confine development economics within the straitjacket of growth economics. Those of us who are development economists or practitioners simply cannot wait until the growth economists are able to incorporate a deeper understanding of the sources of economic development into their models.

## A Postscript on Method

The issues raised in this review of the continuing dialogue between growth economics and development economics are analogous to the more general problem of the relationship between science and technology, particularly the view that advances in scientific knowledge precede and become the source of advances in technology. This view is no longer held by most historians of science and technology. Instead of a single path running from scientific discovery through applied research to development it is more accurate to think of science-oriented and technology-oriented research as two intersecting paths that both lead from and feed back into a common pool of scientific and technical knowledge (Ruttan 2001b, 79–82). I take it as axiomatic that the primary source of demand for advances in natural science knowledge arises from demand for technology development. Similarly, the primary source of demand for social science (including economic) knowledge arises from a demand for policy design and institutional innovation.

The history of growth economics is clearly not consistent with the standard view that advances in science precede development and practice. The major thrust in growth economics has been an attempt to develop formal theory to provide a more coherent interpretation of what was already conventional knowledge in development economics. A major thrust of the neoclassical growth economics, as noted earlier, was to attempt to analyze the properties of steady state growth. Solow's first model was initially extended to incorporate productivity growth in the form of laborsaving (Harrod-neutral) or Hicks-neutral technical change. Attention quickly shifted to attempts to account for and explain the residual.

But growth in partial productivity (output per worker) and total productivity (output per unit of total input) were not new concepts in the mid-1950s. Research on labor productivity had received major attention in the 1930s. Growth in output per unit of total input had been identified by several scholars in the late 1940s and early 1950s (Griliches 1996). Even after the initial Solow study, research on productivity growth, particularly that carried out at the sectoral level, was seldom motivated by the growth economics paradigm. The early research by Kendrick (1961) and Dennison (1962) on trends in productivity and growth owed little or nothing to neoclassical growth theory. Nor did the initial attempt by Griliches (1963) and later by Jorgenson and Griliches (1969) to provide a complete accounting of the growth of output based on changes in the quality of inputs rather than total factor productivity. The theoretical foundation on which much of this work was based was the straightforward Hicksian neoclassical production theory rather than neoclassical growth economics.

The new growth economics is even more explicitly motivated than neoclassical growth economics by an attempt to understand what was already known. Initially it was directed to understanding the perceived failure of substantial convergence of growth rates implied by the neoclassical theory. The issue of convergence was stimulated by a growing body of literature in the late 1970s and early 1980s on the productivity slowdown in the United States and in other developing countries. It was further stimulated by research on long-term productivity and output growth rates by Maddison (1979, 1982), which were interpreted by Baumol (1986) as indicating convergence in productivity and output growth among developed countries and the challenge to the Baumol findings by De Long (1988).

PART III

CHAPTER 6

# Technology Adoption, Diffusion, and Transfer

The diffusion of technology has been a powerful source of economic change since prehistory. It has been an important field of inquiry in anthropology since the late nineteenth century. The adoption and diffusion of technology emerged as an important research agenda in sociology, primarily in rural sociology, in the 1940s and 1950s. It emerged as an increasingly important research agenda in economics during the 1960s. The initial research focused primarily on the adoption decision at the individual or firm level and the diffusion process at the community or industry level. More recent research has focused on the international diffusion of technology.[1]

## The Convergence of Traditions

During the 1950s, strong traditions of diffusion research had emerged in both rural sociology and in medical sociology. By the 1960s these fields were joined by communications, geography, marketing, and economics (table 6.1).[2] The growth of adoption-diffusion (A-D) research has been tracked very carefully in a series of books by an early student in the field, Everett Rogers (Rogers 1962, 1983, 1995; Rogers and Shoemaker 1971). Its development has also been the subject of an important case study in the sociology of science by Diane Crane (1972).

Initially each diffusion tradition emerged as a relatively self-contained intellectual enclave. These enclaves began to converge during the late 1950s (Katz 1960; Katz, Levin, and Hamilton 1963). In a 1960 article, Katz noted that researchers in the field of communications had only recently became aware of the studies of the diffusion of new

---

1. In this chapter I draw substantially on several earlier studies (Thirtle and Ruttan 1987; Ruttan 1996; 2001b, 147–78). I am indebted to Arnulf Grubler and Thomas Murtha for comments on an earlier draft of this chapter.
2. For a bibliography of diffusion studies by discipline see Musmann and Kennedy (1989).

**TABLE 6.1. Major Diffusion Research Traditions**

| Diffusion Research Tradition | Number of Diffusion Publications (% of All Publications) | Typical Innovations Studied | Method of Data Gathering and Analysis | Main Unit of Analysis | Major Types of Findings |
|---|---|---|---|---|---|
| 1. Anthropology | 141 (4) | Technological ideas (steel ax, the horse, water boiling) | Participant and non-participant observation and case studies | Tribes or peasant villages | Consequences of innovations; relative success of agents |
| 2. Early sociology | 10 (—) | City manager government, postage stamps, ham radios | Data from secondary sources and statistical analysis | Communities or individuals | S-shaped adopter distribution; characteristics of adopter categories |
| 3. Rural sociology | 854 (22) | Agricultural ideas (weed sprays, hybrid seed, fertilizer) | Survey interviews and statistical analysis | Individual farmers in rural communities | S-shaped adopter distribution; characteristics of adopter categories; perceived attributes of innovations and their rate of adoption; communication channels by stages in the innovation-decision process; characteristics of opinion leaders |
| 4. Education | 359 (9) | Teaching/learning innovations (kindergartens, modern math, programmed instruction, team teaching) | Mailed questionnaires survey interviews, and statistical analysis | School systems, teachers, or administrators | S-shaped adopter distribution; characteristics of adopter categories |

| | | | | |
|---|---|---|---|---|
| 5. Public health and medical sociology | 277 (7) | Medical and health ideas (drugs, vaccinations, family planning methods, AIDS prevention) | Survey interviews and statistical analysis | Individuals or organizations like hospitals or health departments | Opinion leadership in diffusion; characteristics of adopter categories; communication channels by stages in the innovation-decision process. |
| 6. Communication | 484 (12) | New events, technological innovations | Survey interviews and statistical analysis | Individual or organizations | Communication channels by stages in the innovation-decision process; characteristics of adopter categories and of opinion leaders; diffusion networks |
| 7. Marketing and management | 585 (15) | New products (a coffee brand, the Touch-Tone telephone, clothing fashions) | Survey interviews and statistical analysis; field experiments | Individual consumers | Characteristics of adopter categories; opinion leadership in diffusion |
| 8. Geography | 160 (4) | Technological innovations | Secondary records and statistical analysis | Individuals and organizations | Role of spatial distance to diffusion |
| 9. General sociology | 322 (8) | A wide variety of ideas | Survey interviews and statistical analysis | Individuals | Characteristics of adopter categories; various others |
| 10. General economics | 144 (5) | Technological innovations | Economic analysis | Organizations, individuals | Economics of technological innovations |
| 11. Other traditions[c] | 563 (14) | — | — | — | — |
| Total | 3,890 (100) | | | | |

*Source:* Everett M. Rogers, *Diffusion of Innovations*, 4th ed. (New York: Free Press, 1983), 42, 43.

[a]The exact number of major diffusion research traditions is arbitrary. We choose these because they represent the relatively greatest number of empirical diffusion publications (an exception is the early sociology traditions, which is included because of its influence on certain of the other traditions which developed later).

[b]The rural sociology tradition includes 150 publications by diffusion scholars in extension, whose work is closely related.

[c]Includes public administration and political science (129 publications), agricultural economics (101), psychology (73), industrial engineering (33), statistics (33), and others/unknown (194).

ideas and technology by rural sociologists: "The work of the rural sociologists is of major importance. For the last two decades the latter have been inquiring into the effectiveness of campaigns to gain acceptance of new farm practices in rural communities while taking explicit account of the relevant channels of communication both outside and inside the community" (437). Katz attributed the delay in recognition of the role of interpersonal communication on the part of the mass communication research community to the fact that it drew its intellectual inspiration primarily from psychology rather than from sociology.

In the early 1960s Katz and several colleagues attempted to integrate the conceptual frameworks that had emerged in the several traditions, which had seemingly been "independently invented" and "scarcely knew of each other's existence" (Katz, Levin, and Hamilton 1963, 240). A definitive effort to catalog the literature and to synthesize and evaluate the theories and research findings on the diffusion of technical innovations was made by Everett M. Rogers in his now classic *Diffusion of Innovations* (first published in 1962 and updated in 1971, 1983, and 1995). "The research tradition that can claim major credit for initially forming the intellectual paradigm for diffusion research, and that has produced the largest number of diffusion studies, is rural sociology" (Rogers 1983, 51).

### Learning from Hybrid Corn

A 1943 study by Bruce Ryan and Neal C. Gross (1943, 1950) on the diffusion of hybrid-seed corn "more than any other study, influenced the methodology, theoretical framework, and interpretations of later students in the rural sociology tradition, and in other research traditions as well" (Rogers 1983, 54; Valente and Rogers 1995). Hybrid corn (maize) was first released to farmers by the Iowa State Agricultural Experiment Station in 1928 and promoted by the Iowa Agricultural Extension Service and by the commercial seed companies that marketed the seed.[3] It had a yield advantage, relative to open pollinated corn, in the 15–20 percent range. By 1940 it had been adopted by most Iowa corn growers. In 1941 Ryan, a professor of rural sociology at Iowa State University, obtained funding from the Iowa Agricultural Experiment Station to conduct a study of the spread of hybrid seed to

---

3. For a brief discussion of the invention of hybrid corn see Hayami and Ruttan (1985, 217–19).

Iowa farmers. Sponsorship by the Iowa Agricultural Experiment Station was based on the presumption that a better understanding of the hybrid corn diffusion process would be useful in designing more effective efforts to diffuse other innovations developed by the station. The study involved interviews with 259 farmers living in two small communities in central Iowa.

The Ryan-Gross study attempted to answer a series of questions that dominated most subsequent diffusion research until well into the 1970s. "What variables are related to innovativeness? What is the rate of adoption of an innovation, and what factors explain this rate? What role do different communication channels play at various stages in the innovations-decision process?" (Rogers 1983, 56). The pattern of diffusion that they observed was the now classic sigmoid or S-shaped adoption curve and the symmetric bell-shaped curve describing the distribution of adopters over time (fig. 6.1). They interpreted the slope of the diffusion curve as a consequence of the interpersonal network of information exchanges between those individuals who had already adopted the innovation and those who would be influenced to do so. The sociopsychological research paradigm created by the Ryan and Gross investigation became the academic template that was to be adopted first by other rural sociologists in agricultural diffusion research and then by almost all other diffusion research traditions.

One of the remarkable aspects of the technology diffusion studies by rural sociologists was how rapidly the results of the research were utilized by practitioners. Agricultural extension workers and other change agents found the diffusion model, particularly the five-stage categorization of the distribution of adopters over time—(1) innovators, (2) early adopters, (3) early majority, (4) late majority, and (5) laggard—and the social-psychological and communication characteristics associated with each category, exceedingly useful in their educational programs with farmers. During the late 1950s and early 1960s the Farm Foundation distributed over 80,000 copies of a 1955 report on the status of diffusion research prepared by a committee of Midwestern rural sociologists. During this same period Professors George M. Beal and Joe M. Bohlen of the Department of Rural Sociology at Iowa State University utilized a visual presentation illustrating diffusion processes and concepts in over 160 meetings with change agents including extension workers, advertising agency personnel, sales workers, and managers (Beal and Bohlen 1957).[4] It was this immediate operational value

---

4. I personally recall the enthusiasm with which a Beal-Bohlen presentation was received at a meeting to "train the trainers" that I attended at Purdue University in the mid-1950s.

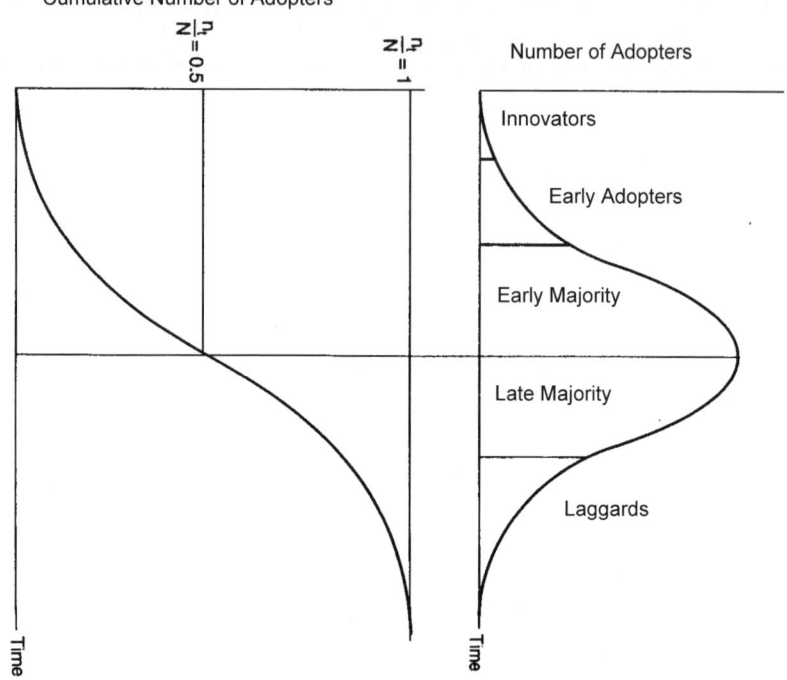

Fig. 6.1. The epidemic model. (From Colin G. Thirtle and Vernon W. Ruttan, *The Role of Demand and Supply in the Generation and Diffusion of Technical Change* [London: Harwood Academic, 1987], 81.)

of the insight gained from diffusion research for health campaigns, technical assistance programs, and other areas that led to the rapid diffusion of diffusion research.

One effect of the convergence of the several diffusion traditions during the 1960s was an attempt to locate diffusion research within the broader field of communications research. This was a major thrust in the second edition of Rogers's book on diffusion of innovations (retitled *Communication of Innovations*). The learning behavior of individuals was identified as the underlying source of the symmetry of the diffusion process. The elementary S-M-C-R communication model in which a source (S) sends a message (M) via certain channels (C) to receiving individuals (R) had, however, been implicit in much of the early diffusion research. Diffusion research, by tracing communication patterns over time, enriched the understanding of communications researchers in the dynamics of the communication process and of the differential impact of the several communication channels at different stages of the communication process (Rogers and Shoemaker 1971, 250–66).

Research on spatial diffusion of technology was initiated with a remarkable series of studies by the Swedish geographer Torsten Hägerstrand (1952, 1953) of the University of Lund.[5] Hägerstrand pioneered a simulation approach to spatial diffusion in which the tendency for an innovation to spread from one adopter to another was expressed as a probability function that decreased with distance from the previous adopter. Hägerstrand's work was largely ignored, even by geographers, for over a decade. Since the mid-1960s a number of quantitative geographers and economists have utilized Hägerstrand's simulation approach and modified it to incorporate insight from the rural sociology and communications traditions (Brown 1981).[6] Sociologists, however, have been slow to adopt the more formal modeling approaches employed by Hägerstrand (Fliegel 1993, 58). As of the late 1970s the trend toward convergence of traditions in research on technology adoption-diffusion remained largely confined to subfields within sociology and communication (Cottril, Rogers, and Mills 1989).

### Challenges to the Diffusion Research Traditions

Prior to the early 1960s most diffusion research by rural sociologists had been conducted within the United States and Western Europe. The 1960s, however, witnessed a dramatic increase in the number of diffusion studies in developing countries. These studies were stimulated by the emphasis on technology development and transfer by the U.S. Agency for International Development (USAID) and other development assistance agencies and by the return to developing countries of students who had received graduate training in departments of rural sociology at American universities. During the 1960s Rogers and several graduate students from Ohio State University conducted research

---

5. The 1953 study was translated and published in English as *Innovation Diffusion as a Spatial Process* (1969). In a postscript, the translator, Alan Pred, traces the intellectual origins and impact of Hagerstrand's work (299–324). Hagerstrand was familiar with the Ryan and Gross hybrid corn studies. The major influence on his work, however, was the strong tradition of quantitative geography in Sweden.

6. For critical reviews of diffusion research in geography see Blakie (1978); Blaut (1987); and Yapa (1991).

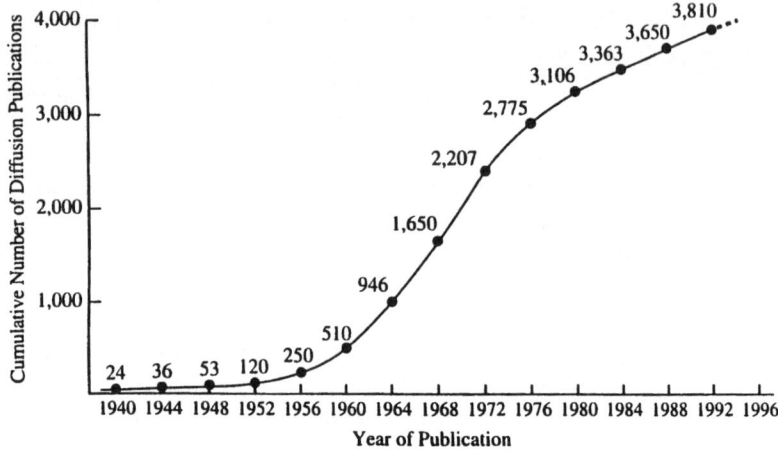

Fig. 6.2. Cumulative number of diffusion research publications by year. (From Everett M. Rogers, *Diffusion of Innovations,* 4th ed. [New York: Free Press, 1995], 45.)

on the diffusion of agricultural technology in Colombia (Rogers and Svenning 1969).

By the late 1950s, however, the number of diffusion publications by rural sociologists in the United States and Europe had begun to decline (fig. 6.2). This was followed by a decline in diffusion studies by rural sociologists in developing countries in the mid-1960s (fig. 6.3).[7] By the mid-1970s, rural sociology had lost its dominance, both relatively and absolutely, as a source of diffusion studies. "Up to 1964, 423 of the 950 diffusion publications were in the rural sociology tradition.... Since 1974 only 8 percent, 45 of 578 publications of all diffusion studies, are in rural sociology" (Rogers 1983, 51).

The slacking of the pace of diffusion research in rural sociology in the 1960s was followed by a rising body of criticism of both theory and method. Writing in the early 1980s, Rogers noted, "Until the past decade, almost nothing of a critical nature was written about this field; such absence of critical viewpoints may have indeed been the greatest

---

7. Rogers's tabulation of diffusion studies in developing countries appears to be incomplete. For example, Dasgupta (1989) reports more than twenty adoption diffusion studies per year in India during the peak years of 1965 through 1974.

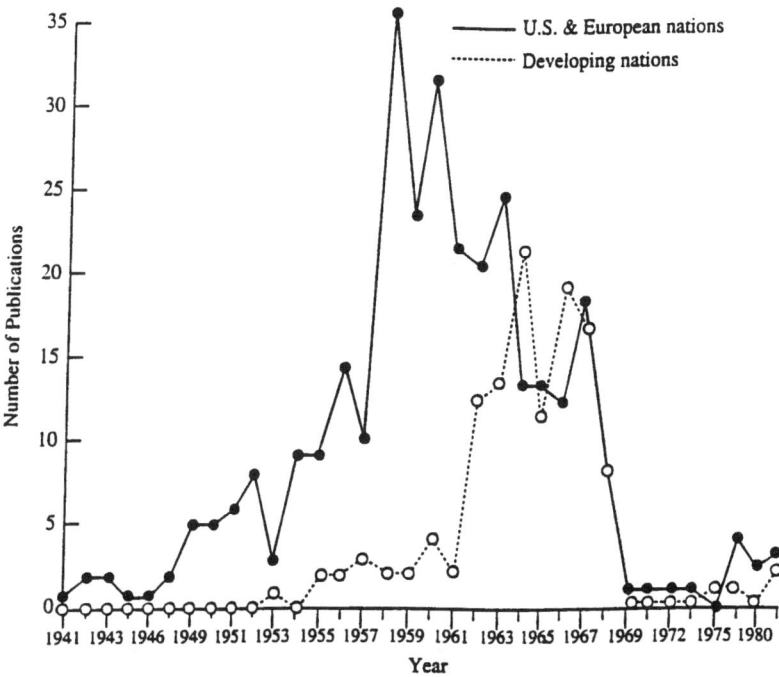

Fig. 6.3. Number of diffusion publications by rural sociologists by year. (From Everett M. Rogers, *Diffusion of Innovations,* 4th ed. [New York: Free Press, 1995], 58.)

weakness of all in diffusion research" (1983, 91). In making this comment he neglected to mention the vigorous and extended argument between Zvi Griliches and several rural sociologists over methodological and empirical issues involving the diffusion of hybrid corn.

A study of diffusion of hybrid corn in the mid-1950s by Zvi Griliches has occupied a role in economics similar to the Ryan and Gross study in sociology (Griliches 1957, 1958). Griliches's contribution was to simultaneously explain (1) the origin of the innovation and timing of diffusion across areas (the supply of innovation) and (2) the rate of diffusion within areas (the demand for innovation) in terms of the profitability of the innovations to hybrid seed producers and farmers. His assertion that the variables considered by the sociologists tend to cancel themselves out, leaving the economic variables as the major determinant of technical change, set off a vigorous and extended controversy.[8] In retrospect, it appears that the economic and sociological explanations were more complementary than competitive. The disagreements were in part semantic. Terms such as *compatibility* and *congruence* can be translated into economic variables with appropriate consideration of imperfect information and risk preference.

Criticism of adoption-diffusion research also emerged from within the rural sociology tradition. Even in the 1960s some rural sociologists had complained that there was a tendency for the methodology employed in diffusion studies to be excessively empirical—a "search for variables" approach (Lionberger 1960). Inadequate attention was given to specification of dependent and independent variables—what economists refer to as the "identification problem."

There have also been several more substantive criticisms. These included (1) a "pro-innovation" bias in diffusion research and a lack of attention to resistance to innovation and the sources of failure of unsuccessful innovation; (2) a tendency to blame the farmer or the peasant for failure to adopt rather than question the appropriateness or profitability of the innovation; (3) inadequate attention to the interrelated processes involved in innovation generation and utilization; and (4) failure to develop methodology for the analysis of an interrelated complex or package of technologies (Rogers 1995, 96–130).[9] The development of more rigorous theoretical and methodological approaches did not, however, reverse the decline in the growth of diffusion research by rural sociologists, either in the United States or abroad.

By the 1970s the research agenda on the diffusion of technology was becoming considerably broader than it had been in the 1950s and 1960s. Initial research had focused primarily on the adoption decision by indi-

---

8. See Bradner and Straus (1959); Griliches (1960); Rogers and Havens (1962); Babcock (1962); Klonglan and Coward (1970); Arrow (1969). Arrow points out that "the economists are studying the demand for information by potential innovators and sociologists the problem of the supply of communication channels" (33). It is of interest that Rogers did review the controversy between Griliches and the rural sociologists in the first edition of *Diffusion Innovations* (1962, 136–42) but omitted the review in later editions.

9. The simultaneous and interdependent development of an innovation and its diffusion had earlier been emphasized by Rosenberg (1972). Fliegel (1993, 71–99) discusses the difficulty of analyzing the adoption of conservation technologies within the classical communications-oriented adoption-diffusion model. He also refers to the difficulty in applying the traditional model when the adopter is involved in the development process as in the case of the many agricultural innovations.

viduals and its diffusion over time and space. The uncritical use of the epidemic model and the logistic curve and the presumed linear relationship between status and adoption were challenged by both sociologists and economists.[10] By the mid-1960s rural sociologists and medical sociologists were beginning to give attention to the modification or "re-invention" of technologies and practices by users (Rogers 1983, 178–84). Economic historians and agricultural economists pointed out that the diffusion of new technology had often been slowed by improvements in the older technology (Rosenberg 1976; Knudson and Ruttan 1988). Resource economists began to explore the factors that constrain adoption during the latter stages of the diffusion process (Fuglie and Kascak 2001). Greater attention was given to the adoption decision process and to how the characteristic or attribute of the innovation affects adoption and diffusion. Sociologists became concerned with how differentiated social structures affected the diffusion process. "In person to person diffusion (as contrasted to constant source diffusion) the models assume that the numbers of contacts between the haves and the have-nots are proportional to the number in each. . . . But when this assumption is not true (as it never is in human populations), serious bias may be introduced" (Coleman 1964, 493).

Different models were developed to track individual and group adoption. "One describes diffusion (i.e. contagion) of a disease throughout a population as the rate of propagation (or in stochastic models, the probability of infection in a given short period of time) proportional to the product of the number of infectives and the number of susceptibles. The other describes contagion within households, and utilizes a chain of binomial distributions to describe the probability of . . . infections in the households stemming from an initial infection" (Coleman 1964, 493).

New Paradigms in Sociology

Several reasons have been advanced for the decline of adoption and diffusion studies within the sociological tradition. It is tempting to infer that research on diffusion simply followed the classical sigmoid adoption curve. The only rural sociologists who are still conducting

---

10. Gartell and Gartell (1985). Dosi (1984) argued that the diffusion curve provides "an ex-post rationalization of the conditional probability of a non-adopter to become an adopter" (286).

diffusion studies are the laggards!¹¹ This explanation would, however, be too mechanistic. A more convincing explanation for the decline is related to a shift in social theory away from modernization and toward dependency and other class-based perspectives (chap. 3, this vol.).[12] Mounting food surpluses turned the attention of agricultural research administrators in the United States away from concerns about the diffusion of technology. Technical change in agriculture began to be viewed by populist critics not as a source of prosperity, but as a source of inequity and as destructive of rural communities and culture. Industrial development was not seen as a source of opportunity for more productive employment and of products leading to improvement in the quality of life, but as a source of exploitation, environmental degradation, and pathological growth of urban populations. Many sociologists and historians began to view modernity not as an emancipation from tradition but as the destruction of tradition (chap. 2, this vol.). Thus diffusion of technology became part of the problem rather than the solution. They began to view their work as designed to illuminate the "thoughtless" consequences of research leading to technical change in rural areas.[13]

---

11. Rogers notes that since the mid-1970s scholarly attention of U.S. rural sociologists studying diffusion turned to conservation and other ecology-related innovations (1985, 61). The findings of these studies have typically been published in journals such as the *Journal of Alternative Agriculture* and the *Journal of Soil and Water Conservation* rather than in disciplinary journals in sociology. See, for example, Nowak (1992) and Fuglie and Kasac (2001). The only article dealing with the diffusion of technology that appeared in *Rural Sociology* during 1990 through 1995 dealt with an ecological issue (Thomas, Ladewig, and McIntosh 1990). During 1966 through 1975, 8.2 percent and between 1976 and 1985, 5.4 percent of all articles in *Rural Sociology* dealt with diffusion (Christianson and Garkovich 1985).

12. In the third edition of *Diffusion of Innovations* Rogers (1983, 371–413) calls for those working within the diffusion tradition in sociology to give greater attention to the consequences of technical charge, particularly the impact on income distribution and social structure. He also noted that in 1968 only 38 of the nearly 1,500 diffusion publications then available dealt with the consequences of innovation and that the imbalance had not changed much in the intervening period (376, 377). Between the mid-1970s and mid-1980s there was an intense debate among economists and other social scientists on the distributional effects of the green revolution (the seed-fertilizer technology) in developing countries in the late 1970s and early 1980s (Hayami and Ruttan 1985, 329–62).

13. For a review of the "new" sociology of agriculture literature see Buttel, Larson, and Gillespie (1990, 73–125). Much of this literature was packaged with the trappings, but without the substance, of scholarship. For one of the more egregious examples see Pearse (1980). Among the more substantial studies were those by Friedland and Benton (1975) and Friedland, Benton, and Thomas (1978). See also the several articles by Brumeister, Koppel, and Grabowski in Koppel (1995). For a review of the controversies associated with the introduction of improved green revolution crop varieties in Asia see Hayami and Ruttan (1983, 336–45). For the literature by historians on resistance to technology see Mokyr (1992) and Isaacman (1995).

These concerns were reflected in an exceedingly critical article by Lawrence Busch (1978). Busch called on the diffusion research community to develop a much deeper understanding of the role of communication by incorporating insights from interpretive and postmodernist theory (box 2.1). He contrasted the adoption-diffusion (A-D) tradition with what he termed Hermeneutic-Dialectic (H-D) tradition. Busch argued that the insights of the H-D tradition into the process of understanding suggest a number of new avenues for adoption-diffusion research: (1) Researchers should take the life-worlds of other cultures seriously, rather than naively assuming the superiority of "scientific" rationality; (2) more attention should be paid to the metaphors and similes used by those who communicate new knowledge; (3) attention should be given to the motives and interests of the institutions and individuals who are the sources of innovations—in whose interest the innovations are being introduced; (4) efforts should be made to understand how adopters and nonadopters understand the potential innovations; (5) more attention needs to be given to how adoption and adaptation modify social structure.

The perspective outlined in the 1978 *Rural Sociology* article provided the theoretical background for an important program of research that Busch conducted, with William Lacy, of the U.S. agricultural research system (Busch and Lacy 1983). In that study Busch and Lacy "examine the factors that influence the choice of research problems by agricultural scientists. Put simply: Why do agricultural scientists do what they do?" (1). The authors conducted in-depth interviews with scientists at several institutions; analyzed mail questionnaires from over 1,400 scientists; and reviewed official government and disciplinary research. They were particularly concerned about the extent to which the research agendas of agricultural scientists are influenced by external factors—the federal, state, and private organizations that fund and manage research and the producer, agribusiness, or resource interest groups that support agricultural research politically. They were also concerned how commitments to professional norms by individual scientists influence their choice of research agendas.

Finishing up this brief review of diffusion research in rural sociology, I found myself surprised in several respects. I had expected to find much stronger convergence among the several diffusion research traditions. The move toward convergence during the 1950s and 1960s was limited to the several diffusion research traditions within sociology.[14]

---

14. There is some evidence that the process of convergence had been substantially weakened even within rural sociology by the early 1970s. "In the rural sociology areas ideas

But the sociologists have been only marginally responsive to the diffusion research in geography and even less to the diffusion research in economics.[15] The next section traces the rapid diffusion of diffusion research among economists and technologists since the mid-1970s.

## The Diffusion of Diffusion Research

The decline in adoption-diffusion research by sociologists coincided with rapid expansion of research on adoption-diffusion and technology transfer by economists.[16] Research on diffusion of agricultural technology by agricultural economists traces back to the Griliches hybrid corn study of the late 1950s (1957, 1958). Griliches's econometric analysis of adoption and diffusion proceeded in two steps. The first involved the collection of data on diffusion of hybrid corn in different areas and the mapping of the (logistic) diffusion path. The second step involved the identification and estimation of the importance of the variables that explained differences in invention adoption, speed of diffusion, and the adoption ceiling among areas. Research on diffusion of industrial technology was initiated by Edwin Mansfield in a series of studies of interfirm and intrafirm diffusion of technology in the early 1960s (1961, 1963a, 1963b). The Mansfield research was substantially influenced by Griliches's earlier research. The effect of Mansfield's research was to emphasize, in addition to the importance of profitability and the proportion of firms already using a new technique, the size of the investment required (relative to the size of the firm), differences in interindustry structure, and differential profitability in determining differences in the rate of interfirm and interindustry diffusion of new inventions.[17]

---

appearing during later stages of growth were less frequently utilized than ideas produced early in its history. . . . Groups of collaborators within the rural sociology area became increasingly unreceptive to each other's ideas" (Crane 1972, 83; see also Rogers 1983, 38–86).

15. Rogers identifies five variables that determine the rate of adoption of innovations: (1) type of innovation-decision; (2) communication channels; (3) nature of the social system; (4) change agents' promotional effort; and (5) perceived attributes of an innovation. But he does not explicitly mention either the spatial dispersion of potential adopters (from geography) or potential profitability (from economics) (1983, 233; 1994, 207).

16. In the 3d edition of *Diffusion of Innovations,* Rogers (1983) did not list economics among the diffusion research traditions. In the 4th edition Rogers (1985) lists 144 studies by economists (table 6.1, chap. 6, this vol.).

17. For a review of other research in the Griliches-Mansfield traditions see Nasbeth and Ray (1974) and Davis (1979). The Davis study is particularly useful for its formalization of the diffusion-adoption model.

Interest by agricultural and industrial economists on research in diffusion of technology did not take off, however, until well into the 1970s.[18] By the early 1980s, at the time when research on technology diffusion by sociologists had fallen off, research on diffusion by agricultural, industrial, and marketing economists and by technologists began to experience explosive growth.[19] The research by economists on diffusion of technology drew substantial inspiration, at least initially, on the diffusion research by the rural sociologists. Research by technologists was, however, much less self-conscious about drawing on the earlier sociological literature. By the mid-1980s, an increasing share of the literature on diffusion was being produced in Western Europe at institutions such as the Science Policy Research Unit (SPRU) at Sussex in the United Kingdom, the International Institute for Applied Systems Analysis (IIASA) in Austria, and the Maastricht Economic Research Institute on Innovation and Policy (MERIT). Just as sociologists have been little influenced by the research on diffusion by economists, recent research by economists and technologists has progressed with little reference to the earlier research by sociologists.

As research on diffusion evolved, however, economists engaged in an increasingly intensive theoretical and methodological dialogue. Explicit attention was given to recent theoretical developments and to the formalization of production, human capital, market, and spatial relationships. Attention was also given to issues of congruence between the implications of the formal theory and the statistical or econometric models used for testing and estimation (Feder, Just, and Zilberman 1985; Silverberg 1991). Economists also became concerned with additional dimensions of adoption such as intensity of use and complementarity among closely related innovations (Smale, Heisey, and Leathers 1995). They extended the reinvention concept to include the continuous flow of improvements by the suppliers of technology

---

18. The delayed interest by agricultural economists in research on technological diffusion is in contrast to the rapid growth of studies of rates of return to research. In his research on hybrid corn in the late 1950s Griliches gave approximately equal attention to diffusion and returns to research. Research on rates of return expanded rapidly from the 1960s through the 1970s. For an early listing, see Ruttan (1983). For more complete lists see Echeverria (1990) and Alston et al. (2000).

19. For reviews by technologists, see Grübler (1991); Marchette (1991). For reviews by economists see Gold (1981); Feder, Just, and Zilberman (1985); Thirtle and Ruttan (1987); Metcalf (1982); Soete (1985); Karshenas and Stoneman (1995); Stoneman (2002). Stoneman is particularly critical of the continuing lack of congruence between the theoretical and empirical literature on the economics of diffusion.

and the development of unified models of the invention-diffusion process (Thirtle and Ruttan 1987; Knudson and Ruttan 1988). Economists also began to give increasing attention to the international diffusion of technology (Vernon 1966, 1979; Hayami and Ruttan 1985; Soete 1985).

In spite of some cross-disciplinary fertilization in the 1960s, sociologists, geographers, and economists have continued to follow alternative rather than convergent agendas in diffusion research. The convergence, which Katz noted in the late 1950s, has occurred primarily within the several subdisciplines of sociology. Several suggestions have been made for the failure of further convergence. Kenneth Arrow's comment on the exchange between Griliches and several sociologists referred to earlier in this chapter remains apt: "The economists are studying the demand for information by potential innovators and the sociologists the problem of the supply of communication channels" (1969).

Since the late 1970s there has emerged a second generation of diffusion research by economists and technologists that departs substantially from the initial Griliches-Mansfield approach. This research falls into two broad classes: (1) research that employs conventional microeconomic equilibrium models and (2) research that employs a newer set of evolutionary models (Thirtle and Ruttan 1987, 108–24; Grübler 1988; Lissoni and Metcalf 1994; Karshenas and Stoneman 1995).

Equilibrium Models

In the equilibrium models diffusion is seen as a transition between equilibrium levels, defined by changing economic attributes (e.g., costs, prices) and a changing environment (e.g., differences in market structure). Diffusion is seen not so much as a learning phenomenon, but as a result of the interaction of changes in the innovation and adoption environment (i.e., the interaction between suppliers and customers of an innovation) (Grübler 1988, 26).

An important second stage innovation has been the further adaptation of the sociological diffusion models by marketing economists to forecast consumer demand. Bass modeled and estimated demand for consumer durables as a function of (1) a coefficient of innovation, which measures the number of initial innovators buying a product, (2) a coefficient of imitation, which is intended to capture the behavior of the population assumed to be imitating the behavior of the rest of the population, and (3) an index of market potential to reflect the ceiling on market penetration. In the Bass model potential adopters are influenced by mass media and word of mouth. Individuals adopting the new product because of mass media are concentrated in the early postintroduction period (Bass 1969, 1980; Rogers 1995, 79–83).

Another important innovation of the second stage research within the equilibrium tradition has been the conceptualization by technologists, building on the earlier research by Mansfield (1963a, 1963b), of the process of technological change as a replacement or substitution phenomenon. The focus in this research has been on the changing relative share of competing technologies (fig. 6.4). In models where several technologies are competing, each technology undergoes three distinct phases in terms of market share—logistic growth, nonlogistic saturation, and logistic decline. Substitution models have been developed most fully and applied most extensively by Nakićenović and several colleagues at the International Institute for Applied Systems Analysis (IIASA).[20] A primary motivation has been the use of such models in technology assessment and forecasting.

The more radical applications of the equilibrium approach completely abandon the communication model. Diffusion takes time not because information is imperfect (or because contigation takes time) but because the new technology is initially not superior to existing technology for some potential adopters or uses. A second departure is that firms are assumed to behave optimally. Thus firms that have not adopted are not interpreted as ill informed or behaving irrationally but as simply waiting for the optimal timing of adoption (Lissoni and Metcalfe 1994; Chari and Hopenhayn 1991). Intellectual links with the sociological origins of diffusion research were completely severed.

In spite of substantial advances in both the theoretical and methodological sophistication, Grübler notes that the equilibrium models continue to be criticized on several counts: (1) the mathematical properties and economic rationale for the application of the logistic model to the diffusion process; (2) the binary nature of most diffusion models and the static assumption on size of potential adopters; and (3) a narrow definition of the group of influencing variables (1991, 14). I find it rather striking that Grübler's criticisms of the diffusion research by industrial economists and technologists as well as the earlier criticism of diffusion research by agricultural economists by Feder, Just, and Zilbeman (1985) are remarkably similar to the early criticisms by rural sociolo-

---

20. Substitution models for two technologies (old and new) were developed by Fisher and Pry (1971). For an extension of the Fisher-Pry model to incorporate multiple substitution see Marchette and Nakićenović (1979).

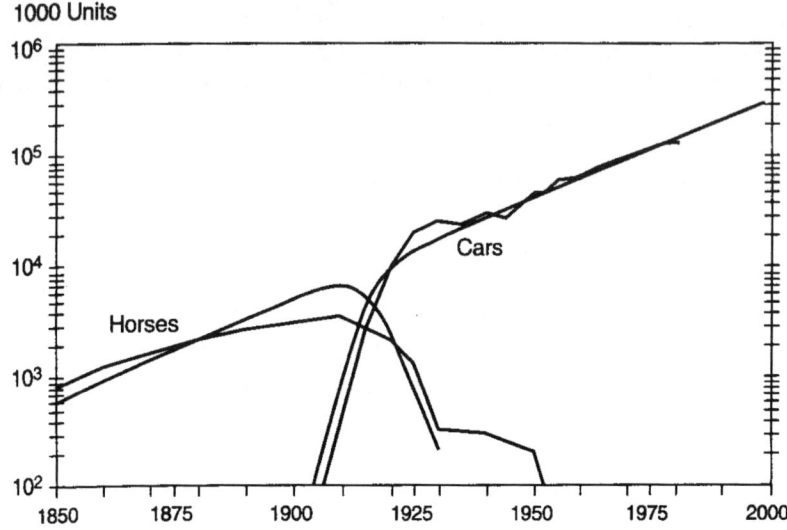

Fig. 6.4. Number of horses and cars in the United States, 1850–2000. (Reprinted from *Technological Forecasting and Social Change,* 39, Arnulf Grübler, "Diffusion: Long-Term Patterns and Discontinuities," 162. Copyright © 1991, with permission from Elsevier Science.)

gists of the methodological limitations of the sociological literature (Lionberger 1960). I also find it significant that, in spite of the criticism that use of the S-shaped logistic curve lacks adequate grounding in either microeconomic or communication theory, efforts to dislodge it have been unsuccessful. It has remained remarkably robust as a description of both the diffusion and the substitution processes.[21] Attention continues to be focused, as in the early Griliches study, on explaining the parameters of the diffusion curve.

### Evolutionary Models

The evolutionary models attempt to remedy a number of the perceived limitations of the equilibrium models. They describe diffusion as an

---

21. For an attempt to develop a more secure analytical foundation for the S-shaped diffusion curve see Henrich (2001). Henrich argues that diffusion models based only on "payoff" relevant innovation evaluation information are not capable of generating the S-shaped diffusion curve. The innovation evaluation information must be supplemented by what he terms *cultural-biased information transmission,* information not directly related to costs and benefits, such as the social status of the initial adaptor to generate an S-shaped diffusion curve.

evolutionary process under conditions of uncertainty, diversity of economic agents, and disequilibrium dynamics. And they try to model the complex feedback mechanisms at work at the microlevel among economic agents. In these models, the structural changes induced by the diffusion of an innovation are regulated by changing technological and behavioral diversity, learning, and selection mechanisms. Together with their interrelated feedbacks, these generate continuous adjustment processes to a changing technological environment and result in ordered evolutionary paths at the macro (industry) level. The basic task in the development of an evolutionary (or a self-organizing) model consists in representing the feedback loops between the structure of an industry, the behavior of firms, and the evolution of the industry (Grübler 1991, 8, 29; Lissoni and Metcalfe 1944, 120–26; Katz and Shapiro 1986; Arthur and Lane 1994; Dosi, Orsenigo, and Silverberg 1986; Silverberg, Dosi, and Orsenigo 1988).

Giovanni Dosi and his colleagues at IIASA have outlined an exceedingly ambitious agenda for the evolutionary modeling of the innovation-imitation-diffusion process. They accept the challenge that earlier research had given inadequate attention to the interactions among technology development and diffusion. A fully developed evolutionary model should be capable of incorporating: (1) the characteristics of each technology (sources of basic knowledge; degrees of appropriability; tacitness of innovation; complexity of research, production, and products; existence and role of various forms of economies of scale; and cumulativeness of technological learning); (2) the degrees and forms of diversity between economic agents (including their technological capabilities and variety of search procedures and behavioral rules); and (3) the endogenous evolution of incentives, constraints, and selection mechanisms (including the evolution of relative profitabilities of different technologies, firm sizes, cash flows, and market shares) (Dosi, Orsenigo, and Silverberg 1986, 8, 9). In practice, substantial compromise, when compared to the preceding objectives, has been necessary. Nevertheless, Dosi and his colleagues have been able to develop simulation models incorporating capital stock innovations that generate industry-level diffusion curves similar to those that describe historical industry diffusion patterns.

In his evaluation of recent progress in the development of evolutionary models, which he regards as more promising than the equilibrium models, Grübler notes that the outcome of a simulation run is the result of the particular technological, market, and behavioral variables assumed. He argues that it is premature to discuss the results of a simulation in terms of how well it represents reality. Instead the model demonstrates only the dynamic implications of the assumptions used in constructing a self-organizing evolutionary model. The main lesson to be learned is that the dynamic interaction between the macro- and microlevels in such a system leads to the emergence of spatial and temporal patterns that are driven, rather than dissipated, by microlevel diversity. From this perspective, regularity in evolutionary paths at the macrolevel is not a contradiction but rather a consequence of the diversity of technological expectations, designs, dynamic appropriability, and behavior of economic agents (Grübler 1991, 37, 38). Chari and Hopenhayn (1991) have suggested that the next step in the development of equilibrium models should be a model in which technical innovation and adoption are jointly and endogenously determined. My sense is that the development of such a model would eliminate most of the defining distinctions between the equilibrium and evolutionary models.[22]

**International Technology Transfer**

By the mid-1960s increasing attention was being given to attempts to understand the role of the "technology factor" in international trade (Vernon 1966, 1979; Gruber and Marquis 1969). In 1966 Raymond Vernon of the Harvard Business School proposed a "product cycle" model to explain why new consumer durables tended to be first invented and produced in the United States, why foreign production facilities were established by the innovating firms, and why this was followed by exports back to the United States.

In the first several decades after World War II the U.S. market consisted of a large number of consumers with incomes higher than in any other national market. Thus, the United States offered a unique market opportunity for products responsive to the wants of high-income consumers. The United States was also characterized by high (and rising) labor costs and highly developed capital markets. Vernon noted that although these two factors help explain why new consumer durables designed for high-income consumers were initially developed in the United States, they did not explain why the new products were

---

22. I do not attempt to discuss here the recent development of "information cascade theory" that attempts to explain herd behavior ranging from fads and fashions to crowd behavior. While the problems addressed by information cascade theory are similar to those addressed by diffusion theory, there has been almost no intellectual contact between the two research traditions. See, for example the review of information cascade theory by Bikhcandiani, Hirshleifer, and Welch (1998).

initially produced in the United States rather in lower-wage locations. His explanation was in terms of the evolution of product technology and design in the early stages of the product cycle: "The product itself may be quite unstandardized for a time; its inputs, its processing, and its final specifications may cover a wide range. Contrast the great variety of automobiles produced and marketed before 1910 with the thoroughly standardized product of the 1930s, or the variegated radio designs of the 1920s with the uniform models of the 1930s" (1966, 195). If Vernon had been writing in the late 1990s he might have made a similar comment about the evolution of personal computers.

During the initial stage of the product cycle, feedback from the market about consumer preferences is exceedingly important. Engineers highly skilled in product design and a labor force that can efficiently implement new product technology are also important. These considerations argue for a location in which communication between all those involved in the success of the new product—engineers, research and development personnel, suppliers, financial institutions, and consumers—is rapid and effective. As demand for a product expands, a "technological trajectory" becomes established. Designs become standardized, the need for flexibility declines, mass production becomes feasible, and concerns about costs of production, particularly labor costs, become more important. If the product has a high income elasticity of demand, or is an effective substitute for labor, demand will grow in other countries in which per capita income is converging toward U.S. levels. This demand from abroad will first be met by exporting from the United States and later, as demand continues to grow, by establishing subsidiaries or by entering into joint venture arrangements with firms in other advanced countries. A third stage occurs when production capacity is established in low labor cost countries. Production in these countries may initially include only labor-intensive components. But the establishment of such capacity often results in substantial human capital investment and learning by doing that lead to the establishment of firms that are able to compete with producers in the innovating countries.[23]

The issue of technology transfer has also been an important focus of the new growth economics literature (chap. 5, this vol.). In an influential book published in the early 1990s Gene Grossman and Elhanan Helpman drew on the new growth economics to advance a "new trade

---

23. Hayami and Ruttan (1985, 260–62) have employed a somewhat similar model involving three phases in the process of the transfer of agricultural technology: (1) material transfer, (2) design transfer, and (3) capacity transfer.

theory." They constructed a series of models in which freer trade serves as an instrument of technology transfer. More open trading relationships enable smaller countries in the South to acquire technology developed in advanced countries and to achieve equilibrium growth rates that are higher, even in the long run, than under more restricted trading arrangements (Grossman and Helpman 1991a, 237–57). They analyze several ways that more open trading relationships affect a country's growth performance: (1) more open channels of communication; (2) international competition; (3) international integration; and (4) reallocation of resources. Their analysis suggests that countries with a large high technology sector (and labor with a high human capital component) may experience long-term gains relative to countries with abundant supplies of raw labor (Grossman and Helpman 1991a, 237–38).[24]

But can the transfer of technology from the developed countries of the North enable the developing countries of the South to achieve levels of productivity and income per capita comparable to those of the North? The microeconomic induced technical change literature suggests that transfer of technology to countries with different factor endowments and different relative factor prices will generate fewer gains than in the countries in which the technology was initially developed (Ruttan 2001b). To the extent that there was substantial disequilibrium associated with barriers to trade, freer trade can be expected to result in a narrowing of differences in productivity and income levels between the North and the South. In a recent article, Datta and Mohtadi (2001), building on the Grossman-Helpman work, have shown that as long as the North remains the source of technological innovation, growth rates (but not income levels) will, after an initial catch-up, continue to diverge. But there is an important caveat to this conclusion. The transfer of technology to the South may, as a result of learning by doing and using, formal research and development, and accumulation of human capital, result in a transition from imitation to innovation. Imitation can become the ladder to innovation in the South in a manner not unlike that of the product cycle (Vernon 1966, 1979).[25]

---

24. Grossman and Helpman (1991a, 207–33) have also constructed a model in which a country that is behind in technology can be driven by freer trade to specialize in traditional goods and experience a reduction in its long-run growth rate.

25. Jones and Marjit have suggested an extension of the Vernon consumer durables product cycle model to include labor-intensive "fragments" of formerly vertically integrated production systems that "can be located in countries in which factor prices are well matched to the factor intensities of the particular fragments" (2001, 363). The location of labor-intensive software production in Banalore, India, is an example.

The analytical results obtained by the new trade theorists represent an important contribution to the analysis of trade and growth. The history of Japan over the last century and the more recent history of Taiwan, Korea, and Chile have been interpreted as lending support to the view that trade can represent an important first step in the process leading from imitation to innovation. In spite of the plausibility of the results suggested by the endogenous trade models, however, the implications of the analysis rest on a very weak empirical foundation. Until recently much of the very large, and inconclusive, empirical literature on the impact of trade on growth had failed to address the issue of the effects of trade on technology transfer. A series of studies published since the early 1990s have shown that although technical knowledge has become more global in the last several decades, its generation and diffusion remain surprisingly local (Jaffe et al. 1993; Audretsch 1998; Keller 2001, 2002). The generation of new technical knowledge is highly localized. The technological spillovers of technical knowledge decline with distance. Technical knowledge spreads initially through personal contacts among people engaged in specialized research, development, and production in a limited number of urban innovation centers (World Bank 1993).

The escalating pace of international knowledge transfer lends new urgency to efforts to better understand the role of newly emerging industries in economic development. A recent study of new industry creation by Murtha, Lenway, and Hart (2001) challenges traditional concepts such as the transition from innovation to technical maturity in the product cycle model. In the case of flat panel display (FPD) technology, for example, new generations of technology began to emerge in a matter of a few quarters rather than years in the late 1990s. Successful FPD producers were part of a network of Japanese and U.S. firms located in Japan that were caught up in a rapidly changing knowledge stream in which the concepts of invention and diffusion were no longer distinct. The faster a technology evolves, the greater is the proportion of knowledge accumulation that remains tacit. This means that geographic proximity and social interaction become more important as the speed of technology invention and transfer accelerates (Murtha 2002).

**Resistance to Technology**

Much of the recent technology adoption-diffusion-transfer literature reviewed here has employed an implicit assumption that new technology will be adopted if it is able to pass the market test of profitability. Delays in diffusion, captured in the S-shaped adoption relationship, are assumed to be the result of structural or economic considerations, such as imperfectly competitive factor or product markets, that might delay, but not prevent, adoption.

In contrast to the literature on diffusion of technology, there is a remarkable paucity of serious literature on resistance to technology (Bauer 1995).[26] A continuing theme in economic history has been the source of the large and persistent differences in output per worker among countries in the textile industry (Clark 1987). There is a tradition, running back to the beginning of the industrial revolution, of resistance to laborsaving technology on the part of workers who feared their jobs would be displaced by machines (Randall 1991). Protests by workers against the introduction of textile machines in England between 1811 and 1816 gave rise to the term *Luddite,* after the legendary leader of the protests, as a generic term of denigration for antitechnology protests. It is somewhat surprising that explanations have remained sui generis. Mokyr (1992, 1998; 2002, 218–83) has provided useful accounts of the sources of resistance to technology. But an analytical framework for the interpretation of resistance comparable to the adoption-diffusion model has been slow to emerge. In part, this may be because resistance to new technology (and new institutions) often does not take the form of organized or spontaneous protest movements. James C. Scott (1985, xvi) has explored what he terms "everyday forms of resistance" to authority by workers and peasants such as foot-dragging, dissimulation, absence and desertion, feigned ignorance, sabotage, and pilfering.

In a remarkable new study, *Barriers to Riches,* Stephen Parente and Edward Prescott (2000) have developed a formal analysis of the sources of resistance to the international transfer of technology. They also address the question of why, when technology is transferred, the productivity level achieved with the new technology often remains lower than anticipated. Parente and Prescott insist, somewhat counterintuitively, that "countries are not poor because incentives to develop new technology are lacking. The technologies have been developed in other countries, and it is just a matter of using the technology that is best, given factor prices, and using that technology efficiently" (Parente and Prescott 2000, 142). Parente and Prescott arrive at this con-

---

26. The literature on the diffusion of institutional innovation is even more limited. For examples see Gray (1973), Grübler (1998, 56–58), and Ruttan (2001b, 447–52).

clusion after carefully examining other sources of differences in the level of per capita income. They first show that differences in savings (and investment), including investment in both physical and human capital, as a share of national product are capable of accounting for only a small share of the differences in per capita incomes across countries. They then go on to show, within the framework of a neoclassical growth model, that differences in total factor productivity (output per unit of total input) does account well for differences in per capita income across countries. Relatively small differences in total factor productivity (TFP) across countries (in the range of three times) can give rise to relatively large differences in the level of per capita income (in the range of fifteen to twenty times).[27]

But why do TFP levels differ across countries? Parente and Prescott's response is that it is not differences in the availability of technologies but the differences in the institutional and policy constraints on the use of available technologies that account for the differences. These include constraints on the establishment of new firms and on the modification of work practices. They also include constraints that have the effect of dampening the efficient use of existing technology and that protect the vested interests of domestic monopolies and oligopolies. The importance of the constraints are supported by a series of simulation exercises in which the coefficients are calibrated to reflect the historical experience of contemporary economies such as Japan, Switzerland, and the United States.

The Parente-Prescott analysis provides substantial empirical support for the argument that policy and institutional constraints are important sources of productivity differences across countries. I am, however, uncomfortable with the assumption of a "common world technology" (Parente and Prescott 2000, 84). They finesse this concern by using an augmented production function, which they define to include policy and institutional constraints that can differ across countries. My own sense is that defined this broadly the "common world technology" is only partly available to the entrepreneurs and managers, or even the agronomists and engineers, in poor countries.

Transfer of technology often requires substantial adaptation and reinvention to become viable in locations with institutional arrangements or relative factor and product prices that are different than in the location in which the technology was initially invented. Ronald

---

27. Parente and Prescott insist "that relative income levels rather than growth rates are the key to understanding the problem of development" (2000, xv).

Findlay has insisted: "While the book of blueprints, in some abstract sense, may be open to the world as a whole, one may have to pay a stiff price to look at some of the pages" (1978, 1). The Hungarian economist Janos Kornai, a leading student of the transition from socialism, observed that it is not possible for a society to choose its economic policies and institutions as if it were "pushing a shopping cart down the aisles of an institutional supermarket and select those that have been successful in different countries" (2000, 40n).

There are important sources of resistance to technology that are not fully captured by the Parente-Prescott model. Since the industrial revolution there has been an intellectual tradition that is critical of the impact of technical change on civic culture, social organization, and human and natural environments. One strand in this criticism traces to the perspective advanced by Karl Marx and his followers that economic and political polarization is an inevitable consequence of capitalism and modern technology (Marx 1967, 1968; Lenin 1964). A second strand in this criticism has its origin in the romantic criticism of industrial culture in both England and the United States. In this literature "The Machine in the Garden" has come to serve as a metaphor for the ills of industrial civilization (Marx 1964; Goodheart 1980). These two strands have combined to support a perspective that technology is responsible for the cataclysm of war, destruction of cultural values, and environmental degradation.[28]

Recent literature on the effects of information asymmetry has opened up the possibility of a more rational understanding of resistance to the introduction of new technology. The emergence of formal social structures such as regulatory regimes and of informal institutions and cultural attitudes that slow the introduction of technology can be interpreted as a response to the asymmetry of information availability to the suppliers of technology relative to workers and consumers.[29] Such concerns are only partially resolved by the introduction of analytical techniques such as risk-benefit calculations.

Over the longer term, resistance to technology has seldom been successful. The Luddites have usually lost! Nuclear power is one of the few modern examples of a major on-line technology that has been success-

---

28. For a dramatic example of change in perspective it is useful to compare the early and later work of Lewis Mumford. His 1920s writings, exemplified by *Techniques and Civilization* (1934), presented an optimistic view of the impact of technical change. His later work, exemplified by *The Pentagon of Power* (1964), was deeply pessimistic about the social, cultural, and political effects of technical change.

29. For the seminal information-theoretic literature see Akerlof (1970) and Stiglitz (1989).

fully challenged by protest movements (Rucht 1995). It is an open question, as this chapter is being written, whether agricultural biotechnology might become another example. But even unsuccessful protest movements have often led to important institutional innovations. Among the more important have been labor market reforms, regulation of food and drugs, and legislation designed to limit the negative health and environmental spillover effects of agricultural and industrial intensification. When such protests have been successful it is usually because political entrepreneurs have been able to mobilize substantial political resources. The elements of a model that might provide greater insight for understanding the resistance to technical (and institutional) change have been suggested (in a somewhat different context) by Saleth and Dinar (2002 draft). They insist that there has been an overemphasis on the role of information and an underemphasis on mental models. Those who resist adoption of new technology or new institutions may reject the information-theoretic models employed by sociologists and economists in favor of different models of how the world works, or should work.

**The Divergence of Traditions**

As I completed this chapter I found myself confronted with several puzzles. Why did the convergence of traditions that appeared so promising to students of adoption-diffusion research in the early 1960s falter? And why did research on the diffusion of technology emerge as a major research agenda by economists and technologists just as it was faltering in sociology?

There have been several attempts by sociologists to confront the puzzle. Everett Rogers has argued that part of the answer is to be found in the maturing of the adoption-diffusion research agenda. "The decline of interest in diffusion research by the rural sociologists resulted from the success of the paradigm in answering the major theoretical questions. . . . The paradigm was not found inadequate in its explanatory power but rather it became stale as the main research questions were answered" (1995, 60). Frederick Fliegal (1933) has provided an alternative interpretation. He argued that commitment to survey research methodology deprived rural sociology of the capacity to interpret the more dynamic aspects of the diffusion of increasingly complex technologies. The richness of the invention-adoption-diffusion process in which several closely related components of a complex technology are involved or in which the distinctions between invention, adoption, and diffusion are no longer as distinct as in the case of hybrid corn could not be captured by the classical model or research methods. Both Rogers and Fliegal are partially correct. An implication is that the theoretical and methodological commitments of the first generation adoption-diffusion "invisible college" became a barrier to confronting the larger issues of spatial diffusion and transfer of technology.

But why did the sociologists remain confined within the sociopsychological model that had initially proven so productive? Hägerstrand, as noted previously, had advanced a probability model of spatial diffusion. Griliches had developed an economic interpretation of the spatial diffusion of the hybrid corn invention and adoption decisions. Mansfield had incorporated structural aspects of industrial organization in his analysis of intrafirm and interfirm diffusion. It appears to me, at least in retrospect, that the contentious argument between economists and sociologists over the priority of sociological and economic interpretation played an important role in the failure to develop a model of invention, adoption, and transfer that would draw on knowledge from both sociology and economics.

I suggest three primary explanations for the growth of research on the diffusion of technology by economists, often conducted under the rubric of technology transfer. One is that there is a continuing demand by both domestic and international development agencies for policy-relevant knowledge of the sources and diffusion of technical change (World Bank 2003). A second is that mainstream economists have typically remained skeptical toward the argument that technology is the problem rather than the solution. Agricultural economists in particular have continued to stress that invention and adoption of agricultural technology have been induced by broad economic forces, primarily changes in relative resource endowments and growth in demand arising out of population and income growth (Hayami and Ruttan 1985, 73–114, 255–98; Ruttan and Hayami 1995a, 1995b).

An even more important factor has been the intellectual challenge of explaining the failure of convergence of productivity and income growth between rich and poor countries (Baumol 1986; De Long 1988; Baumol and Wolff 1988; Dollar and Wolff 1993) This issue became the subject of intense intellectual debate in the mid- and late 1980s. Attempts to interpret this failure of convergence in income and productivity have given rise to a new growth economics in which investment in human capital and institutional constraints have occupied a central role in the diffusion of technology (chap. 5, this vol.).

Failure to maintain the momentum of the convergence of research traditions that seemed so promising in the early 1960s clearly remains an obstacle to a richer understanding of the diffusion of technology. Advancement of understanding of knowledge about the sources, processes, and consequences of the diffusion of technology is inherently an interdisciplinary project. It will be necessary to draw on psychology, sociology, economics, geography, technology, and history in order to advance the project. It is important that sociologists who value the research tradition that established the adoption-diffusion paradigm as a field of research contribute to this broader research agenda.

CHAPTER 7

# Social Capital and Institutional Renovation

During the 1970s, in the most environmentally degraded region of Burkina Faso, one of the poorest countries of sub-Saharan Africa, farmers constructed earthen dikes to harvest water and retain soils across an estimated 60,000 hectares (150,000 acres). In the following decade, finding that porous stone dikes were more effective and durable, they covered thousands of hectares with stone dikes—despite an estimated labor input of up to two hundred person-hours per hectare (Sanders, Nagy, and Ramaswamy 1990; Critchley 1991).[1]

Although the labor cost of dike construction to individual farmers might seem prohibitive, the water-retention technology enhanced the private profitability of a recommended technical package of improved sorghum seed and modest levels of fertilizer (ICRISAT 1985; Sanders, Nagy, and Ramaswamy 1990). The stone dikes, a traditional technique that was improved by Yatenga farmers and members of the nongovernmental organization Oxfam, were in large part diffused by the *Groupements Naam* (Harrison 1987). The *Groupements Naam* are mutual assistance groups derived from the traditional age-set associations of Mossi society.

The contour dike construction by the *Groupements Naam* illustrates the role of social capital in economic development.[2] The traditional

---

1. For an earlier version of this chapter see Smale and Ruttan (1997, 182–99). I am indebted to Klaus Deininger, Karlyn Eckman, David Feeny, Elinor Ostrom, John Sanders, Earl Scott, Robert Tripp, and Fred Zimmerman for comments on the earlier version and to Habib Fetini for introducing me to Bernard Ouedraogo and the *Groupements Naam*.

2. Social capital is the shared knowledge, understandings, institutions, and patterns of interactions that a group of individuals brings to any activity (Ostrom 1997, 156–60). "Physical capital is wholly tangible, being embodied in observable material form; human capital is less tangible, being embodied in the skills and knowledge acquired by an individual; social capital is less tangible yet, for it exists in the *relations* among persons" (Coleman 1988, S101). For discussion of the role of social capital in social development see chapter 3, this volume. For other case studies of the role of social capital in economic development see Montgomery and Inkeles (2001).

cultural endowments of the Mossi were the foundation on which the social capital of an indigenous development organization was built. In the case of the dikes, membership in the indigenous development organization provided incentives for farmers to implement a costly technical innovation that enabled them to raise the yields of their staple food.

The story suggests several hypotheses that are related to the role of cultural endowments and the accumulation of social capital in the process of economic development. First, specific cultural endowments can facilitate the formation of indigenous development organizations. Although culture is frequently viewed in the economic development literature as an impediment, cultural endowments such as the tradition of the strongly disciplined, self-reliant *naam* groups may facilitate cooperative arrangements. An important theme in the New Institutional Economics is that, while policies and incentives matter most in determining the path of economic development, the incentives faced by individuals are strongly conditioned by cultural endowments (Clague 1997, 13–36). Cultural norms can, for example, affect the time it takes to develop the set of mutually agreed upon and enforceable rules that serve as the foundation for social capital (Ostrom 1997, 2000).

Second, such indigenous development organizations can create incentives for the adoption by farmers of new technologies. In the example presented here, the rules of conduct and function in the indigenous *Groupements Naam* movement facilitated the construction by communities of costly water-retention infrastructure. This infrastructure, in turn, enhanced the private profitability to individual farmers of a seed-fertilizer innovation. This hypothesis bears on the study of collective action and its applications. Third, foreign institutions are less likely to be successfully transferred than "renovated" institutions.[3]

The following sections outline the cultural endowments of the Mossi that were recognized by the founder of the *Groupements Naam* movement as instrumental to their formation; summarize how economists, political scientists, and anthropologists explain the existence of similar mutual assistance groups; describe features of the institution-building process and how the *groupements* may have facilitated technical innovation; and discuss issues affecting the continuity and replicability of

---

3. Fetini (1993) defines three archetypal models of institutional development: innovation, adaptation, and renovation. Renovation consists of building a new institution from an existing—but no longer functional—institution. Fetini characterizes some aspects of the farmers' group movement in the Sahelian region of Africa, and in particular that of the *Groupements Naam*, as institutional renovation.

the movement and implications of the *Groupements Naam* experience for donor assistance.

### The *Kombi-Naam* Cultural Endowment

Those cultural endowments most clearly instrumental in the formation of today's *naam* groups are expressed in the operation of the *Kombi-Naam*, one of the traditional mutual assistance organizations of the Mossi who inhabit the Yatenga province of Burkina Faso.[4] Here, the term *cultural endowment* refers broadly to the dimensions of culture, including religion and ideology, that have been transmitted from the past.[5]

The *Kombi-Naam* was a temporary association of young men and women from the same age group, reestablished spontaneously each year when they gathered to choose leaders and designate annual group activities. They worked in common fields, provided labor in the fields of those who demanded or were in need of special assistance, and were remunerated in kind according to the nature of the labor and the means of the household employing them. Certain other vital community tasks were collectively and inexpensively completed by the youth association. At the close of the year, to mark the dissolution of that particular realization of the *Kombi-Naam*, they organized a festival to which the youth of other villages were invited.

Several features of the *Kombi-Naam* may be said to compose the basis of the social capital upon which B. L. Ouedraogo sought to build the *Groupements Naam* movement. First, membership and internal leadership cut through the hierarchy to include individuals from all socio-occupational sets, including minorities and servile groups. Strict sexual mores were enforced, to encourage young men and women to learn mutual respect in a cooperative work environment. For a moment in time, all villagers of that generation were "equal." The second special feature was the method for selecting leaders. In contrast to Western definitions of democracy as majority rule, Ouedraogo describes the traditional election as an exercise in "qualitative democracy." Consensus was achieved when unanimity emerged from a long

---

4. The description of the *Kombi-Naam* is drawn from Ouedraogo's (1990) account, which he based on interviews with elders who had been courtiers of the last king of Yatenga. Ouedraogo is the founder of the *Groupements Naam* movement. A similar description is found in a manuscript by Skinner (n.d.), who is a scholar of Mossi society.

5. There has been a substantial evolution of the concept of culture in the anthropology literature (see appendix, this vol.)

process of discussion and mutual concessions. A third special feature is that the *Kombi-Naam* leadership included some dignitaries who represented institutions of the greater traditional society. The link between youth and elders mediated intergenerational tensions in a society in which age confers status. The content of the cooperative work activities, such as cleaning the village mosque, increased positive intergenerational contact.

The signal feature of the *Kombi-Naam,* for purposes of this study, is the tradition of rigorous discipline and the denial of opportunism to which the young people freely consented, given that unanimity had been achieved for certain key decisions. As late as the 1960s, Hammond (1966, 91) observed that "the young people set to work without supervision after the leader and farmer have agreed upon a price . . . work is performed collectively and profits shared equally."

Other related facts about the historical evolution of the *kombi-naam* may also be of importance. The concept of *naam* is a dominant leitmotiv in the lives of the Mossi of Burkina Faso (Skinner 1964; Izard 1985a, 1985b). Literally, *naam* means chieftainship (Ouedraogo 1990), sovereignty or power (Izard 1985b), or the power first possessed by the ancient founders (Skinner 1964). Philosophically, the Mossi refer to it as that "force of God that enables one person to dominate others" (Skinner 1964, 13). Many of the key economic, political, religious, and social institutions of the Mossi are patterned around *naam* and are designated by titles derived from that word (Izard 1985b; Skinner 1964).

What limited the potential of the *Kombi-Naam* as a development organization? First, the experience in equality, self-reliance, community service, and mutual assistance lasted only one year. Membership in each realization of the *Kombi-Naam* was only temporary. There was no time for individuals to reap long-term benefits from a current investment in mutual assistance. Youth carried into adulthood the sense of having participated, but the titles and active solidarity were embodied only in the *Kombi-Naam* institution. Servants then undertook their hereditary occupation as servants, and young women prepared to become wives and mothers in a patriarchal society. Second, the *Kombi-Naam* accumulated no capital. At the close of the year, its earnings were exhausted in one ceremony.

**Understanding the *Kombi-Naam***

Mutual assistance groups and solidarity networks are a common feature of many traditional societies. Anthropologists, political scientists, and economists have long debated why such groups exist and how effective they are. In the 1970s this debate centered around the "moral economy." Scott (1976) interpreted the existence of mutual assistance and solidarity networks in Southeast Asia in terms of the "subsistence ethic" of peasants, which reflects, in a "moral economy," the overriding common need to organize against food crises. Opposing Scott's view, Popkin (1979) argued that collective action often fails because of the opportunistic behavior of peasants who have limited incentive to pursue cooperative strategies. Posner (1980) attempted to reconcile the views of Scott and Popkin by arguing that mutual assistance systems can be sustained in the long run by the existence of a lasting relationship between self-interested members. Similarly, anthropologists have long recognized that gift giving is a form of commodity exchange that, as compared to an impersonal exchange of goods in a market, can establish a durable personal relationship between donor and recipient (Mauss 1954; Gregory 1982). Posner's argument has been formalized in work by Kimball (1988). Related literature, models, and arguments are exhaustively and critically reviewed by Platteau (1991, 2000) and Fafchamps (1992).

Political scientists and economists have advanced a game-theoretic interpretation as a formal explanation for the existence of mutual assistance groups and cooperation. They solved Olson's free-rider problem (1965) by showing that in repeated, rather than one-shot games, there is no dominant strategy. Wealth- or utility-maximizing individuals find it worthwhile to cooperate with other players when the game is repeated, when they possess information about other players' past performance, or when there is a small number of players. Building on the extensive literature and expanding the game-theoretic approach, Fafchamps (1992) demonstrated how key observed features of mutual assistance groups and solidarity networks can be explained through the theory of repeated games. Using a different approach with game theory, Ostrom (1990, 2000) demonstrated simply that a contract enforced by unanimous approval of the rules, such as in Bernard Ouedraogo's "qualitative democracy," can result in a cooperative equilibrium in which agents share costs and returns from common property.

The use of game theory in this context has also been criticized. Fafchamps (1992) has argued that the theory of repeated games does not explain how specific organizational structures or mutual assistance associations are chosen. The elements needed to build such a theory would include culture, political institutions, and historical events. Fafchamps argues, in effect, that though game theory can be used to demonstrate why a particular institution assures a stable equilibrium, it cannot explain why that institution evolved historically. More recently Elinor Ostrom (2000) has employed a unique combination of political theory, field investigations, and game theory–based public good experiments to distill a set of design principles for long-surviving, self-organized resource management regimes. These design principles are listed in chapter 4 (box 4.1) in this volume. They are effectively illustrated in the *Groupements Naam* experience.

Some mutual assistance groups and solidarity networks can be viewed as special cases of risk-pooling insurance mechanisms in which villagers organize vertically (as in patron-client relationships) or horizontally (as in the original *Kombi-Naam*) to protect themselves against famine and other disasters. In land-abundant, semiarid countries with simple technology, a high covariance of risk in crop or livestock output helps to explain the lack of formal insurance mechanisms and the need for geographically extensive social institutions or private capital accumulation in the form of grain stocks or livestock (Binswanger and McIntyre 1987). For example, analysis by Nugent and Sanchez (1993) highlights the role of tribes in the transhumant (long-distance herding) activities of semiarid regions. As population density increases, however, the propensity to rely on such extensive social institutions as insurance mechanisms can be expected to decrease because of the reduced cost of the infrastructure and access to economic activities with less covariation (Binswanger and McIntyre 1987; Ruttan and Bogerhoff-Mulder 1999). Although the *Kombi-Naam* were organized horizontally, the only risk they were designed to cover, or could cover effectively, was uncorrelated and specific, such as providing labor to assist the sick or needy.

### Renovating the *Kombi-Naam* as a Development Organization

Bernard L. Ouedraogo, the charismatic leader of today's *Groupements Naam* movement, initiated a process of building a community development institution from the traditional *Kombi-Naam*.[6] In his own words, he and his coworkers sought "development without damage" (*développer sans abîmer*) (1990, 13). The process of renovation occurred in several stages. First, public education programs designed to provide basic instruction to village children were reorganized as training programs for young farmers (*Formations des Jeunes Agriculteurs,* or

---

6. Much of this section is drawn from Ouedraogo (1990).

FJAs). The FJAs were then broadened to include postschool farmer youth groups (*Groupements des Jeunes Agriculteurs,* or GJAs). In 1967, the GJAs were then "grafted" onto the *Kombi-Naam* by redefining rules of function and organization.

As in the *Kombi-Naam,* the GJA-*Naam* groups maintained close ties with village elders. To gain acceptance in a village, the GJA-*Naam* groups provided assistance to village elders, retaining an honorary presidential post for elders (usually a former soldier who had traveled) and in return obtaining land-use rights from elders, borrowing their plows and oxen. As in the *Kombi-Naam,* membership in the GJA-*Naam* was open to individuals of all ethnic and socio-occupational groups. Leaders were selected through a consensus process similar to that used in the *Kombi-Naam,* although the leadership functions were defined in new ways. The general functions of the GJA-*Naam* groups were newly defined as educational, political, recreational, and economic.

Ouedraogo recognized that temporary association and an inability to accumulate capital were inherent features of the *Kombi-Naam* that needed to change before the traditional institution could be used as a development tool by the community. Solving the capital accumulation problem in the economic function of the GJA-*Naam* entailed convincing members that the earnings, formerly exhausted at the annual festival, could be more usefully directed toward collective savings and investment. According to Ouedraogo, this was no easy task. The effort to prolong the life span of the *naam* groups seems to have been reinforced by concurrent changes in resource endowments. Those who joined the GJA-*Naam* in the 1960s remained members well into adulthood in the 1970s. Ouedraogo and others (Gentil 1986; Pradervand 1989) contend that the 1967–85 period of low rainfall, which included the severe droughts of 1967–73 and 1982–84, contributed to the proliferation of the *naam* groups. To reclaim degraded land through the construction of dikes, dams, reservoirs, and other public works, villagers needed cooperative, collective action. In terms of the induced innovation framework, a change in resource endowments (the drought) increased the demand for an institution that supported collective action. Cultural endowments (the *Kombi-Naam*) facilitated cooperative arrangements by reducing the cost of achieving consensus.

The *Groupements Naam* of today are a federation of associations of varying size and composition, including young and old, men and women. Crudely, the "renovation" of the *Kombi-Naam* extended the payback period for investment in mutual assistance and diversified its portfolio.

## *Groupements Naam* and Technical Innovation

In the Sahelian and Sudanian climatic zones of sub-Saharan Africa, the principal production constraints are lack of soil moisture and low soil fertility. Land degradation has caused even further deterioration in soil quality. High capital and maintenance costs have deterred investments in conventional, large-scale irrigation projects over the past two decades (Matlon 1987). Other water conservation or retention techniques are potentially important, particularly in degraded Sudanian regions like Yatenga. In such areas, rainfall is low and irregular, and soil encrustation leads to infiltration problems. Water retention techniques can reduce runoff and help exploit rainfall by increasing the effectiveness of nutrients, especially when combined with improvements in soil fertility.

Sorghum and millet are the staple cereals in the Sahelo-Sudanian zones of Burkina Faso. Based on information from nearly a decade of research on new technologies in Burkina Faso, Sanders, Nagy, and Ramaswamy (1990) concluded that only with increased soil moisture and moderate fertilizer application (organic or inorganic) are improved sorghum varieties more attractive than traditional varieties in terms of yield potential, profitability, and risk. Increasing soil moisture when the nutrient levels remain low does not generate large increases in yield; applying fertilizers without an assured water supply is risky since the response to fertilizer depends on the availability of water at critical stages of plant development. As these scholars portray it, technical change in the Sahel is a staged process, depending on initial soil conditions and expected rainfall; the adoption of technologies to improve water retention and soil fertility is the precondition for the productivity increases associated with change in plant varieties (see also Sanders and Ramaswamy 1992).

Using the induced innovation framework, Sanders, Nagy, and Ramaswamy (1990) also argue that, with the Sahelian drought, changes in resource endowments spurred the demand for technical innovation. As soil resources declined through depletion and erosion, and as nonfarm employment opportunities (even through migration to the south) grew less rapidly than the rural population, farmers were induced to adopt yield-increasing, labor-intensive technologies such as contour dikes and organic fertilizer. Sanders, Nagy, and Ramaswamy (1990) report that the region of most rapid adoption of water retention and soil fertility technologies was Yatenga, even though the estimated rates of return to the technologies were higher in other, less degraded regions. They explained this anomaly by hypothesizing that, unless farmers are pressured by soil deterioration and falling labor-land price ratios, the implicit returns on these large labor inputs are too low to interest farmers. In other words, changes in relative returns to factors, rather than the absolute level of returns, encouraged the adoption of new technologies.

One of the more effective dike (alternatively, *diguette,* bund, contour ridge, or contour line) techniques had its genesis in Mossi methods. Based on fieldwork in western Burkina Faso, Savonnet (1958) identified and described in detail four techniques for soil erosion and water control used by the Mossi and other ethnic groups. These and other techniques used historically by the Mossi are also reported in Reij (1989). Some traditional techniques had fallen into disuse. The efficiency of stone and earthen lines was limited. Contours were not accurately measured, and other construction details such as stone placement and line spacing needed improvement.

Harrison (1987) called the process of dike improvement in Yatenga "barefoot science." When Oxfam began work on an agroforestry project in Yatenga in 1979, one of their fieldworkers brought the concept of water harvesting from a visit to the Negev desert in Israel. In emphasizing that their priority was food production rather than agroforestry, farmers helped shift the design of the project. The farmers and the Oxfam project director set to work improving the simple stone contour line. One improvement involved placing large foundation stones preceded by smaller stones to act as a permeable water filter (Critchley 1990). But accurate contours are impossible to gauge by eye on the slight slopes (2–3 percent, according to Reij, Mulder, and Begemann 1988) in Yatenga. Aside from questions of spacing, breadth, and depth, the key was the development of a simple water level made of plastic hose pipe that cost six dollars to make, could be mastered by illiterate villagers in a day or two, and ensured correct alignment of the contours (Wright 1985; Harrison 1987). The *Groupements Naam* groups subsequently adopted the technique and diffused it rapidly (Younger and Bonkoungou 1985; Harrison 1987).

In the short term, the contour dikes improve crop yields in two ways. First, controlled rainfall runoff improves infiltration, increasing water absorption by crops and land. Second, fertilizer (usually manure) and other organic material applied behind the dikes is much more effective since it is less likely to be washed away. The organic matter also attracts termites, which bore into the ground and aerate the soil. Over the longer term, the dikes control erosion and increase soil quality.

Evidence concerning the economic rate of return to contour dikes is positive but patchy and inconclusive. According to Sanders and Ramaswamy (1992), the yield effects of contour dikes combined with organic fertilizer placed behind the dikes were "small but the diffusion is impressive" (249). In 1985, ICRISAT staff evaluated a package of stone dikes, tied ridges, a low dose of fertilizer, and an improved sorghum variety. On farmers' fields, where only the package was evaluated, the yield difference was 67 percent in the first year in the Sahelian (less favored) zone. ICRISAT calculated that a break-even sorghum yield increment of only 155 kg would assure a return of 15 percent on labor and cash investment. This increment was exceeded by over two-thirds of farmer participants in the Sahelo-Sudanian zone. On researcher-managed fields in the Sahelian zone, the yield increment from stone dikes alone was 40 percent in the first year. Oxfam's experiments were less controlled, but their data generally showed statistically significant yield increases when stone contours were present on farmers' fields, with the largest difference occurring in dry years (Wright 1985).

Using ICRISAT's cost data and some of the lower yield differentials in Wright's data, Younger and Bonkoungou (1985) estimated an internal rate of return to the Oxfam project (excluding the diffusion outside of the project zone by the *Groupements Naam*) of about 40 percent. For newly diked land they calculate a rate of return of 147 percent (excluding sunk R&D costs).

The existence of institutions such as the *Groupements Naam* clearly affects the calculated rates of return to investment in technologies such as the contour dikes and other water retention infrastructure. As North (1990a, 5) argues, institutions affect the costs of exchange and production by reducing transaction and transformation costs. Rate-of-return calculations do not capture the external effects of dike construction by one farmer on other farmers' fields or the possible cost-reducing effects of construction by the *Groupements Naam* as compared to construction by either an individual or a construction firm. Dikes are a divisible but lumpy technology. The only portion of the benefit that is measured in conventional rate-of-return analyses is the private benefit to the farmer applying fertilizer and planting an improved cultivar behind the dike.

**The Continuity and Replicability of the *Groupements Naam***

Supporters, critics, and Ouedraogo himself questioned the potential for continuity of the *Groupement Naam* movement and its replicability in other regions of sub-Saharan Africa. Movements of this type are widespread in the Sahel, particularly in Senegal and Burkina Faso. Gentil (1986) cites a number of factors that have been related to the evolution of successful farmers' movements in West Africa. The first is the openness of the political climate. The governments of Senegal and Burkina Faso have been relatively tolerant of trade union activity, multipartyism, and a free press. He notes, however, that most successful groups serve at least some government interests. For example, the economic activities of farmers' groups in Senegal have never threatened the cash crops that are the lifeline of the Senegalese economy. Charismatic leaders also seem to play a crucial role. A unifying ideology, a "hostile" external force, or a major obstacle to overcome can also to help rally support and enforce cohesion.

Both economic and political issues affect the potential for continuity in the *naam* movement. Critics most often question its economic viability. Bernard Ouedrango indicates that basic organization of savings and credit in *naam* groups is simple: all earnings are deposited in a bank account and distributed in three unequal parts, depending on the needs of the group. The first part is seed money for a revolving fund or for repaying bank loans. The second, smallest, part is an expense account for funding annual harvest festivals. The third part is for expansion of activities through purchase of factors of production or for livestock and tool replacement. The drawing of individual loans from the account, such as consumption credit, is prohibited.

To secure the necessary seed capital for *naam* groups, its creators insisted that, in the first phase, capital should be self-generated, "so that potential supporters and detractors of the organization could take it seriously" (Fetini 1993). In 1976, to meet the challenges of training farmers in both technical spheres and project development, and to relieve key constraints on scientific know-how (human capital) or physical capital in projects they identified, Bernard Ouedraogo and a colleague (Bernard Lecomte) established an umbrella organization, the 6-S (*Se Servir de la Saison Sèche en Savane et au Sahel*), that included the *Groupements Naam* and other local and international nongovernmental organizations across the Sahel. Although the 6-S expanded the capital base of the *naam* movement, a large proportion of its activity involved the provision of reimbursable loans.

Central to the financial functioning of the 6-S was the notion of *fonds souples* (flexible funds), as contrasted to project aid. Initially, modest funds are made available to the *naam* federation without earmarking them for specific end uses. The funds were distributed only when a group showed sufficient initiative and were increased only as the group demonstrated management ability. When a group's project became profitable, the funds were reimbursed and the money was recycled to the project of another group. At the level of the 6-S, the funds were disbursed as grants, but they are used as a 70 percent reimbursable loan by *naam* recipients. That 70 percent, when repaid, was placed in a revolving fund to be extended to poorer *naam* groups. Funding diminished or ceased when groups demonstrated enough efficiency to attract other donors (Fetini 1993).

As *naam* groups proliferated and as the range of their commercial activities expanded, some groups and some activities were clearly more successful than others. Ouedraogo (1990) stressed the significance of the villagers' acceptance of the revolving fund concept, toward which they were originally suspicious. He related how the concept was explained to villagers and how, in the ideology of the *naam* movement, they were constantly reminded of how their actions relate not only to their own group but also to the potential of other groups. The concept of repayment with interest and the prohibition on giving credit or food to the needy from the group account were difficult problems, but usually not insuperable. In some cases, more flexible arrangements were developed.

*Naam* groups engage in a portfolio of activities, only some of which generate cash income. Their commercial activities included the purchase, installation, and operation of grain mills, the constitution and regulation of village grain banks, livestock production, artisanry, vegetable production on irrigated plots operated by the collective, and petty trading. Some observers (Beaudoux and Nieuwkerk 1985; Gentil 1986) expressed reservations about the economic performance of some of these activities.

The very breadth of *naam* activities may also be an important key to their success in adopting and diffusing agricultural technology. Many failed efforts at community action in similar projects (agricultural experimentation and terrace building) were organized to carry out only a limited set of tasks. The transaction costs involved in carrying out limited specific tasks were too high to be sustainable. The lesson to be learned may be that, in encouraging one community-based activity, it is better to base the work on existing institutions and to ensure that the institutions carry out a broader set of functions (R. Tripp, personal communication).

But Ouedraogo (1990) and others further emphasized that commercial activities are in some sense of lesser importance to the long-term vitality of the group than the community and social activities that lend the group its fundamental cohesion. Ouedraogo regarded the water retention work as one of the principal activities of a *naam* group, although he considered it a community rather than an economic activity. In his survey of several hundred groups, Buijsrogge (1989) was startled to find that, among the identified future intended activities of these groups, most were public works. He expected greater interest in profit-making enterprises, for which incentives are obvious. He explains his finding by recognizing that, for many of the villages beset by drought and youth out-migration, community action is a question of survival. To maintain viable communities, it is necessary to retain the population. Questions of environment and soil fertility, for example, were the foremost concerns.

Over the long run, political viability seems likely to represent a more serious constraint than economic viability. Assembling data from across West Africa, Gentil (1986) described farmers' movements as consisting of three types of groups: (1) groups promoted by a nonstatal apparatus, such as a church or nongovernmental organization; (2) groups initiated by farmers but linked closely to a charismatic, nonfarmer leader; and (3) groups initiated and sustained by farmers and farmer-leaders. He claimed that a large proportion of the groups were created and dissolved within a short period of time. He described many of the groups as artificial and entirely dependent on external organizations.

The *Groupements Naam* have clearly been initiated by, and derive their strength from, village communities. Pradervand (1989) refers to the 6-S as the first international organization run by peasants. Harrison wrote that in the 6-S, professionals and experts assumed the proper roles of training and technical and financial backup, but that "the heart of the *naam* is . . . in the villages" (1987, 282). But the unique qualities of groups like those of the early *naam* movement have often succumbed to externally driven development fads and fashions. Certainly the *naam* federation of today is likely to include more "collective opportunists" than in the early years before government acceptance and the establishment of the 6-S. The *naam* movement was officially recognized in 1978 by the government of Burkina Faso, and the *Groupements Naam* were then federated and linked to a broader national structure. Most of the groups in Buijsrogge's survey (1989) were formed around 1974 through 1977, when the government began to pursue specific community development policies.

A remarkable aspect of the *naam* movement is how its leaders have been able to work effectively with revolutionary regimes in Burkina Faso as well as in other parts of West Africa (Skinner n.d.). A more philosophical question is, however, the extent to which the views of farmers, rooted in their own experience and localities, can continue to mesh with those of national politicians and international development organizations. Ouedraogo himself (1990) concluded that the future of the federation in the 6-S, in terms of retaining a sense of village base and control, was uncertain. As the scale of the *Groupements Naam* bureaucracy and the involvement of its leaders with national politics and donor culture increase, how will the institution evolve?

Another interesting dynamic in the development of the *naam* movement is its relation to the national government and what that means for the structure of political power. Through the *naam* movement, a substantial portion of the labor-intensive work involved in constructing infrastructure and of the costly work of on-farm research and demonstrations was transferred from the state to its poorest citizens. One might hypothesize that some "enlightened" governments may choose to use farmers' groups to shift the burden of development costs back onto rural communities. Some governments may find it in their interest to co-opt successful farmers' groups to consolidate their own political bases. The case of farmer groups in Bakel, Senegal, detailed by Adams (1981), demonstrated that at least some associations, perceiving government recognition as potentially exploitative, have resisted assimilation.

**Conclusions**

The water retention work of the *Groupements Naam* of Yatenga, Burkina Faso, represents a specific example of the relationship between cultural endowments and the formation of social capital. Culture affects economic outcomes by influencing the formation and composition of the social capital that conditions economic behavior. In the case of the *Groupements Naam,* specific cultural endowments of the Mossi in Yatenga appear to have facilitated the formation of a development organization through the renovation of a traditional institution. The rules of conduct and function embedded in the social capital of the new institution seem to have provided special incentives for communities to undertake costly public works in water retention infrastructure. This infrastructure, in turn, enhanced the profitability and risk characteristics of a recommended technical package of improved sorghum seed and fertilizer relative to traditional varieties. Cultural endowments, therefore, appear to have played a role in raising the potential for technical change in staple food production.

The implications for donor assistance are not so clear. One obvious implication is that donor, and national government, involvement would need to be much more sensitive to indigenous institutional arrangements and cultural endowments. Another, suggested by the success of the flexible funds approach, is that project-based funding may be damaging for the development of indigenous social capital. Conventional rate-of-return analyses are too limiting and project life cycles are too short to serve as a bases for evaluation. The story also suggests that donors should avoid assuming that institutions must be transformed or built anew. Donors should show greater sensitivity to the possibility of traditional institutions as a potential foundation for economic development.

One hypothesis about how donors might encourage the development of similar institutions is suggested only peripherally in this account. The *Groupements Naam* have survived several regimes in Burkina Faso and are part of a transnational federation of farmers' groups that spans the Sahel. Government recognition of the rights of indigenous communities to organize has been exceedingly important. Central governments that have been insecure in their command over political resources have often been intolerant of successful ethnic or communal-based organizations. In the case of Burkina Faso the *Groupements Naam* have flourished precisely because the national government saw it as a means of consolidating its rural political base.

CHAPTER 8

# Religion, Culture, and Nation

The closing decades of the twentieth century witnessed a sharp transition in the grand themes that had provided the context for development thought and practice during the previous half century. The process of political decolonization was followed by concerns about the implications of the globalization of economic activity. The fall of the Berlin Wall, which signaled the end of the Cold War, eliminated tension between the United States and the USSR as a source of discipline in international relations. A new set of issues, including the rise of militant fundamentalist religions, the role of cultural endowments in economic and political development, and the ambitions of the constituent nationalities in multinational states, forced themselves onto the development agenda.[1]

These issues have not been easy for social scientists to address. Students of the sociology of religion have found it easier to study the evolution of religious thought and practice than to agree on the significance of fundamentalist religion for political and economic development. The analysis of the political and economic consequences of national culture no longer commands credibility among anthropologists. The disintegration of multinational states has challenged academic thought on the processes of nation building. These issues have largely been neglected by development economists. In this chapter I attempt to confront some of their implications for political and economic development and practice.

**Fundamentalist Religion and Economic Development**

To a world that viewed itself as becoming modern, or even "postmodern," the emergence during the late twentieth century of religious

---
1. I am indebted to Robert T. Holt and Timur Kuran for comments on an earlier draft of this chapter.

movements that were "intense, impassioned, separatist, absolutist, authoritarian and militant" has been difficult to comprehend (Marty 1996, 24).[2] At the same time that the "death of God" was being celebrated by liberal theologians (Cox 1969), the United States was experiencing a prolonged religious revival (Gilbert 1997, 13). Fundamentalist religious movements, previously regarded as fossilized remnants of premodern culture, were emerging with renewed vitality to challenge the assumptions of both democratic and authoritarian order.

But are there sufficient common elements in the diverse movements that have arisen from Christian, Islamic, Jewish, Hindu, Sikh, and Buddhist religious traditions to qualify for a fundamentalist (or fundamentalist-like) appellation? One criterion is that all of the fundamentalist movements have defined themselves in opposition to modernity. They are vigorous critics of what they regard as the compromise of traditional religious institutions and the erosion of fundamental religious principles, in response to the corruptive forces of modernization. A second common feature is the rejection of the liberal principle of the separation of the sacred and the secular—the separation of church and state. Another is that most fundamentalist movements have arisen as a reaction to modern secular pluralistic social development in which the cultural constraints and support networks of the institutions of traditional preindustrial society were undergoing serious decay.[3]

---

2. This section was originally written in the summer of 1998 as a review of studies on fundamentalism and economic development conducted as part of the Fundamentalism Project directed by Martin E. Marty, Fairfax M. Cone Distinguished Service Professor of Modern Christianity at the University of Chicago, and R. Scott Appleby, Director of the Cushwa Center for the Study of American Catholicism and Associate Professor of History at the University of Notre Dame. The Fundamentalism Project was sponsored by the American Academy of Arts and Sciences and funded by the MacArthur Foundation. The results of the project, initiated in 1987, are reported in Marty and Appleby (1991b, 1993a, 1993c, 1994, 1995).

3. The appropriate application of the term *fundamentalism* has been vigorously debated. A case can be made for limiting its use to its initial manifestation in North American Protestantism where it originated from the Great Awakenings of the eighteenth and nineteenth centuries and from the biblical scholarship, initially centered at Princeton Theological Seminary, that attempted to provide a rational defense of the Bible as the authoritative and inherent revealed word of God. The term has been extended to include other evangelical Christian movements, although it rests uneasily on the twentieth-century Pentecostal and other charismatic churches that continue to seek direct divine revelation. A strong case has also been made for extending the term to include antimodernist movements within Judaism and Islam, the other religions of the Abrahamic tradition, who draw inspiration from a divinely inspired sacred text. Another argument has been made for including the religionationalist Adi Granth movement in Sikhism because of its foundation in a presumed divinely revealed scripture. Finally, the case for extension to the militant religiocultural movements in Hinduism and Buddhism has been made in the Fundamentalism Project on the basis that their inclusion

Almost all fundamentalisms are grounded in an absolute truth that generally, but not always, draws on a particular sacred scripture. With few exceptions, such as Judaism, they also share a common missionary zeal to reform and convert society to their way of life. While they tend to be characterized by an enclave culture—a "wall of virtue"—designed to shield their members from the morally defiled culture of the "world," they do not practice an ascetic monasticism nor do they attempt to avoid the accumulation of material wealth. In effect they embrace much of the technology and the materialism of the modern world while rejecting many of its institutions.

## Fundamentalism and Society

In retrospect it is apparent that the several fundamentalisms, for the most part, have had greater impact on family and social institutions and behavior than on political and economic institutions and policy. "Fundamentalists have expended the greater part of their energies, and have enjoyed greater success, in reclaiming the intimate zones of life in their own religious communities than in remaking the political or economic order according to the revealed norms of traditional religion" (Marty and Appleby 1993b, 5).

Marty and Appleby identify three reasons for the preoccupation and the success that fundamentalists have achieved in the area of personal and family life. One is that the religious traditions from which they emerged have placed strong emphasis on the regulation of personal conduct, family life, and education according to divinely revealed norms. A second is that it is in the realm of domestic relations, in the more intimate zones of life, that fundamentalists can most easily succeed in shaping behavior according to revealed norms and traditional patterns with little resistance from the external society. A third

---

provides a wider base from which to draw inferences concerning the emergence and behavior of the larger universe of fundamentalist and fundamentalist-like movements (Ammerman 1991, 1–65; Marty and Appleby 1991a, 814–42; Oberi 1995, 96–114; Almond, Sivan, and Appleby 1995, 399–424). The characterization of fundamentalism employed in the Fundamentalism Project has been criticized as both too broad and too ethnocentric (Bendroth 1996). For a response to critics see Marty and Appleby (1993a, 18). For an attempt to provide a psychological rather than an institutional interpretation of the fundamentalist movements see Boyer (2001, 265–96). Hart and Negri (2000) argue a causal linkage between past modernism and fundamentalism. They have both arisen not only at the same time but in response to the same global economic and political changes. "Postmodernist discourses appeal primarily to the winners in the process of globalization and fundamentalist discourses to the losers" (150).

explanation for the fundamentalist preoccupation with personal and family life is found in the conviction that the moral and spiritual development of the young is the most certain path to the transformation of society.

Fundamentalists have not, however, attempted to avoid responding to the apparent contradictions between religious and scientific interpretations of natural phenomena. The fundamentalist challenge to science rests on the insistence, particularly by some Christian, Jewish, and Muslim fundamentalists, that all knowledge must be subordinated to a sacred text or texts. In American Protestant fundamentalism this concern has given rise to "creation science"—the natural knowledge required to validate the biblical story (or stories) of creation (Moore 1993a, 50). Among Muslim fundamentalists insistence that the Qur'an is absolute and not subject to revision has given rise to a movement to "de-Westernize" scientific knowledge.[4] The leaders of the Jewish ultraorthodox Hasidic movement insist that "whenever contradiction is to be found (between the literal meaning of the biblical text and scientific conceptions) the scriptural passage should be read in its clear literal fashion" (Ravitzky 1994, 307).

Most fundamentalists are, however, quite comfortable in adopting the instrumentalities and artifacts of modern technology. Contemporary Christian, Jewish, and Islamic fundamentalists have aggressively employed the instruments of modern technology, particularly communication technology, in their efforts to disseminate their messages and mobilize their constituencies. Furthermore, when radical or militant fundamentalists in the developing world have found themselves (as in Iran) faced with concrete problems of health care, population growth, adequate useable water, food production, modern industries, participation in international trade, or the requirements of securing borders and maintaining modern armed forces, then modern science and technology has been embraced as the means by which social problems may be effectively addressed.

Any effort to confront the implications of fundamentalist and fundamentalist-like movements for economic development must address the implications of fundamentalism not only on the adoption but also on the generation of scientific knowledge and on technology development. Most individuals have a remarkable ability to compartmentalize

---

4. Islamic fundamentalist authors tend to display considerable ambiguity regarding Western science. Modern science is often portrayed, correctly, as having deep roots in early Islamic scientific knowledge. Its modern diffusion to the Muslim world is viewed as "repossession" rather than borrowing.

contradictory beliefs or different spheres of action (Kuran 1998, 254–57). Commitment to fundamentalist Christianity did not prevent Faraday from making seminal contributions to the understanding of the principles of electromagnetic induction and related electrical phenomena. Nor has commitment to biblical literacy prevented contemporary physicists from advancing knowledge in the field of space science and technology or contemporary biochemists from advancing knowledge in molecular biology or biotechnology.

I remain concerned, however, that pervasive diffusion of fundamentalist religion might, over time, result in cultural changes that could dampen interest and capacity for advancing scientific knowledge. There is a tendency for members of fundamentalist groups to avoid careers in the sciences. In the United States, the schools established by fundamentalist religious movements have made only limited investments in scientific education and research. Their emphasis has generally been on sacred rather than secular knowledge and on religious texts and practices rather than on things scientific or material (Mendelshon 1993, 331). This emphasis may, however, be more a reflection of an enclave fundamentalism than the priorities that would be pursued by a society committed to fundamentalist belief. In the Islamic world, fundamentalism has not been a significant barrier to education in medicine, engineering, and the applied sciences.

Fundamentalism and Economy

Economists working in the field of development have, in general, given remarkably little formal attention to the linkages between religion and economic development (Kuran 1997; Lal 1998).[5] Even those whose work has been influenced by the New Institutional Economics have given little attention to the role of religion in economic development (chap. 5, this vol.). I find this rather surprising in view of the continuing controversy among social scientists about Weber's thesis on the role of Protestant religion on the rise of capitalism (chap. 1, this vol.). The impact of fundamentalist religion on economic doctrine and policy is explored in some depth in five chapters in Marty and Appleby's *Fundamentalisms and the State* (Kuran 1993a, 289–301; Kuran 1993b, 302–41; Iannaccone 1993, 342–66; Keyes 1993, 367–409; Lal 1993, 410–26).

---

5. There is, however, a substantial new literature on the economics of religious behavior and of religious organizations (Iannaccone 1998, 2002).

The economic agendas pursued by adherents to the several fundamentalisms appear, at least on the surface, to have little in common. In India Hindu economic concerns have been driven by a desire to keep India's traditionally closed economy highly protected from foreign competition. In Thailand the articulated economic agenda of Buddhist fundamentalists has been to liberate individuals from the shackles of materialism while protecting the socioeconomic status of the Buddhist monks. Islamic economic reform movements have given major attention to innovation in financial institutions, particularly Islamic banking, designed to replace interest by profit sharing. Fundamentalist Christians generally share with libertarians a distrust of government intervention in the marketplace.

Kuran insists, however, that these disparate agendas reflect a common conviction. "Fundamentalist economists all believe that the ills of modern civilization are rooted in moral degeneration" (1993a, 291). Every variant of fundamentalist economics places emphasis on behavioral reform through moral uplift. "Each encourages people to bring social interest into their calculations, promotes the display of generosity, and insists on the individual's obligation to refrain from waste" (291).

For Kuran a fundamentalist economist is not simply an economist who draws inspiration from, or who is a member of, a fundamentalist religion. "When I use the term 'Buddhist economist,' I mean an economist who subscribes to Buddhist fundamentalist economics, not just any economist of the Buddhist faith or one influenced by Buddhist teachings" (1993a, 290). But what does it mean to adhere to fundamentalist economics? Fundamentalist economics includes a cluster of doctrines "that are formally unrelated but share a professed, though not necessarily exercised, opposition to secular ideals and practices—an opposition based, moreover, on a common aspiration to ground economic prescriptions in normative religious sources" (Kuran 1993a, 290).

Defined in these terms, what has been the impact of fundamentalist economics on economic behavior or economic policy? I attempt to respond to this question by examining the impact of U.S. fundamentalist economics and Islamic fundamentalist economics.

*Fundamentalist Economics in the United States*

Within the American fundamentalist and evangelical traditions it would be almost correct to assert that no distinctly fundamentalist economics has emerged. Adherents to the more strict fundamentalist doctrines have not sought to articulate a unique understanding of the workings of the U.S. economy or to critique its workings from a fundamentalist perspective. As of the mid-1990s none of their larger colleges—Bob Jones University, Tennessee Temple, Baptist Bible College of Missouri, and Baptist Bible College of Pennsylvania—had an economics department or offered an economics major, and only Tennessee Temple offered even an introductory course in economics. Even using a broader definition that includes evangelical as well as strictly fundamentalist colleges and universities, there has been relatively little attention to economics. The small economics department at Jerry Falwell's Liberty University in Lynchburg, Virginia, did offer an economics major and teach a full range of economics courses (Iannaccone 1993, 347). It is also the site of a public policy center, the Contemporary Economics and Business Association (CEBA). The center publishes a journal, *Christian Perspectives: A Journal of Free Enterprise,* which promotes the conservative economic doctrines of the New Christian Right (NCR).

Iannaccone argues, drawing on opinion surveys, that the economic attitudes of rank-and-file fundamentalists and evangelicals are much more diverse than the views articulated by Jerry Falwell, Pat Robertson, and other leaders of the Christian Right. He argues that, unless one equates fundamentalism with the New Christian Right, it is inappropriate to claim that fundamentalists, or even all evangelicals, are ardent supporters of laissez-faire capitalism or other conservative economic policies. Fundamentalists and evangelical leaders with similar theological beliefs and social traits subscribe to a wide range of economic views (Iannaccone 1993, 352). Iannaccone interprets the identification of the NCR with conservative economic policies in terms of political coalition building designed to advance the economic agenda of secular conservative activists and the social agenda of the New Christian Right.

Iannaccone goes on to argue, mistakenly in my view, that in the United States the fundamentalist and evangelical communities have had no measurable impact on economic policy—that they have not been interested in economic policy and have not tried to change economic policy. An obvious area of some impact is in U.S. economic assistance policy. Objection to use of abortion as an instrument of population policy has been used by the NCR as a red herring to mobilize efforts to limit support, not only for family planning, but for the entire bilateral and multilateral economic assistance budget (Ruttan 1996). I agree with Iannaccone, however, that whatever impact the fundamentalist movement has had on U.S. economic policy has been largely indirect—through support by the NCR for policies advanced by secular conservatives.

## Islamic Fundamentalist Economics

In contrast Islamic fundamentalists and fundamentalist economists have taken a proactive interest in economics and economic policy. Attempts to develop a distinct Islamic economics date from the middle of the twentieth century. A strong motivation has been to defend Islam against external political and economic domination (Kuran 1995a). Among the diverse proponents of a distinct Islamic economics there is a common concern with the perceived moral failures and economic injustices and inefficiencies of the national economic systems in Muslim countries. "In capitalism interest promotes callousness and exploitation, while in socialism the suppression of trade breeds tyranny and monstrous disequilibria" (Kuran 1993b, 304). The two major policy thrusts of Islamic economic discourse have been reform of financial institutions (Islamic banking) and equity in income distribution (zakat).

### ISLAMIC BANKING

The purpose of Islamic banking is to prevent the perceived inefficiencies and injustices associated with interest-based lending in which the risk of loss falls entirely on the borrower. Islamic banks, at least in principle, accept only *transaction deposits,* which are risk free but carry no interest, and *investment deposits* based on profit sharing, which carry both the risk of capital loss and the promise of a variable return.

Banks claiming Islamic identity were, by the early 1990s, in operation in most of the countries of North Africa, the Middle East, and South Asia. There is a substantial professional literature, including several specialized economic journals that focus on Islamic banking and finance. In Iran and Pakistan all banks are required to conform to Islamic banking principles. Even in Islamic countries in which conventional banks continue to operate, Islamic banks sometimes hold as much as 10 percent of commercial deposits.

What has been the impact of institutional innovation of Islamic banking? The banks do provide an institutional arrangement that meets the demand of individuals and firms whose religious convictions lead them to hold strong preferences against interest. Kuran argues, however that the lending practices of the Islamic banks do not conform to the stipulations of Islamic economics (1993a, 311). Various subterfuges (detailed by Kuran) have been developed that, while formally consistent with the requirement of profit sharing, are functionally equivalent to interest. While there continues to be a consensus among both the theorists and practitioners of Islamic economics that interest is sinful, the practice of Islamic banks continues to be vigorously debated. One of the issues that has been difficult to confront is how closely the practices employed by Islamic banks conform to the ideology of Islamic economics.

ZAKAT

Islamic economic policy has also devoted considerable effort to the reform of zakat, a traditional Islamic institution for the redistribution of income and wealth. The zakat is an annual tax assessed on income and wealth that was traditionally devoted to assistance for "the poor, the handicapped, orphans, travelers in difficulty, debtors, dependents of prisoners, to free slaves, and to assist people serving the cause of Islam (including the zakat collectors themselves)" (Kuran 1993b, 318). In Arabia, during the early years of Islam, zakat was collected and distributed by the state. In more recent history it has been a voluntary contribution administered at the local level where it has often been subject to considerable abuse. A common objective of Islamic economic reform has been to modernize the base for zakat contributions and to recentralize the collection and distribution of zakat revenues.[6]

Malaysia has instituted modest reforms, and Pakistan and Saudi Arabia have attempted to implement more extensive reforms in the zakat system. Administration of the reforms has been characterized by substantial evasion in collection and corruption in administration. As a percentage of GNP, in Saudi Arabia zakat revenue collection has ranged from 0.01 to 0.04, and in Pakistan it runs in the 0.35 range. Grants to the poor, even when administered effectively, have been too small to have any measurable impact on income distribution. The tradition of zakat, and of Islamic communalism more generally, would

---

6. In seventh-century Arabia, zakat was levied on the traditional sources of income and wealth—agricultural production, livestock, minerals, and precious metals. The rate varied from 2.5 percent on precious metals to 20 percent on mining income. Islamic economists generally favor broadening the collection base to reflect contemporary sources of income and wealth and to direct disbursement to redress income inequality (Kuran 1993b, 319).

seem to provide essential ideological support for the institutional innovations associated with the modern welfare state. Kuran argues, however, that substantial renovation of the zakat will be necessary if it is even to begin to live up to the promise held out by Islamic economists (1993b, 324–25).

Cultural Change

It is somewhat remarkable that so little attention was given by the Fundamentalism Project economists to the impact of the institutional and cultural changes that have been set in motion by the several fundamentalist movements for longer-term economic development. This is particularly surprising in view of the attention that sociologists and historians have given to the role of the rise of early Protestantism, particularly Calvinism, on the emergence of modern capitalism. The Protestant contribution consisted of the idea of vocation, in which ordinary work was invested with religious significance (from Lutheranism), and an ascetic and activist approach to life (from Calvinism) (Weber 1958; Tawney 1962; Swedberg 1998, 108–45). "The importance of Protestantism lay in the fact that it brought rational asceticism into the everyday world and provided it with religious sanctions" (Gellner 1982, 532).

Several Fundamentalism Project authors did draw attention to the puritan-pietistic-evangelical roots of Protestant fundamentalist and evangelical movements in Latin America—to their effect on incalculating the virtues of diligence, thrift, sobriety, prudence, and responsibility (the "domestication of husbands") in family life (Deiros 1991, 142–96; Stoll 1994, 99–123; Levine 1995, 155–78). "The cultural impact of the new evangelicalism is unquestionable. . . . The fundamentalist message for the family offers alternative models of relationship as well as new roles for women and men that are adjusted to the exigencies of contemporary society. . . . This theologically informed domestic conservatism provides a greater measure of security to women and children and establishes new guidelines of affection, responsibility and leadership for men" (Maldonado 1993, 234). It is worth noting that in Weber's view these qualities were not viewed as instrumental in sixteenth- and seventeenth-century Calvinist thought. The "Spirit of Capitalism" was an entirely unintended (a latent rather than manifest) consequence of (irrational) religious conviction (Weber 1958, 45).[7]

---

7. Weber did not argue that the new economic ethic associated with Protestantism was the only source of the rise of modern capitalism. He also argued that in the modern world, where

To what extent can the several fundamentalisms and fundamentalist-like religionationalist movements in developing countries be viewed, not simply as a revolt against modernity, but as a stage in the modernization of culture, somewhat analogous to the role attributed by Weber to the Protestant Reformation in Europe? It may not be too difficult, as noted previously, to draw such inferences about the impact of evangelical Protestantism in Latin America. But does the analogy carry over to such movements as the Rashtriya Swayamsevak Sangh (RSS) and Vishva Hindu Parishad (VHP) promotion of militant Hindu ethnonationalism in India? Peter van der Veer has argued that the embrace of nationalism and capitalist development is sufficient to characterize these movements as "fundamentally modernist" (1994, 656).

The authors of the Fundamentalism Project chapters failed, however, to directly confront the issue of the extent to which contemporary fundamentalist movements were induced by changes in material culture. They have also failed to confront the extent to which the fundamentalist movements have become a source of change in material culture and impacted economic growth and development.

Regardless of the implications of fundamentalist and fundamentalist-like movements for longer-term economic development it would be hard to argue that their short-term impact has been other than regressive in the states in which they have been most successful in achieving some measure of political power or influence. The short-term impact on political development has been even more regressive than on economic development. The intimate relationship between the sacred and the secular has been conducive to the perpetuation of unstable autocratic forms of governance. My own sense is that it would be premature to base any conclusions about the long-term impact of contemporary fundamentalist or fundamentalist-like movements on economic or political development. It is not, however, too soon to conclude that the civil disorder associated with attempts to bring about a transition to

---

the economic implications of capitalism are fully established and economic activity has become highly rationalist, the personality characteristics associated with the Protestant Reformation play a less important role than they played in the past (Swedberg 1998, 108–45). I avoid the argument as to whether the rise of capitalism in late medieval Western Europe gave rise to the Protestant Reformation or whether the Reformation, particularly its Calvinist variant, was the source of the "Spirit of Capitalism." The issue of whether change in institutional structure induces change in culture or whether change in culture induces change in structure is a long-standing subject of debate in the social sciences. The answer is that the relationships between changes in cultural endowments and in institutional structure are dialectical (Samuelsson 1961; Thompson, Ellis, and Wildavsky 1990; fig. 1.1, chap. 1, this vol.).

fundamentalist ideology and polity have generally had at least a short-term negative impact on economic growth and political development.

## A Fundamentalist Economics?

There has been a reluctance in the chapters of the Fundamentalism Project that were written by economists to challenge the presumption that a separate fundamentalist economics is possible, or even a separate Christian or Muslim economics. I insist, however, that there is no need for a separate Muslim or Sikh microeconomics to understand the effects of price and income change on rice production and consumption in either the Pakistani Punjab or the Indian Punjab. Of course, one might expect to find, from empirical analysis, that the parameters of the supply and demand relations are affected by cultural considerations.

It is possible, however, for different societies to have distinctly different economic policies that do draw on fundamentalist (or other) religious convictions. The renovation or reform of zakat to achieve more effective income distribution is an example of such a policy. But the analysis of the effects of a zakat tax on the investment, production, and consumption behavior of those who pay the tax, those who administer the tax, and those who are its beneficiaries does not require a separate economics. It does require a rigorous application of modern economic (and social) analysis within the context of the particular economic structures and institutions in each Islamic country in which the reforms are to be implemented (Kuran 1995a).

## Culture and Development

Next I turn from the specific impact of contemporary innovations in religious belief and practice to the broader issue of differences in cultural endowments on economic development. I focus specifically on the role of cultural endowments on economic development in East Asia, South Asia, and the Near East.[8] The failure of the Chinese, Hindu, and Muslim civilizations to emerge as a seedbed for an industrial revolution has been a continuing puzzle to historians and social scientists. The Weber thesis that China and India lacked the ideological resources—an active worldly asceticism—to produce a capitalist

---

8. I do not attempt to address the even more difficult issue of the relationship between cultural endowments and economic growth in Africa. For a very useful introduction see Herbst (2000). The Herbst book has been reviewed by Robinson (2002).

spirit has been widely debated (Gellner 1982). By the late twentieth century, China had experienced over two decades of extraordinary growth, and India seemed no longer trapped in what Raj Krishna referred to as a low-level "Hindu Growth Rate" (Krishna 1980). But why was the Middle East, which experienced several centuries of dynamic cultural, scientific, and economic development associated with the diffusion of Islamic religion, so slow to respond to Western dominance? In addressing these puzzles I am forced to draw primarily on literature from political science and economics. Anthropologists have largely withdrawn from the kind of macrosocial anthropology associated with the work of Ruth Benedict (1946; chap. 2, this vol.).

East Asia

The rapid emergence of the East Asian economies—first Japan, then Korea, then the several Chinese economies—as the only area of the world that has been able to mount an effective challenge to the economic dominance of the Euro-American societies has triggered renewed interest in the role of cultural endowments in economic development. Much of this interest has focused on the role of Confucian thought and practice.[9]

The issues posed in this recent literature can be grouped under two broad headings, one emphasizing the historical sources of Chinese backwardness and the second the sources of the contemporary spurt in Chinese economic growth. John C. H. Fei, one of the pioneers in the study of economic development, articulated the historical question as follows: "If all the cultural values essential to the modern growth epoch (MGE) can be detected in Confucianism, the natural question . . . is, why was it that the modern epoch, based on science and technology, did not first occur in China? Why was it that industrial capitalism spread historically from England to China rather than the other way around?" (1989, 277). The questions raised by Fei seem particularly relevant when one recalls the dynamic period of cultural, technical, and economic change in China during the Sung dynasty (960–1279) (McNeill 1982).

There is fairly broad agreement that there are a number of behavioral and institutional characteristics of East Asian societies that are conducive to modern economic growth. The behavioral characteristics

---

9. See, for example, the articles in Chung-Hua Institution for Economic Research (1989); American Academy of Arts and Sciences (1993); Peter L. Berger and Hsin-Huang Michael Hsiao (1988).

include high savings and investment rates and a strong emphasis on education. These behavioral characteristics are reinforced at the institutional level by the pragmatic, materialist, and secular character of Confucian philosophy; a positive attitude toward private property and market exchange; and a strong emphasis on meritocracy and effective performance in the management of public sector institutional infrastructure.

But are these behavioral characteristics the product of endogenous cultural development? To what extent are they the product of the exogenous economic and political forces that have impinged on East Asian societies since the middle of the nineteenth century? I turn now to a survey of the responses to these questions by several students of East Asian culture and economic development.

*Savings and Investment*

Almost all students of contemporary development in East Asia have emphasized the high rates of saving and the high rates of investment in both human and physical capital as important elements in the growth of the East Asian economies. But is the high rate of saving and investment a response to new opportunities for productive investment, or is it a response to missing financial markets? Or is it merely a continuing cultural endowment? Many students of East Asian culture have emphasized the role of "filial piety" in Confucian teaching and practice as a source of high savings rates. The high savings rates and high rates of human capital investment have persisted, at least during first and second generation descendants of overseas Chinese (Tai 1989, 199–236). But then why didn't the forces that led to high levels of saving and investment in human and physical capital in the past lead to modern economic growth?

*Technical and Scientific Innovation*

By the fourteenth century China had achieved a level of agricultural and industrial sophistication that placed it on the threshold of a scientific and industrial revolution.[10] In retrospect it is clear that by the

---

10. The definitive work on the history of Chinese technology is by Joseph Needham and a series of collaborators. Needham's research program was initiated in 1948. The first volume of his work, *Science and Civilization in China*, was published in 1954. More than a dozen volumes have since been published. For an introduction see Jones (1988, 73–84).

eighteenth century China was falling behind the West because it failed to make the shift from the learning-by-doing system of invention to a science-based system of technology development (Meyers 1989; Lin 1995). The answer that John Fei gave to this puzzle is that traditional Chinese culture was deficient in its capacity to reason about the physical world (1989). Lin argues that the incentive structure provided by the Chinese bureaucratic system "diverted the intelligentsia away from scientific endeavors, especially from the mathematization of hypothesis about nature and from controlled experimentation" (Lin 1995, 284). Government service, in which entry was controlled by highly competitive examinations, was the primary vehicle for upward mobility. Chinese society failed to develop the institutions that could provide the economic support and autonomy from the political process that the practitioners of science and science-based technology need to acquire and expand scientific and technical knowledge (Pye and Pye 1985, 182–214).

*Psychological Constraints*

Meyers provides a psychological explanation—a heightened sense of moral responsibility—in which individuals are continuously evaluated and esteemed in terms of Confucian ethical guidelines that have acted to narrow the scope of economic cooperation. In the Chinese economy, households and firms relied on personal networks in factor markets to obtain and allocate resources. In the commodity and service markets they behaved as highly competitive units, reluctant to engage in market integration and strongly preferring to deal across markets through brokers. "Avoidance of impersonal ties and strong preference for personal ties to limit transaction costs is not only a salient feature of the Chinese firm's behavior, but flows as well from specific Confucian values and social behavior" (Meyers 1989, 288).

But this psychological constraint, formerly viewed as a barrier to Chinese economic development, is viewed much more positively by recent observers (Jones 1988, 100). It is this very cultural endowment that has enabled Chinese firms in Taiwan, Hong Kong, and now south China to successfully practice what Danny Kin-Kong Lam and Ian Lee have referred to as "guerrilla capitalism." Overseas Chinese, primarily in North America and Europe, became a conduit to channel technical knowledge and market opportunities to their relatives back home. They also provide Chinese multinational firms with the managerial expertise needed to enter new markets (Drucker 1994).

## The Corporate State

Reviewing recent literature on East Asian culture and economic development, there are two puzzles left. What happened to convert the Confucian "social man" into the "economic man" that seems to characterize contemporary economic behavior in East Asia? Why is the continuing aversion to impersonal cooperation, which seems to be a persistent feature of Confucian culture, no longer the serious obstacle to economic development that it appeared to be in the past?

In a paper presented to the Conference on Confucianism and Economic Development in East Asia, Marion J. Levy (1989, 555–64) suggested several answers to these questions. Levy approaches the first puzzle by asking a question: Is there a viable substitute in twentieth-century East Asia for the personality characteristics that Weber identified with the Protestant ethic—the worldly religious asceticism that resulted in the internalization of the personal drive for material betterment? His answer is the demonstration effect of a rising level of material prosperity in the West. It is not that the demand for material betterment was new in East Asian culture. It was the fact that the possibility of its realization has been demonstrated in Western Europe and North America.

Levy's answer to the second part of the puzzle is the emergence of the modern corporate state. The Chinese state has been able to build on cultural endowments associated with traditional Confucian teaching and practice. The ideology of a virtuous government, managed by an educated elite in the public interest, has been particularly important as an ideal, if not always in practice, in the Chinese economies of East Asia and in Korea and Japan. With the creation of a modern corporate state, capable and willing to override the disintegrative force of "filial piety," the Confucian ethic was able to provide a supportive environment for economic development. "With familism no longer threatening the primacy of the national state as a focus of organization in the face of exponentially increasing interdependency, people striving in this world in the interest of their families was a healthy antidote to individualism. It placed a higher emphasis on cooperation and conformity in the pursuit of family interest that was probably a healthier base for economic development for latecomers than an adversarial individualism" (Levy 1989, 563–64).

## External Intervention

But were the forces leading the modern corporate state due to exogenous or endogenous forces? The Chinese societies (but not Japan) were the inheritors of states, as Wittfogel (1957) emphasized, with substantial bureaucratic capacity. But much of this capacity had atrophied by the middle of the nineteenth century.

One can hardly avoid the conclusion that exogenous political and economic intervention played an exceedingly important role in the reemergence in East Asia of modern states capable of mobilizing human and material resources to achieve political and economic objectives. The economic and political intrusions by the West into East Asia in the nineteenth century induced a dramatic effort to reform and modernize political and economic structures—first in Japan and later in China. Japanese intervention in Korea, Taiwan, and mainland China reinforced the insult of Western intervention. The emergence, in 1949, of a modern corporate state in China must be viewed as an institutional innovation induced, at least in part, by more than a century of political and economic decline (Binswanger and Ruttan 1978, 407–13). And in both Taiwan and South Korea the economic and political challenges from Communist China and North Korea have been an incentive to implement policies designed to stimulate development (Pye and Pye 1985).

What implications can be drawn from this review of the literature on the role of cultural endowments in contemporary economic development in East Asia? Clearly, cultural endowments differ among the several Asian societies. In addition to China itself, we are confronted with the "state Confucianism" of South Korea, the folk Confucianism of Taiwan, the implicit Confucianism of Japan, and the efforts by the government of Singapore to introduce Confucian culture to an indifferent population. Yet one cannot avoid being struck by the similarities in behavior and organization among the several societies of East Asia (Pye and Pye 1985).

But how decisive have these common elements been in the emergence of the region as a major challenge to the economic dominance of the West? John Wong and Arline Wong have argued that the common success of the East Asian newly industrializing countries (NICs) "can simply and adequately be explained by means of standard economic analysis" (1989, 509). Lin, Cai, and Li have shown that the common feature of the East Asian NICs' success has been a comparative advantage strategy—"better utilization of their comparative advantage at each stage of their development" (1996, 92). At the superficial level of standard economic modeling, it is hard to disagree. At a deeper level, however, we have clearly not progressed very far in our efforts to understand the sources of modern economic development in the Chinese economies of East Asia beyond Weber's suggestion that Confucianism emerged and developed in a political milieu that was not conducive to the invention of industrial capitalism. But with the diffusion of industrial capitalism in the nineteenth and twentieth centuries, Confucianism provided a fertile soil for the assimilation of capitalist behavior (Weber 1958).

Confucianism has been a less fertile soil for the emergence of democratic institutions. "Classic Chinese Confucianism and its derivatives in Korea, Vietnam, Singapore, Taiwan, and in diluted fashion, Japan emphasized the group over the individual, authority over liberty, and responsibility over rights" (Huntington 1991a, 36). Where democracy has emerged in East Asia it was initially characterized by a dominant-party form of democracy that emphasized political stability. The perspective that Deng Xiaoping brought to the reforms that he presided over in China in the late 1970s was that social and political order and stability, imposed by a centralized authoritarian regime, provided the essential climate for economic liberalization and economic growth. This perspective has protected China from the economic and political disintegration that was associated with the transition from communism in the former USSR. But the Deng rationalization does not explain why the economic reforms that initiated the rapid spurt in economic growth beginning in the late 1970s resulted from the spontaneous actions of peasant communal farmers in the relative backward province of Szechwan rather than being initiated or even guided by agencies of the Communist party or the central government in Beijing.[11]

## South Asia

Religion has traditionally played a more pervasive role in the economic and political culture of South Asian than East Asian societies. In exploring the impact of religion on economic development in South Asia I draw heavily on the work of Deepak Lal, the James S. Coleman Professor of International Development Studies at the University of California—Los Angeles. Lal's research on the impact of Hindu reli-

---

11. The agrarian reforms, termed the Household Responsibility System, that began in the mid-1970s involved the disintegration of the communal farming system. As late as September 1979 the Central Committee of the Communist Party declared the system illegal except for some forms of sideline activities and for households in remote or isolated areas. A year later, confronted with continued rapid diffusion and evidence of a dramatic positive impact on agricultural production, the Central Committee conceded that the government should support the system (Lin 1988; Zhou 1996; chap. 4, this vol.).

gion and the caste system on economic development in South Asia is one of the most convincing analyses of the effects of religion, or more broadly, of cultural endowments, on economic development (1988, 1998).[12] Lal's search for the cultural constraints of Hindu society on economic growth stemmed from dissatisfaction with his earlier attempts to advance the understanding of economic growth within the framework of neoclassical growth economics. His frustration with his own work, and growth economics more generally, was similar to that expressed earlier (chap. 5, this vol.): that it addresses only the proximate sources of economic growth.

The caste system is central to any attempt to understand the role of Hindu religion in economic development in South Asia. Two issues have dominated the social science literature on the caste system: its origin and its persistence. The caste system consists of hierarchically structured castes and subcastes. The subcastes involve the reciprocal flow of goods and services among castes.[13]

Lal provides an induced institutional innovation interpretation of the origin of the caste system (see chap. 1, this vol.). The caste system, in his view, represented a response by the nomadic Aryan invaders of the Indian subcontinent to the problem of securing a stable labor supply for an intensive system of agriculture in order to exploit the rich agricultural resources of the Indo-Gangetic Plain. The great puzzle that has confronted historical and social science research is why the caste system has persisted for over two thousand years in spite of internal turmoil, foreign invasion, and economic change. One answer is found in the cosmological beliefs that are the foundation of the Hindu social system. The central concepts, reincarnation and ritual purity, bind a Hindu to his caste and determine his location in the social system upon reincarnation. Lal, quoting Gellner (1998, 121), notes that a Hindu Robinson Crusoe "would be a contradiction. He would be des-

---

12. See also Kuran (1995a, 128–36, 196–204). Efforts have been made to provide more formal analytical interpretations of the caste system (Akerlof 1976; Scoville 1996). I find the Akerlof model too "thin" to capture the essential features of the caste system. Scoville employs a richer Leontieff-type model to analyze the reciprocal flow of goods and services among castes and an extension of the classical growth model to explore intergenerational change in occupational and marriage arrangements.

13. "The social system consists of numerous endogamous hierarchically ranked occupations and often region specific sub-castes (jatis). These were subsumed under the . . . four broad varnas (castes): Brahmins (priests), Kshatriyas (warriors), Viashy (merchants), and Shudra (workers and peasants). This scheme merely provided the broad theoretical framework for Hindu society. The interweaving of the hierarchically arranged occupational sub-castes were the real fabric of the Indian caste system" (Lal 1998, 27).

tined for perpetual pollution: if a priest, then his isolation and forced self-sufficiency would oblige him to perform demeaning and polluting acts. If not a priest, he would be doomed through his inability to perform the obligatory rituals" (Lal 1998, 38).

A second response emphasizes the role of the autarchic village community. The combination of the caste structure and associated cosmological beliefs reinforced the viability of "a decentralized system of control that did not require any overall (and larger) political community to exist for its survival and ensured that any attempt to start new settlements outside its framework would be difficult if not impossible" (Lal 1988, 29). The local administration structure inherited by any new political overload could be taken over with little disruption to fund their courts, their armies, and their imperial ambitions.[14]

The third source of stability within the village was the joint family. In addition to its occupational specialization, the joint family rather than the individual or the nuclear family was the basic unit of the social system. The joint family often lived under one roof, shared property rights, and were linked through cosmological belief to both ancestors and descendents (Lal 1988, 32). Occupational specialization assured mutual dependence. Economic considerations, particularly the relatively high per capita income compared to other "organic societies," sustained the system until well into the twentieth century (35).

It is relevant to ask, as in the case of China, Why didn't the industrial revolution begin in South Asia? The answer that Lal provides is that the Hindu equilibrium constrained the emergence of individualism. Lal insists that individualism has been the essential instrument of modern (Promethean) growth in the West. Like Weber he identifies the development of individualism with Western Christianity.[15] It is the

---

14. "The great strength of the social and economic system set up by the ancient Hindus . . . was its highly decentralized nature, which provided specific incentives to warring chieftains to disturb the ongoing life of the relatively autarkic village communities as little as possible. . . . The tradition of paying a certain customary share of the village output as revenue to the current overlord meant that any new political ruler had a ready and willing source of tribute in place" (Lal 1988, 309–10).

15. Lal extends the source of individualism, as an unintended consequence of Western Christianity, back almost a millennium before the Protestant Reformation where it was located by Max Weber. He argues that the two revolutions initiated by Pope Gregory I in the institutions of marriage (in the sixth century) and by Pope Gregory VII in the creation of modern legal institutions for the administration of the Church and the papal state (in the eleventh century) were the critical institutions leading to the rise of individualism. And the instrumental rationality that was induced by the rise of individualism was in turn responsible for the emergence of the industrial and scientific revolutions. "Max Weber was right, he just got his dates wrong!" (Lal 1998, 174). In both cases Lal argues that the reforms were not moti-

repression of individualism that has been responsible for stagnation in China and India.

Lal makes a convincing case that the institutional structure and cultural endowments that characterized the "Hindu equilibrium" have constrained the path of South Asian economic development. The Hindu equilibrium model may also provide the missing link to issues that have long puzzled historians of Indian economic development such as the source of the large and persistent difference in labor productivity in the British and Indian cotton textile industry even when the same technology was employed (Clark 1987; chap. 6, this vol.).

However, Lal joins Levy in insisting that once the instruments of growth were discovered and tested in the West they can be, and have been, diffused worldwide to very different societies "that do not have to adopt the cosmology and beliefs that led to their creation" (Lal 1998, 175).[16] There is clear evidence that during the last decades of the twentieth century the constraints associated with the Hindu equilibrium have become less compelling.[17] Public sector investment had provided India with a heavy industry base with substantial capacity to sustain industrial growth in other sectors. Indian peasant producers became eager adopters of the new green revolution seed-fertilizer-irrigation technology. The Indian industrial labor force has become increasingly differentiated and skilled. Indian entrepreneurs and technicians led the way to the development of a dynamic computer software industry. The annual rate of growth in per capita income has risen from the Hindu growth rate of less than 1.5 percent to over 4 percent (Adams 2001). The Hindu equilibrium mold has been, if not broken, at least severely fractured.

---

vated by the internal logic of Christian ideology but rather were the unintended consequences of the concern of the Church to advance its own material interests.

16. At an even deeper level Lal argues, correctly I believe, that emotions play an important role in internalizing the moral codes embodied in the cultural endowments that are the source of economic development. Shame and guilt, unlike the other emotions (disgust, sadness, anger, and fear) play particularly important roles in the development of a culturally acquired "moral sense" (Lal 1998, 13). A sense of shame is derived from being seen by others as involved in inappropriate behavior. A sense of guilt is derived from being seen by oneself as unworthy. Lal insists that shame and guilt were cojoined in the cultural development of the West. In the West the emotion of guilt arose from the Christian doctrine of original sin. This sense of guilt was used by the Church to check the individualistic passions that greed had unleashed. In China and India a sense of shame was an internalized response to inappropriate behavior in hierarchical society and has been sustained by civic culture. But a sense of shame was not accompanied by a sense of guilt.

17. Osborne (2001) notes, however, that as caste membership has become less important with respect to the division of labor, it has become more important in Indian politics.

## The Middle East

The economic decline of the Muslim countries relative to Europe has been a continuing puzzle to both scholars and reformers. The puzzle is confounded by the fact that from the tenth through the thirteenth centuries the Muslim world occupied a predominant leadership role in scientific and technical innovation. "During the first few centuries of Islam regions under Muslim rule, including the Middle East, North Africa and Spain flourished economically" (Kuran 1997, 48). The economic integration of the trading worlds of the Mediterranean and Indian Oceans under a common language and culture stimulated economic growth through both the larger market it generated and the exchange of scientific and technical knowledge (Lal 1998, 54). The West did not achieve economic dominance until at least the seventeenth century, and the Ottoman empire represented a continuing military threat in Eastern Europe into the eighteenth century.[18]

Timur Kuran has critically examined a series of explanations that reject the role of Islamic religion as a source of economic backwardness or that attribute the decline of Islamic countries to European imperialism. He insists that explanations for why the Islamic world experienced a relative decline and why the decline lasted so long must be sought within Islamic society itself (1997). Kuran gives greater credence to explanations that emphasize (1) the static nature of the Islamic worldview; (2) the failure of Islamic society to understand the sources of change in Western culture, society, and economy; and (3) the persistence of communal forms of social organization while Western Europe was turning increasingly individualistic. While he regards each of these explanations as partially valid, Kuran insists that they do not provide a fully adequate response to several fundamental questions. "Why . . . did the Islamic world produce no diagnosis of the growth retarding effects of its cultural beliefs? Why did ambitious Muslims . . . not recognize the ideological sources of their disadvantages? Why did rulers, threatened by the rise of Western Europe, not recognize the economic disadvantages of their communalist cultures? Why, in the face of mounting competition from Europe, did Muslim merchants fail to generate solutions to their economic losses?" (64).

In more recent work Kuran has explored in considerable depth a series of institutional constraints on economic development in the

---

18. I feel much less confident in commenting on the role of institutional and cultural change in economic development in the Islamic world than in East and South Asia. In this section I draw on an important series of studies by Timur Kuran (1997, 1998, 2001a, 2001b).

Muslim world. He explores, in particular, the failure to develop the financial and commercial institutions that might have enabled Muslim trading firms to compete with modern Western firms in Mediterranean trade. Kuran argues that Islamic partnership law remained essentially unchanged during several centuries during which Western commercial partnership law, which had originally been based on principles drawn from Islamic law, evolved to include more advanced institutions such as the modern corporation. At the time when Islamic commercial law took shape it played an important role in facilitating the expansion of trade in the Mediterranean region and the Middle East. Even prior to the nineteenth century, when the Middle East and North Africa came under European colonial administration, the Muslim role in trade with Western Europe had slipped into insignificance. Perhaps the most striking failure in the premodern Islamic world was the failure to develop institutions capable of mobilizing large-scale resources for commercial and industrial development (2001b).

Kuran also explores the failure of traditional private "pious foundations" (*waqf*), which provided public services such as soup kitchens for the poor or the provision of public infrastructure such as lighthouses, to evolve into organizations such as municipal or industrial corporations. A primary motivation in the initial emergence of the *waqf* was to provide a decentralized institution for the delivery of public services. In this respect it represented an important institutional innovation that compared very favorably to the institutions that serve similar purpose in other contemporary societies. In retrospect, however, there were several major deficiencies. In principle the manager of a *waqf* was required to use its resources in a manner stipulated by its founder. This deprived the *waqf* of the flexibility to respond to changes in the demand for public services. In practice the primary motive of many *waqf* founders was to shelter their wealth against taxation or opportunistic expropriation. As the resources controlled by the *waqf*s expanded to include substantial shares of agricultural land and even of commercial property, the resources available to sustain government became increasingly constricted. Kuran argues that the Islamic *waqf* had the potential to evolve into a rich source of "social capital" on which to build the modern governance and property rights institutions that would have provided more effective safeguards against opportunistic expropriation and taxation (2001a). In effect, however, the Islamic societies failed to evolve the institutions of civil society that could become dynamic sources of economic growth (chaps. 3 and 4, this vol.).

Kuran argues that by the thirteenth century those who were in a position to challenge the constraints on institutional innovation tended to conceal their views lest they be accused of harboring designs against Islam. The effect was a failure on the part of educated Muslim observers to recognize the cultural and institutional sources of the growing economic and military strength of the West.[19] This interpretation draws on the theory of preference falsification that Kuran developed in his widely acclaimed *Private Truths, Public Lies* (1995b): "The book shows how inefficient social structures can survive indefinitely when people privately supportive of change refrain from publicizing their dispositions. The motivation for such preference falsification is to avoid the punishments that commonly fall on individuals who enunciate unpopular public positions" (Kuran 1997, 65). One of its by-products is the corruption of public discourse. A second is what Kuran terms "collective conservatism" that constrains any impulse to challenge existing institutional arrangements.

The fragmentation of the Ottoman Empire at the end of World War I led both to a dramatic change in perception on the part of the Turkish elite of the role of Turkey in the world system and to the establishment of a secular Turkish republic. It is possible that the recent intensification of the external confrontation with the West could become the shock to the political system of the Arab nations of the Middle East that would lead to rapid political and economic reform (Kepel 2000).[20] Kuran insists that if, and when, such revolutionary changes occur they will almost certainly be unanticipated by many observers and many of the consequences will be unintended (1995b, 289–309).

A definitive assessment of the role of cultural endowments in economic development has yet to be constructed. It does seem clear, as Levy emphasized in his discussion of East Asian economic develop-

---

19. This was in sharp contrast to the Japanese experience where a vital school of "Dutch" studies contributed to deepening Japanese understanding of Western culture, science, and technology even before Japan was confronted with the demands of Western imperialism represented by the arrival of Admiral Perry and his "black ships" (Keene 1969).

20. Kepel (2002, 361–76) presents a class-based interpretation of the troubled history of political and economic development in East Asia and North Africa. Only Turkey has come close to being able to resolve the tensions between the urban poor, the pious middle class, the militant Islamist clericals and intellectuals, and a nationalist secular elite. Kepel argues that by the beginning of the twenty-first century the Islamic movement was losing its momentum, thus opening up the possibility of constructing a more open Muslim version of democratic society.

ment, that the cultural endowments that were associated with the emergence of economic growth in the West have been less essential for the developing economies of East and South Asia and the Middle East. "The development of individualism as an ideal was almost certainly essential for the first comers. It is almost certainly not essential for the latecomers to the process who, if they are to be successful with the process, require higher levels of coordination and control" (Levy 1989, 561). It is no longer necessary to look only at Japan to find an Eastern economy that has been capable of generating a modern rate of economy growth.

In general the countries of East and South Asia and the Middle East have achieved less success in political than in economic development. It has been difficult to overcome the belief that power flows downward from the center to the base. In China this conviction has resulted in the establishment of large-scale heavy industry and massive resource development programs that have been destructive of economic resources. In India it has taken the form of a highly sophisticated central planning capacity combined with weak implementation capacity. In the Middle East, institutional arrangements for the transfer of political leadership remain, with few exceptions, unresolved. Political power, conceived as "power over," continues to constrain the capacity of "power to" (chap. 4, this vol.).

It is possible that a new global civic culture based on the norms that are emerging from the international diffusion of the institutions of capitalism is emerging (Gellner 1999). These include not only fewer constraints on market transactions and greater social mobility, but also others such as a decline in nepotism and rent seeking, that are appropriate in a global economy characterized by Promethean growth. The predatory state has been an insecure incubator for the development of technology and science and an inhospitable cradle for the growth of the institutions of capitalism. Markets and commercial culture have existed under a wide range of traditional and modern political systems. Even in Stalin's Soviet Union they could not be completely suppressed. But the public and private investments necessary for sustainable accumulation of physical, human, and social capital have occurred primarily in political systems in which the consent of the governed is required to raise revenue for the state and in which market behavior is governed by the rule of law (chap. 4, this vol.).

Taking changes in cultural endowments into account is analogous to taking changes in resource endowments into account. In the induced innovation model (see chap. 1, fig 1.1), changes in resource endowments, cultural endowments, technology, and institutions were treated as recursive—as both independent and dependent variables. Changes in cultural endowments induce changes in resource endowments, technology, and institutions. But changes in resource endowments, technology, and institutions also induce changes in cultural endowments. The theory of induced innovation suggests that a widening of the disequilibrium between economic potential and performance will induce changes in culture and institutions that lead to narrowing the disequilibrium.

The great unanswered question for the three societies discussed in this section—Confucian, Hindu, and Muslim—is why the disequilibrium widened and persisted not just over decades but centuries. One answer is that the very high transaction costs of institutional and cultural change locked these societies into inefficient development trajectories (North 1990b). A second is that the "collective conservatism" resulting from preference falsification delayed the transition to a more dynamic or Promethean development path. My own sense is that both transaction costs and preference falsification represent important components of a more general interpretation. But a more general interpretation has not yet been constructed. We are fortunate, however, as the experience of East Asia has already demonstrated, and as the experience of South Asia is beginning to show, that it is not necessary to wait for a fully satisfactory interpretation of historical experience to begin to design the institutional and policy changes that can lead to a breakout from historical path dependence.

### Nationalism and Nation Building

Nationalism is a political ideology that insists on congruence between the state and the nation (Gellner 1983, 1–3). In Europe prior to the French Revolution and in the preindustrial world generally, neither language nor culture served as a basis for national political boundaries. State formation was generated by conflicts of the patrimonial distributional coalitions that characterized late feudalism. Cultural differences in the form of estates, guilds, and castes defined status within society. People were expected to know their place! A shared culture was not available to serve as a plausible basis for the formation of political units (Gellner 1999, 192). Nationalism emerged in Western Europe as a major political movement following the Napoleonic Wars. It has been one of the defining institutional innovations of the nineteenth and twentieth centuries.[21]

The relationship between culture and the state has been a contentious issue for students of political development (chap. 4, this vol.). For idealists, cultural factors are viewed as independent variables in the formation of national consciousness. For materialists, national culture is a dependent variable—molded by the state rather than a source of state formation (Steinmetz 1999, 1–49). In the induced innovation model (fig. 1.1, chap. 1) the relationship between state formation (institutional innovation) and cultural change is viewed as recursive—running in both directions.

Nation building became a major theme in the post–World War II political development literature (Connor 1972). In that literature it was generally assumed that the object of nation building in the new postcolonial states of Asia and Africa should be the strengthening of the national state (Huntington 1965; chap. 4, this vol.). The nation appeared to be the necessary vehicle for political and economic modernization (Hard and Negri 2000, 133). To the extent that geographically based linguistic, religious, and ethnic differences were considered, they were viewed as a potential threat to the political stability of the state and as obstacles to be overcome during the nation-building process.[22] But it was generally anticipated that the endowments of social capital embedded in ethnic identity would erode with the

---

21. Isaiah Berlin argues that nationalism "was the one movement which dominated much of nineteenth century Europe and was so pervasive, so familiar, that it is only by a conscious effort of imagination that one can conceive of a world in which it played no part" (1980, 337). Berlin defines nationalism in terms of four characteristics: (1) the belief in the overriding need to belong to a nation; (2) belief in the organic relationships of all the elements that constitute a nation; (3) belief in the value of our own simply because it is ours; and (4) faced by rival contenders for authority or loyalty, belief in the supremacy of its claims (1980, 345). In Berlin's view, nationalism is distinct from mere national consciousness or patriotism. Until well into the twentieth century, most political and social theorists regarded nationalism as a passing sentiment or ideology. After the liberation of the constituent nationalities of the multinational empires it was expected to decline in favor of ideologies such as socialism designed to overcome economic oppression.

22. The typology of state formation in the political science literature is less than precise. In this section I will refer primarily to three types. One is the *ethnically homogeneous state* where a nation has its own state such as Japan or Poland. A second is the *immigrant state* in which the population may be derived from a variety of ethnic immigrant streams and accompanied by substantial acculturation and assimilation such as in the United States or Argentina. A third is the *multinational state* consisting of two or more geographically based ethnic homelands such as the former Czechoslovakia or Nigeria. These categories do not completely exhaust the typologies that might be identified (Connor 1987, 208–10).

progress of modernization. Modern societies would be characterized by patterns of status and identity that were achieved rather than ascribed (chap. 3, this vol.). Secularization would erode the religious foundations of ethnic identity, and rational self-interest would lead to class-based rather than ethnic-based political and economic commitments (Deutsch 1953; Lerner 1958).[23]

Nation building in states with geographically based ethnic nationalities is my primary focus here. I do not address the issues of conflict among the culturally defined civilizations emphasized by Samuel Huntington. Huntington argues that in the twenty-first century ethnic conflicts will continue to occur within civilizations, but the most dangerous cultural conflicts will occur along the fault lines between civilizations (1996, 28). Huntington's generalization rests on weak historical foundations.[24] From the mid–nineteenth to the mid–twentieth centuries many of the bloodiest conflicts were the civil wars that occurred within Western civilization—the U.S. Civil War and the two great European civil wars (World War I and World War II). In the late twentieth century the most brutal conflicts were interethnic civil wars (Rosecrance 1998). I also do not attempt here to deal with external assistance for the purpose of nation building.

### Why Nationalism?

Since the French Revolution the unitary state has represented the preferred model of the nation-state for the architects of the nation-building project. But in France political unity was established well before the emergence of modern nationalism.[25] In Eastern Europe, however, ethnic national communities were sustained within the framework of the multiethnic Austrian and Turkish empires. For these countries the German model, shaped by language and history as a nation before it became a unified state, represented an alternative to the French model.

---

23. See Connor (1972) for a listing and review of the major post–World War II literature on national integration and nation-building theory. This is essentially the same literature reviewed by Holt and Turner (1975) in their critique of the Social Science Research Council–sponsored political development literature (chap. 4, this vol.).

24. Both Huntington's *Foreign Affairs* (1993) article and his book (1996) have been widely reviewed and debated in the foreign policy literature and in the popular press. Reactions have ranged from polemical critiques to warm appreciation. Several reviewers have characterized the article and the book as "persuasive discourse" designed to counter the triumphalist "end of history" theme advanced by Francis Fukuyama (1989).

25. Even in France regional ethnic differences persisted well into the twentieth century (Weber 1976).

The new states of Eastern Europe created in the aftermath of World War I, the new postcolonial states that emerged after World War II, and the states created by the disintegration of the Soviet empire were in many respects more suited to following the German than the French model. But their intellectual and political leaders have aspired to the French model of a unitary nation-state. Ethnolinguistic-based regional autonomy, even within a federal system, was viewed as a threat to the nation-building project. Each new state wanted to blend its diverse peoples into a new national identity to constitute a unified nation-state rather than preside over a federal state with substantial autonomy for its constituent nationalities.

Jeffrey Herbst has argued that the problem of nation building has been substantially different in the new postcolonial states of Africa than Europe. In Africa a fundamental problem, conceived by the new governments of the formerly English, French, and Portuguese and Belgian colonial territories, was how to achieve control over the diverse people who lived in the former colonies. One of the few areas of common agreement among postcolonial governments was that the territorial boundaries that they inherited should not be subject to external challenge. Thus, freed from the external threats that had confronted the builders of nation-states in Europe, they were able to direct what limited force was available to them to suppress internal challenges to central authority (Herbst 2000).

The dismemberment of Yugoslavia has forced a painful reassessment of what went wrong in what had once appeared to be a successful nation-building effort (Pavlowitch 1994; Somer 2001). The serious student cannot avoid recognizing that ethnic consciousness continues to represent a vital political force even in the older developed countries that have long been governed by hierarchical central political institutions such as Spain and Great Britain. And the prospect of further fragmentation of the former Soviet Union or the prospect that national political boundaries may have to be substantially redrawn in sub-Saharan Africa is a source of uneasy contemplation in the corridors of the U.S. State Department, the World Bank, and the Organization for African Unity.[26] The idea that "any people, simply because it considers itself to be a separate people has the right, if it so desires, to create its own state, remains a revolutionary idea" (Connor 1987, 215).

---

26. In the *World Development Report: The State in a Changing World* (World Bank 1997) only a few paragraphs were devoted to the problems of multinational states. In those paragraphs the emphasis was on managing multiethnic societies (113) and decentralization (120–22).

At a time when nation-states have acquired greater capacity to limit external aggression they continue to repress the legitimate political interests of their constituent nationalities (Montgomery 1998, 78).

Social scientists have been slow to articulate a coherent theory of the rise of nationalism. An important exception has been the advancement by anthropologist Ernest Gellner of a materialist theory of European state formation (1998, 1999). He starts by asking, Why has the application of Enlightenment rationality been so successful in the area of science and technology and why has it achieved such limited success in the area of political and moral development? Industrial development, a product of Enlightenment science and technology, has resulted in the dramatic economic transformation of much of the world over the last two centuries. Gellner goes on to argue that industrialization was successfully co-opted by a romantic Counter-Enlightenment that identified national culture with the state. He traces the political motivation for the Counter-Enlightenment to the uneven diffusion of the industrial revolution and to the disparities in the growth of economic and political power associated with the industrial revolution.[27] The nation-state came into existence, he argues, in an attempt to manage and protect the sociocultural conditions necessary for the advance of industrialization.

But what were the forces, beyond the rise of industrialization, that induced the rise of nationalism in nineteenth-century Europe? The rise of nationalism in Germany has been described as a reaction against the political and cultural hegemony of revolutionary and post-Napoleonic France. But the insult of foreign domination was not a new phenomenon. It was the coincidence of industrial and democratic change that energized a sense of insult from foreign domination and induced the emergence of nationalism as a dominant political ideology. The democratic and industrial revolutions eroded the political and economic resources available to the hierarchically based feudal dynastic military and clerical leadership and transferred both political and economic resources into the hands of new men.

The institutions that had ordered society, whether established by

---

27. The convergence of economic and political power associated with the industrial revolution is illustrated by the invention of the Watt-Boulton steam engine. Progress on the steam engine was delayed until the invention, by ironmaster John Wilkinson, of a boring machine capable of drilling a straight hole through a solid bloc of metal for the manufacture of an accurate cannon. The same technology was used to bore the cylinder necessary for an efficient steam engine. Wilkinson was, in turn, one of the first purchasers of a Watt-Boulton steam engine that he used to operate the bellows of his blast furnace (Ruttan 2001b, 73–74).

nature, ordained by God, or enjoined by hereditary political leadership, began to lose their authority. The resulting sense of loss of cohesion and order created an opportunity for a new class of intellectual and political leaders to emerge (Schumpeter 1950). The emergence, particularly in the German lands, of a view that the collective will of the people provided the essential base for the ordered political life of the nation provided an opportunity for political entrepreneurs to practice the symbolic politics of ethnic nationalism. It became their role to create by deliberate social action the modern equivalents for the lost cultural, political, and religious values on which the old order rested. "The conception of the political life of the nation as the expression of the collective will is the essence of political romanticism—that is, nationalism" (Berlin 1980, 349).

In retrospect it is clear that the nation-building project has imposed exceedingly heavy costs on the people involved. Success has been achieved by attempting to convince, or more often to compel, some peoples to adopt one nationality rather than another—French rather than Breton, Turkish rather than Kurdish, Nigerian rather than Ibo, Chinese rather than Tibetan (Levy 2000).

## Toward Postnationalism?

In his work Gellner projected a vision of a postnationalist society in which control over material resource endowments would no longer be the dominant source of national wealth or power. The Marxist class distinctions of industrial society would be eroded in a society in which status depends on the capacity to manipulate symbols rather than materials—where status can conflict with occupational effectiveness! Such a society would be pervaded by a common culture transmitted through universal access to education. "The new society, based on expanding technology, on pervasive impersonal and often anonymous communication by context-free messages, and on an unstable occupational structure, is destined for a standardized, educationally transmitted high culture, more or less completely diffused amongst all its members. Its political and authority structure will be legitimized by two considerations: by whether it can ensure sustained economic growth, and whether it can engender, protect and diffuse the culture of the society" (Gellner 1999, 190). In what appears to me to be an excessive burst of optimism, Gellner argued that the old links between religion and culture, and between ethnicity and culture, will retain their viability at the personal but not at the civic level. "For the first time in the history of humankind a high culture becomes the pervasive operational culture of an entire society" (1999, 187).[28]

But are there any design principles to which one can appeal in constructing institutional arrangements that will meet Gellner's challenge? Most scholars who have addressed the issue of constitutional design, such as Buchanan and Tullock (1962), have avoided the issue of constitutional design in multinational states or the successor states that emerged following the disintegration of multinational states. One exception is the elegant attempt by Jacob Levy to outline a normative political theory of multiculturalism, which addresses the design principles appropriate to preventing such political evils as "violence against an ethnic minority that wishes to retain its own identity; forcible exclusion from citizenship and state protection of small and stigmatized minorities; cruelty directed by community leaders against members who desire to assimilate into a neighboring culture; and the outcast status of those who have been forced to leave ancestral ethnic lands" (2000, 12).[29]

Levy insists that some aspects of life in a multicultural world must be taken as given—"attempts to deny or radically alter them have been bloody failures" (2000, 5). Two institutional constraints on nation building must be taken as given, at least for the foreseeable future. First, ethnic pluralism is an enduring characteristic of most modern states. Second, cultural blending and melding is a continuing process. But rather than being a solution to the problem of life in a multiethnic society, ethnic pluralism is itself a source of new cultural identity. The violence, cruelty, and humiliation that routinely accompany ethnic politics are not avoided by attacking ethnicity, any more than the violence, cruelty, and humiliation of the wars of religion were eroded by convincing people not to be religious! Levy argues that the institutional accommodations and arrangements that make up the separation of

---

28. Gellner proposed a five-stage linear model of the transition from prenationalist to postnationalist society. The stages, using my terminology rather than his, are: prenationalist; irredentism; national self-determination; ethnic cleansing; and postnationalist. He rejects the Marxist-inspired perspective that the transition from a preindustrial society to a capitalist industrial society generated class conflict and the "awakening" of latent nationalism among the constituent "small nations" of multinational states. In chapter 4 I have reviewed stage theories of political development proposed by Rostow and Eckstein. Neither Rostow or Eckstein is as sensitive to the intensity of ethnic conflict as Gellner. Gellner's vision of the future, however, differs dramatically from the vision of a "fortress world" dominated by hostile world civilizations advanced by Huntington (1996).

29. Levy contrasts "the liberalism of fear" with the liberalism of rights and justice. The liberalism of fear emphasized the avoidance of cruelty, humiliation, and political violence. Levy traces the concept of the liberalism of fear to Judith Shklar (1989).

church and state and the protection of freedom of religion are the model that should be followed, at least in spirit if not in particulars, in designing institutions to accommodate ethnic diversity (27). But he insists that there is no single criterion except self-identification that can be used to identify the constituent nationalities of multinational states (97).

Levy devotes the latter chapters of the book to examining a series of cases involving the accommodation of cultural pluralism—the incorporation of indigenous law and the issues of property rights in land—in ethnic politics. Of these, the links between ethnicity and geography are particularly important in the politics of ethnic nationalism. The political movements of indigenous peoples have been about land more than any other issue—about rights to develop or benefit from the development of their traditional lands, about the restoration of lands from which they have been disposed, and about security against future losses (Platteau 2000, 73–188). The conflict often centers around the liberal principle that land rights, like other forms of property, can be alienated. This issue is expressed with particular force in the struggle over land between the state of Israel and the Palestinian Authority: "The problem is that Israel does not distinguish between ownership of land and sovereignty over it. If a certain area is under Jewish ownership, Israel sees it as part of its sovereign territory" (Levy 2000, 217, quoting Greenberg 1997).

If there is any general design principle that emerges from the Levy work it is that the constitutional system or the laws of national states should be open to very substantial modification—other than proscription of any internal laws or rules of constituent nationalities that are violent or cruel—in order to accommodate the civic and economic needs of constituent nationalities. He presents examples, ranging from accommodation of symbolic claims (such as recognition of ethnic holidays) to self-governance. One common principle is the avoidance of the forms of symbolic or physical cruelty that drive latent nationalist sentiment to become overt.

There is a substantial literature that suggests that ethnic diversity, even when it does not lead to national disintegration, imposes a substantial burden on economic growth (Easterly and Levine 1997; Knack and Keifer 1997). Ethnic diversity has been associated with lower investment in education, lower investment in public infrastructure, greater rent seeking and corruption, and lower levels of trust among political and economic elites. In more recent work Easterly (2001a) finds that the negative effects of ethnic diversity have been substantially mitigated in countries that have been able to develop effective formal institutions that assure security of contract, private property, rule of law, and high-quality bureaucracy. Development of effective formal legal and governance institutions can overcome the negative effects of ethnic diversity. This result is not entirely reassuring! Achieving political consensus on the design of economic and political institutions is more difficult in an ethnically diverse than in an ethnically homogeneous poor country.

There is also a modest literature that addresses the economic considerations involved in state integration and disintegration. Alberto Alesina and several colleagues have explored the relationships between openness to trade and country size (Alesina and Spolaone 1997; Alesina and Wacziarg 1998). For example, they show that to the extent that market size influences productivity due to economies of scale in supplying public services or private goods, the costs of autarchic economic policies are smaller in large than small countries—in India as compared to Sri Lanka. A constituent nation that separates from a large multinational state can achieve the benefits of cultural homogeneity while incurring fewer economic penalties in an open trading regime. This is, in effect, the argument made by the Quebec separatists. The cultural gains of separation can, however, be at least partially achieved if the state responds by granting ethnic political autonomy while encouraging economic integration.[30] Even if separation results in lower national income, there may be economic gains to politically dominant groups in the new ethnically based nation (Boulton and Roland 1997).[31] The effect of this literature, as well as the experience of modern nation-building efforts, is to raise substantial questions about international commitments to the preservation of existing national territorial boundaries. Some states would benefit, both economically and politically, if they were to dissolve into constituent nations.

**Perspective**

These issues pose a particularly serious challenge to the contribution of social science knowledge to institutional design. The emergence of fun-

---

30. Krasner and Froats cite the bilateral understanding between Austria and Italy that granted regional autonomy to the German-speaking provinces of South Tyrol and Adige in the late 1960s. Since then, the region has become a model of successful local autonomy and state–minority cooperation (Krasner and Froats 1998, 245).

31. I am reminded of a comment by the Canadian economist Albert Breton, made many years ago, to the effect that nationalism is an ideology used by the middle class to shift the income distribution in its favor (1964).

damentalist religion as a major social movement across a wide spectrum of countries in the late twentieth century was a surprise to both theologians and social scientists. The recent transformation of culture in two of the oldest world civilizations, Hindu and Sinic, to embrace modern science and technology, to accommodate the institutions of capitalism, and to achieve rapid economic growth was not anticipated by development economists. The nation-building design advanced by Marxist political philosophers and Western political scientists has been challenged by the unanticipated strength of ethnic nationalism—by the durability of ethnic social capital.

As noted earlier, there has been substantial disagreement within the social sciences about the role of purposeful institutional design. There is general agreement that institutional change is a product of human action. But those holding an organic or evolutionary perspective, while agreeing that the institutions of civilization have been created by human action, insist that this "does not mean that man must also be able to alter them at will" (Hayek 1978b, 3).[32] A second constructivist, or design, perspective holds that advances in social science knowledge can play an important role in the rational design of institutional reforms and innovations. But the decentralized information concerns that were involved in early arguments about the incentive compatibility problems of market socialism (Hayek 1935; Lange 1938) have only been partially resolved even at the most abstract theoretical level.[33]

I find it difficult, however, to accept the criticisms advanced by conservative political philosophers of the failure of nation-building efforts as evidence that intelligent design of the institutions of governance and economic order is not possible. The unintended consequences of organic or evolutionary institutional innovation have not led to either a more civil society or to greater prosperity in most countries of the world. Social science research has substantially advanced our knowledge of the sources of the political disruptions of the last century. The design principles that were employed by the framers of the U.S. and the Swiss constitutions were based directly on advances in knowledge associated with Enlightenment political philosophy. In Spain the

---

32. The organic view of institutional change is reinforced by a theory of "unintended consequences" that runs through the work of Adam Smith (invisible hand), Max Weber (spirit of capitalism), and Friedrich Hayek (socialism). See Lal (1998).

33. The concept of incentive compatibility was introduced by Hurwicz (1972a, 1998). Hurwicz has shown that it is not possible to specify an informational decentralized mechanism for resource allocation that simultaneously generates efficient resource allocation and incentives for consumers to honestly reveal their true preferences.

highly centralized government of Franco was followed by one that granted substantial autonomy to Spain's non-Castilian nationalities (Connor 1987, 217). India has survived the challenge posed by a wide variety of communal groups with different religious, language, and ethnic identities (Sen 1999, 157). The disengagement between the Czech Republic and Slovakia, described as a "Velvet Breakup," was accomplished in a civilized manner (Saideman 1998, 145). Many ethnonational movements have been prepared to trade off demands for the establishment of a national state for meaningful autonomy within a federal system (Kitschelt 1993; Ghai 2000).

A major challenge to social science knowledge over the next half century will be the design of constitutional systems that will permit geographically based ethnic communities to meet their civic needs while pursuing sufficiently open economic arrangements to remain, or become, economically viable. The alternative is the continuation of the rising tide of barbarism that has been associated with economic and political development over the last two centuries.

CHAPTER 9

# Why Foreign Economic Assistance?

There is general agreement in the economics literature that resource flows among countries in response to market incentives enhance global economic efficiency and welfare in both rich and poor countries.[1] It has also been argued that in the presence of capital market imperfections the transfer of resources from developed country governments to the governments of developing countries on commercial terms, either directly or through multilateral agencies, can generally be expected to be welfare-enhancing in both developed and developing countries (Krueger 1986). These arguments do not apply, however, to official development assistance that involves a substantial grant element.

Two arguments have typically been used in support of transfers that include a grant component. One set is based on the economic and strategic self-interest of the donor country. The second is based on the ethical or moral responsibility of the residents of wealthy countries toward the residents of poor countries. Both sets of arguments have been the subject of continuous challenge.

I argue in this chapter that neither the donor self-interest nor the ethical responsibility argument can be rejected on logical or theoretical grounds. I also insist that the empirical evidence in support of either the economic or the strategic self-interest arguments is exceedingly thin. The ethical responsibility arguments impose less burden on the empirical evidence. They have been subject to continuous challenge by political theorists and moral philosophers.[2]

---

1. In this chapter I draw primarily on Ruttan (1989). See also Krueger, Michalopoulos, and Ruttan (1989) and Ruttan (1996, 2001a).

2. For an exception see Lumsdaine (1993). Lumsdaine insists that the primary reason for foreign economic assistance lies in the humanitarian and egalitarian principles of the donor countries.

## Donor Self-Interest

Donor self-interest arguments tend to assert that development assistance promotes the economic or political interests of the donor country. This argument is frequently made in official and popular pronouncements in defense of developed country aid budgets.[3] It has also been made by the critics from the Left who assert that aid impacts negatively on political and economic development in poor countries (Hayter 1971; Lappe 1980; Escobar 1988). The empirical evidence suggests that donor self-interest plays an important role in bilateral assistance, particularly on the part of the larger donors, while recipient need plays a larger role in multilateral assistance (Maizels and Nissanke 1984; Lumsdaine 1993).

### Economic Interest

Most economic self-interest arguments employ some version of the assertion that aid promotes exports from and employment in the donor country. A crude version of this argument simply draws attention to the presumed gains to the U.S. economy from exports of commodities or services that are subsidized by the assistance program. Producers of food grains in the United States benefit from food assistance, workers in the maritime industry gain from cargo preference provisions, and U.S. engineering firms gain from contracts associated with infrastructure development projects. Programs to protect private overseas investment against economic and political risk have been a prominent component of U.S. and many other nations' assistance programs (Rosen 1974).

Somewhat less obvious appeals to specific interests often emphasize the generalized role of economic assistance for the development of a nation's transportation or communication network in generating future commercial transactions. As the recipient country's infrastructure develops, commercial demand for new and replacement equipment compatible with the aid-assisted investments is expected to widen commercial sales opportunities. Similarly, technical assistance for the development of an LDC grain-milling and feed-processing industry is viewed as enhancing the commercial demand for food and feed grains from the donor country.

---

3. See, for example, Commission of Security and Economic Assistance (1983, 31); Shultz (1984, 142–45).

A more sophisticated argument often made is that if aid is effective in contributing to LDC economic growth the effect will be an expansion of demand for DC goods and services that are characterized by high import demand elasticities (Mellor and Johnston 1984). Agricultural producers in the United States are urged not to become overly concerned about loss of, for example, oilseed markets to Malaysia and Brazil because, as incomes rise, growth in demand for animal proteins will generate demand for U.S. feed grains. Loss of exports by the mature industrial sectors will be more than compensated for by capital goods and high-technology exports (Schuh 1986, 31–42).

The first two arguments rest on relatively weak logical foundations. The use of assistance resources to subsidize domestic suppliers of commodities or services generally reduces the value of a given level of assistance to the recipient country (Pincus 1963; Bhagwati 1970). This concern has generated vigorous discussion about the value and impact of food assistance (Schultz 1960; Isenman and Singer 1977). Other areas, such as tied procurement of such services as technical assistance, have been subject to much less controversy. But there can be little question that the effect of aid tying is to raise the cost to the donor of providing whatever benefits recipients receive from development assistance.

The growth impact argument rests on stronger logical grounds. It should be technically possible to specify conditions under which government-to-government aid transfers involving a grant element could improve welfare in both donor and recipient countries (Krueger 1986, 66). The empirical analysis to support this argument is, however, surprisingly limited. It is not sufficient simply to assert that the transfer of assistance resources may be followed by the growth of exports from the donor and recipients; this growth must be calculated. Empirical estimates of the effects of aid recipients' growth on donor trade balances or on welfare gains and losses have seldom been made (Mosley 1985).

Political and Strategic Interest

The view that development assistance is a useful complement to other elements of donor political strategy—that its primary rationale is to strengthen the political commitment of the aid recipient to the donor county or to the West—has been a consistent and at times dominant theme in the motivation for development assistance (Morgenthau 1962; Sewell and Mathieson 1982). During the Cold War, political considerations in both donor and recipient countries have, however, often made it advisable to cloak the objectives of short-term political or strategic assistance with the rhetoric of economic assistance—hence terms such as "economic support fund" in the USAID budget (Huntington 1971b).

The strengthening of the capacity of Western Europe to resist external aggression and the enhancement of the political appeal of centrist political forces were major motivations for the Marshall Plan (McKinlay and Mughan 1984, 31–34). Strategic concerns were a prominent feature of the Kennedy administration's Alliance for Progress in the early 1960s (Rostow 1985). The Reagan administration's Carlucci Commission report insisted that "the foreign security and economic cooperation programs of the United States are mutually supportive and interrelated and together constitute an integral part of the foreign policy of the United States" (46). The commission urged that efforts be made to enhance the complementarity between U.S. economic and security programs through the creation of a Mutual Development and Security Agency that would bring development, military, and related assistance programs under one agency.

The achievement of donor short-run political and strategic objectives and longer-run political or economic objectives have at times been inconsistent with recipient longer-term political development (Vietnam, Nicaragua, Philippines); their complementarity and conflict need fuller examination. A common assumption in the earlier political development and nation-building literature was that Western-style democratization and bureaucratization would be in the interest of both the donor and the recipient. But the study of political development has provided few guidelines for policymakers or practitioners who would guide political development along mutually advantageous lines (Eckstein 1982; chaps. 4 and 8, this vol.).

## In Summary

There is a disturbing dichotomy in the dialogue about the use of foreign assistance in the pursuit of domestic economic and strategic interests. It is clear that self-interest and security arguments have often represented little more than cynical efforts to generate support for the foreign assistance budget. There have been serious efforts to examine the theoretical foundations of the economic self-interest argument. There have also been increasingly serious attempts to evaluate the economic and social impacts of economic assistance in developing countries. In addition to a large professional literature, the U.S. Agency for International Development has conducted and published an extensive

Project Evaluation Report literature. The World Bank has an Operations Evaluation Department that engages in a major program of project completion evaluation studies.

The security rationale has not, however, been subject to nearly as rigorous theoretical or empirical analysis (McKinley and Mughan 1984). The single background article on the effectiveness of military assistance prepared for the Carlucci Commission asserted a positive linkage between U.S. security assistance expenditures and security interests while admitting that the evidence to support the assertion is "elusive" (West 1983). This is not to suggest that empirical support cannot be provided for the political and strategic self-interest arguments. It is simply to argue that, in spite of Huntington's assertion that the results of security assistance have been at least as successful as efforts to promote economic development (1971a), little convincing evidence has appeared in the professional literature on development assistance.

There is an inherent contradiction in both the economic and the security self-interest arguments. There is no question that donor countries have at times pursued their self-interest under the rubric of aid even at the cost of harm to the recipient country (Ruttan 1996, 253–333). If the donor self-interest argument is used as a primary rationale for development assistance, it imposes on donors some obligation to demonstrate that their assistance does no harm to the recipient.

One effect of the economic and security self-interest arguments has been to clarify that donor governments and assistance constituencies are not indifferent to the form of the resource transfers they make to poor countries. Security assistance draws on a set of ideological concerns that often go beyond a rational calculation of donor self-interest. Food aid taps not only the self-interest of donor commodity producers but also a powerful set of altruistic concerns in donor countries about poverty, hunger, and health in poor countries (Ruttan 1996, 149–202). It is doubtful that these forms of assistance are directly competitive—if food aid were reduced, the resources released would not become available to support the security assistance budget.

**Ethical Considerations**

Efforts to develop an acceptable rationale for development assistance have not been confined to self-interest assertions and rationalizations. There has been an extended argument that development assistance represents a moral responsibility on the part of rich countries, over and above any considerations of self-interest. But neither the advocates nor critics of foreign assistance have adhered to careful distinctions between self-interest and moral responsibility.

The typical criticism of foreign assistance starts with the argument that "foreign aid is neither necessary nor sufficient to promote economic progress" (Bauer 2000, 42). This is often followed by an assertion that the resources devoted to foreign assistance are wasted—that assistance does not achieve its intended objective, either economic or political. This tends to be followed by the argument that, in any event, it is not legitimate within the framework of Western political philosophy for government to forcibly extract resources from its citizens in order to transfer them to foreigners (Banfield 1963; Bauer 1981, 86–134; 2000).

Both the popular and official sponsors of foreign assistance have typically treated the ethical basis for foreign assistance as intuitively obvious (Riddell 1986, 1987). There is, however, a substantial professional literature that attempts to identify a basis in ethical theory or political philosophy for income or resource transfers at the expense of a reduction in welfare in the donor country. In addition, since government-to-government transfers are often involved, there are also attempts to explore the basis for claims for assistance by the recipient country and of the obligation of the donor country to the recipient country.

### Entitlement

An argument frequently put forth during the "New Economic Order" dialogue of the 1970s was that there should be compensation by the rich countries to poor countries for past injustices stemming from political domination and economic exploitation (Singer 1977). A second entitlement argument is based on the uneven distribution of natural resources. It has been argued that natural resources are part of our global heritage and that those areas that are favorably endowed have an obligation to share rents from differential resource endowments with those areas that are less favorably endowed (Beitz 1979, 134–43; Simon 1984, 179–212; Sartorius 1984, 192–224).

The argument based on past injustice, while correct in principle, poses substantial difficulties for translation into contemporary assistance policy. If exploitation occurred and compensation was not made, the effect of compound interest is to magnify the size of the obligation. Much of the development assistance provided by Great Britain and

France, the two major colonial powers, has been directed to former colonies and dependencies. Lenin's model of imperialism, in which capital was exported to low-income staple-producing areas under the direct or indirect political control of the major power and earned enormously high rates of return for a narrow class of investors in the metropolitan country, has not held up even to casual examination (Blaug 1961). It has been difficult to establish the extent to which the imperial relation was, in fact, exploitative (Boulding and Mukerje 1970; Krasner 1974). A more relevant argument in a world of both overt and "voluntary" constraints on the movement of commodities, labor, and financial resources is that developed countries have inadequately "exploited" the human and physical resources of the poor countries.

It is also difficult to decide what weight should be given to the natural resource distribution argument. One can hardly claim that the inhabitants of Kuwait and Somalia "deserve" the differential resource endowments they have inherited. But natural resource endowment differentials do not represent a very powerful factor in explaining differential income levels or growth rates among either developing or developed countries. The difficulty of converting staple exports into a base for sustained national or regional economic development has been a major challenge even in situations that have not been characterized by exploitation (Jacobs 1984). Perhaps the area in which the natural resources distribution issue is of greatest contemporary significance is the debate about the management and distribution of the potential rents associated with the exploitation of the global commons—the ocean and space resources (Cooper 1977, 1986).

## Distributive Justice

Most economists have generally felt fairly comfortable—perhaps too comfortable—with a straightforward utilitarian rationale for foreign assistance. If private rates of return to capital investment are higher in developing countries than in developed countries, investment should flow from developed to less developed countries. Because markets are imperfect, developed country governments should transfer resources to developing countries to assist in physical and institutional infrastructure development. But few economists would be willing to embrace the full implications of the utilitarian income-distribution argument—that rich countries ought to give until the point is reached at which, by giving more, there is a loss in utility in the recipient country or countries.

Most political philosophers, and those economists who adhere to a Hobbesian contractarian view of the role of government, have found it difficult to discover any intellectual foundation for development assistance based on considerations of distributive justice. "Justice has meaning only as a rule of human conduct and no conceivable rules for the conduct of individuals supplying each other with goods and services in a market economy would produce a distribution which could be meaningfully described as just or unjust" (Hayek 1978a, 58). Hayek argues, in effect, that justice is a function of the rules or processes that govern individual and group behavior and not of the outcome generated by the rules. The appropriate role of public policy is rule reform.

The Hobbesian contractarian argument with respect to foreign aid has been forcefully articulated by Banfield: "Our political philosophy does not give our government any right to do good for foreigners. Since the seventeenth century, Western political thought has maintained that government may use force or threat of force to take the property of some and give it to others only if doing so somehow serves the common good . . . government may take from citizens and give to foreigners when doing so serves the common good of the citizens, but it may not do so if . . . all advantage will accrue to foreigners and none to citizens" (1963, 24). This argument reemerged with renewed force in the debate over foreign assistance in the late 1970s and early 1980s (Bauer 1981, 86–133; 1984a, 38–62). It seems apparent that the emergence of social justice as a significant issue on the political agenda, both within nations and in international relations, is due to lack of confidence that the actual behavior of economic markets and political institutions adequately approaches the conditions specified by Hayek, Nozick, and other libertarian political philosophers (MacPherson 1985).

Attempts have been made to develop a contractarian argument drawing on the Rawlsian "difference principle" to establish a moral obligation for foreign assistance. A central thrust of Rawls's theory is that in a just society departures from egalitarian income distribution would be permitted only when differential rewards contribute to the welfare of the least advantaged members of society. Rawls argues that this "difference principle" would be agreed to by rational individuals attempting to design a constitution—given full general knowledge of the political and economic nature of society except the positions that they would occupy by virtue of social class, individual talent, or political persuasion. A Rawlsian constitution does not imply perfect equalization of incomes. If, for example, inequality calls forth economic activity that benefits the least as well as the more advantaged members of society, it is justified (Rawls 1971, 54–192).

Rawls made no attempt to explore the implications of the difference principle for international equality. Beitz and Runge have argued that an intuitively obvious extension of the difference principle to the international economic order is that justice would imply equal access by citizens of all countries to global resources except in those cases where departure from inequality could be justified on the basis of benefits to citizens of the least advantaged countries (Beitz 1979, 141–42; Runge 1977). This argument goes beyond the "past injustice" resource entitlement argument discussed earlier. To the extent that it draws on the Rawls framework, however, it remains vulnerable to the weakness of attempting to derive rules of justice from an "imagined social contract" (Crews 1986). I would personally prefer a stronger behavioral foundation on which to rest convictions about moral responsibility for assistance to poor countries. This preference reflects a skepticism about both the contractarian approach to political philosophy and the public choice approach to political economy that attempt to derive principles for the design of social and economic institutions from primitive assumptions about human nature.

Others insist that both the moral and rational arguments apply only to individuals (or families) and not to collectives such as nations (Cooper 1977). The problem of extending to nations the ethical arguments that have been developed to apply to individuals has been difficult to resolve in spite of the strong popular sentiment that rich nations do have some responsibility for assisting poor nations to achieve adequate levels of nutrition, health, and education. Richard Cooper asserts that "much recent discussion on transfer of resources falls uncritically into the practice of . . . anthropomorphizing nations, of treating nations as though they are individuals and extrapolating to them on the basis of average per capita income the various ethical arguments that have been developed to apply to individuals. This is not legitimate. If ethical arguments are to be used as a rationale for transferring resources, either a new set of ethical principles applicable to nations must be developed, or the link between resource transfers must be made back to the individuals who are the ultimate subjects of standard ethical reasoning" (1977, 355).

## An Implicit Global Contract

A contractarian argument that limits the responsibility of the rich toward the poor to national populations has great difficulty in confronting a world where citizens hold multiple loyalties, where national identity may be wider or narrower than state boundaries, where policy interventions as well as market forces guide the flow of labor and capital and the trade in commodities and intellectual property across state boundaries, and in which the costs and benefits of policy initiatives flow across national borders to generate public "goods" and "bads" (Keohane 1984, 120–24; Cooper 1986, 1–22; Dalrymple 2001). In this world, a "state-moralist" or "political realist" approach to international relations, that would limit the expression of the moral concerns of individual citizens to national governments about the basic political rights or subsistence needs of people in other countries, seems remarkably archaic.

Increased interdependence among nations results in a rise in both political tension and concern about lack of equity in economic transactions. The ethical foundation for a system of development assistance rests on the premise that the emergence of international economic and political interdependence has resulted in an implicit extension of Arrow's argument for redistribution to include the international sphere: "There are significant gains to social interaction above and beyond what individuals can achieve on their own. The owners of scarce personal assets do not have substantial private use of these assets; it is only their value in a large system which makes these assets valuable. Hence, there is a surplus created by the existence of society which is available for redistribution" (1983, 188).

The growth of global and political interdependence implies a decline in the significance of state-established boundaries (chap. 8, this vol.). Since boundaries are not coextensive with the scope of economic and political interdependence, they do not mark the limits of social obligation in the sharing of the benefits and burdens associated with interdependence (Beitz 1979, 143–61). An open international economy increases the value of the natural, human, and institutional resources of the developed countries and makes part of this surplus available for redistribution.

**Lessons from Experience**

Acceptance of ethical responsibility by the citizens and the governments of rich countries does not resolve the question of what level of assistance is appropriate. It was noted earlier that the utilitarian or consequentialist argument seems to be based on equating marginal utilities—the rich countries ought to give until the point is reached at which, by giving more, the loss in utility in the donor country would exceed the gain in utility in the recipient country or countries. However, the actual level of aid allocations by donor countries seems to reflect the much weaker moral premise that, if it is possible to contribute to welfare in poor countries without sacrificing anything of moral or economic significance in the donor country, it should be done. There seems to be an implicit moral judgment among the citizens and governments of the rich countries that the moral obligation to feed the poor in Ethiopia is stronger than the moral obligation to help sustain a 3 percent rather than a 1 percent per year rate of growth in Ethiopia's GNP (Huntington 1971, 175; Singer 1977; Pogge 1986).

Neither the commitment to development assistance nor the commitment to a particular level of development assistance provides guidance about who should receive aid. The ethical considerations that support the distributive justice argument imply that assistance should be directed to improving the welfare of the poorest individuals in the poorest countries. But there is also an ethical argument that aid should be directed into uses that produce the largest increments of income from each dollar of assistance—the argument that assistance resources are limited and should not be wasted. There is substantial evidence that assistance resources have, on balance, generated relatively high marginal rates of return—rates of return that are high relative to what the same resources would have earned in the donor countries (Peterson 1989). There is also evidence that donor governments are willing to trade off some efficiency for equity in their aid allocations—that recipient income levels do carry modest weight in the allocation of aid resources (Behrman and Sah 1984). But there is little more than anecdotal evidence on the distributive impacts of development assistance in recipient countries (Bauer 1981, 86–137).

Acceptance of responsibility for assistance does not resolve the question of what form of assistance to offer. The goals of assistance range from attempting to assure immediate "subsistence rights" or basic needs (Shue 1980), to assistance designed to strengthen the capacity of a nation to meet the subsistence requirements of its own people, or to modifying the institutions that influence the resource flows among nations. On some grounds it would seen obligatory to secure some minimum level of subsistence before allocating resources to the other two objectives. But this conclusion is not at all obvious if the effect is to preclude either (1) expansion of the capacity needed to assure future subsistence or (2) reform of the rules of conduct that govern economic and political relationships among nations (e.g., reforming the International Trade Organization [ITO] rules on agricultural trade).

The most important lesson from a half century of development assistance experience is the significance of the macroeconomic environment for economic growth. Understanding of the importance of appropriate trade and exchange rate policies, of fiscal and monetary policies, and of the overall incentive structures of policies and institutions is much more firmly grounded than during the first several decades of development assistance effort. Economic assistance to Korea, for example, was relatively unproductive in the 1950s when the Korean government was pursuing an import substitution industrial policy associated with substantial exchange rate and factor and product market distortions. When export-oriented policies were adopted in the 1960s, growth accelerated despite lower levels of economic assistance. In Ghana overvalued exchange rates, repression of agricultural commodity prices, and proliferation of parastatal industrial enterprises resulted in a decline in per capita income from the level at independence in spite of very substantial external economic assistance. Macroeconomic and structural reforms initiated in 1983 have enabled Ghana to make much more effective use of external assistance and to reverse more than two decades of economic decline (Krueger, Michalopoulos, and Ruttan 1989, 269–302).

Since the early 1980s, the recognition that the effectiveness of assistance in generating economic growth is severely weakened in the absence of appropriate economic policy on the part of the recipient country has led to much greater emphasis on policy-based lending on the part of the World Bank and the International Monetary Fund (as well as a number of bilateral donors). Reform policies have crystallized around what has become identified as the Washington Consensus—macroeconomic stability, domestic market liberalization, openness of international commodity and financial markets, and privatization of public enterprise.[4] It is generally agreed that the effectiveness of policy-based lending depends on policy dialogue leading to a convergence of donor and recipient views regarding policy reform. A number of fac-

---

4. The Washington Consensus term and concept was first articulated by John Williamson (1990) in addressing economic policy reform in Latin America. For the benefit of those critics who identify the Washington Consensus with "market fundamentalism" it should be noted that Williamson's definition included: "A redistribution of public expenditure priorities towards fields offering both high economic reforms and the potential to improve income distribution, such as primary heath care, primary education, and infrastructure" (2000, 252).

tors influence the effectiveness of donor agencies' efforts to engage in constructive dialogue about a recipient country's macroeconomic policies. One is the ability of both donor and recipient to bring substantial professional capacity and experience to the policy dialogue. A second is whether the donor agency is prepared to provide adequate aid during a transition period to support the policy objectives agreed on by donors and recipients. Finally, no amount of policy dialogue or conditionality will be effective unless there is broad agreement and commitment within the recipient country government on the reform objectives. These conditions are rarely met.

By the late 1990s there was a rising chorus of criticism of the effectiveness of policy-based lending on the part of the World Bank and the International Monetary Fund. Much of the criticism focused on the resolve, capacity, and sensitivity of the Bank and the Fund in implementing the principles of the Washington Consensus (Meier and Stiglitz 2001; Easterly 2002; Stiglitz 2002b). What we should learn from this experience is that the Washington Consensus reforms, often advanced or imposed by donors during periods of economic or financial crisis without adequate sensitivity to the technical and institutional capacity to respond to the reforms, have often led to economic regression rather than growth.

Experience also indicates that the macroeconomic setting has a strong bearing on the success or failure of sectoral and project assistance. The economic viability of agricultural and industrial development projects has often been severely weakened by exchange rate overvaluation. Experience also suggests several areas in which assistance efforts have generated very substantial benefits in spite of unfavorable macroeconomic policy or political environments. Assistance efforts in the areas of agricultural research and human capital development have, for example, often generated very substantial benefits even in the absence of favorable economic and political environments (Pardey and Beintema 2001).

Underinvestment by public and private suppliers of agricultural knowledge and technology has been characteristic in both colonial and postcolonial economies. The underinvestment reflected the "public good" characteristics of agricultural research, technology development, and extension education. The economic benefits of technical change in agriculture are characteristically transferred rather rapidly from the public and private sector suppliers of knowledge and technology to agricultural producers and from producers to consumers (chap. 5, this vol.). Since the mid-1950s bilateral and multilateral assistance agencies have made major contributions to enhancing food production in developing countries through their support of a system of national and international agricultural research institutes (Ruttan 1982; Baum 1986).[5]

The rates of return generated by these investments in agricultural research have been high relative to any other investments available to the public or private sectors of developing countries. By 2000 developing country farmers were feeding approximately twice as many people far better than in the early 1960s. Over 90 percent of the increase in production has come from higher yields. Even in Africa, where most countries have lagged in strengthening their agricultural research capacity, about half of the increase in production has come from higher yields (Alston et al. 2000; Pardey and Bentema 2001). These advances in food production have not eliminated food insecurity for the very poor in developing countries. They have strengthened the capacity of farmers to respond to the demands of the poor in the presence of economic and social policies designed to enhance the well-being of the poor.

A more complete discussion of areas in which assistance has been productive, even in the absence of a favorable macroeconomic or policy environment, would include assistance for human capital development, particularly education and health. In addition to the contribution of such investments to economic growth, human capital investment is one of the few instruments available to a poor country to set itself on a sustainable path toward more equitable income distribution.

Both development theorists and assistance agencies were initially slow to recognize the importance of investments in education as an important source of economic growth. By the mid-1960s, however, the major U.S. foundations, the U.S. foreign economic assistance agency,

---

5. The first four international agricultural research institutes were established through the joint efforts of the Rockefeller and Ford Foundations—the International Rice Research Institute (IRRI) in the Philippines (1959), the International Center for the Improvement of Maize and Wheat (CIMMYT) in Mexico (1963), the International Institute for Tropical Agriculture (IITA) in Nigeria, and the International Center for Tropical Agriculture (CIAT) in Colombia (1968). In 1969 the two foundations initiated consultations with the UN Food and Agricultural Organization (FAO), the World Bank, the United Nations Development Program (UNDP), and a number of bilateral assistance organizations that led to the formation of a Consultative Group on International Agricultural Research (CGIAR). During the next two decades, centers were established to conduct research on additional commodities (potatoes, cassava, bananas, plantain, livestock, and livestock disease); resources (soils, water, forest, marine, and genetic); and agricultural, food, and research policies. The centers rapidly became the nodes of a network of a global system for the exchange of scientific and technical information and of genetic material.

and the World Bank were beginning to play an important role in the development and strengthening of higher-level education and research institutions, particularly in the fields of science and technology, in developing countries. Donors were, however, more reluctant to become heavily involved, in spite of evidence of high rates of return, in primary and secondary education (Schultz 1988; Krueger, Michalopoulos, and Ruttan 1989, 317–18).

Assistance for the development of national health systems has also generated very large benefits. Life expectancy in low-income countries has improved by twenty-two years, and mortality rates of children under five years has been halved. Under the leadership of the World Health Organization, smallpox has been eradicated. Polio and leprosy have been nearly eliminated. After substantial initial success toward elimination there has been a resurgence of such mass killers as malaria and tuberculosis and the emergence of several new infectious diseases of which HIV/AIDS is the most dramatic. Reduction of the debilitation and mortality impact of these, as with earlier infectious diseases, will depend on very substantial new investment in laboratory-based biomedical research—most of which must be carried out in developed countries.[6] In addition to laboratory-based health research, effective health systems in developing countries will require the strengthening of national capacity to adapt and produce the knowledge and materials appropriate to the resource and cultural endowments of poor communities and to make that knowledge available to families, particularly mothers.[7]

Programs to achieve significant gains in primary and secondary education and substantial improvements in health indicators will not require massive assistance resources. Many of the resources needed to support institutional development or reform can be mobilized at the local or regional level. Such programs are generally regarded as desirable even by national governments that are too corrupt or do not have the capacity to effectively implement larger-scale development programs.

---

6. It is interesting to speculate, however, about how much more rapid the response to the emergence of HIV/AIDS might have been if there had been a first-class biomedical research facility in East Africa at the time it first emerged. As yet neither the biomedical community nor the development assistance community has recognized the importance of establishing a counterpart to the CGIAR system of international agricultural research institutes in the health field (Bell, Clark, and Ruttan 1994; Jha et al. 2002).

7. For case studies of the remarkable gains in health status in countries or regions that have designed systems to reduce the incidence of illness rather than focusing primarily on illness recovery see Kaseje (1994), Gunatilleke (1994), and Tendler (1997, 21–45).

### A Foreign Economic Assistance Future?

Since the early 1990s the foreign economic assistance effort by developed countries has experienced substantial erosion. In retrospect it is clear that much of the aid effort was motivated more by Cold War tensions than concerns about relieving poverty or generation of long-term growth.

These motivations have not prevented the generation of very substantial development benefits from economic assistance (Krueger, Michalopoulos, and Ruttan 1989). But strategic and political motivation for the allocation of assistance resources has resulted in smaller benefits, measured in terms of poverty reduction or economic growth, than if economic considerations had carried more weight in the allocation of assistance resources. And in some cases, particularly in Southeast Asia, Central America, and the Middle East, it is clear that economic assistance, especially U.S. assistance allocated under the "supporting assistance" rubric, did not measure up well against the criterion that at the very least economic assistance should do no harm (Ruttan 1996). A consequence in the United States, and in a number of other donor countries, has been "aid fatigue." The U.S. economic assistance effort has declined from well above 1 percent in the early 1950s to little more than one-tenth of 1 percent of GNP in the late 1990s.

My own sense is that the deficiency in the U.S. aid effort will not be resolved until a new post–Cold War vision of the kind of world that the United States wants to live in during the early decades of the twenty-first century captures the political and popular imagination. Such a vision can only be perceived dimly at this time. The international environment in which assistance efforts will be conducted in the early decades of the twenty-first century will be vastly different from the bipolar world of the Cold War era. We will continue to be confronted by what Harlan Cleveland has termed "a new world disorder" (1993). Many centralized nation-states are perceived by large numbers of their peoples as increasingly less relevant to their economic and civic needs. This trend is most apparent in large multinational states. It also includes many small multinational states that incorporate geographically based ethnic minorities (chap. 8, this vol.).

International organizations and national governments will be forced to deal more creatively with their constituent nationalities than in the past. The constitutional design challenge of the early decades of the twenty-first century will be how to simultaneously achieve political autonomy and economic viability. On the political side this means a pragmatic search for constitutional arrangements that assures ethnic and other communities sufficient autonomy to satisfy their civic needs. Economic viability will require that political autonomy be achieved within a constitutional framework that permits financial resources, commodities, services, and people to move relatively freely across political borders (chap. 8, this vol.).

What should be the role of U.S. economic assistance in an environment of economic and political disorder? U.S. national security will rarely be threatened by the small conflicts of national liberation or ethnic cleansing. Conventional military responses will not be particularly effective in restoring order. The political development literature offers only limited guidance for the allocation of resources to political development (chap. 4, this vol.). In the case of economic development, both sector development and macroeconomic policy efforts have been able to draw on a powerful body of economic thought—primarily neoclassical economic theory—that provides the analytical tools with which to address the issues of development practice and the design of economic reform. The application of these tools, even when used with skill and sensitivity, has not represented a guarantee against failure in project, program, or policy design (Harberger 1993). There is no comparable body of theory that can serve as a guide in the design of a program to strengthen the institutions of governance or to achieve a liberal political order.

I am quite confident, however, that the liberal impulse that was, at least in part, responsible for sustaining the U.S. development assistance effort over the past half century will reemerge as a compelling source of American foreign policy. The liberal impulse that inspired U.S. leadership in the design of the post–World War II international system and the U.S. bilateral aid program will insist that the United States play a constructive role in responding to the new international disorder. But the implications for the U.S. development assistance program are far from clear. The capacity and the motivation of the U.S. bilateral development assistance program, as we have known it throughout most of the postwar period, have largely eroded. Since the early 1990s the U.S. assistance budget has declined in real terms, and an increasing share of the resources available to it has been devoted to short-term political and strategic objectives. It is doubtful, in the absence of a new vision and a new consensus, that the U.S. assistance agency can survive in its present form.

## In Conclusion

The first conclusion that emerges from this review is the weakness of the self-interest argument for foreign assistance. The individual (or group) self-interest arguments, after careful examination, often represent a hidden agenda for domestic rather than international resource transfers. The political "realists" have not been able, or have not thought it worthwhile, to demonstrate the presumed political and security benefits from the strategic assistance component of the aid budget. Rawlsian contractarian theory does provide a basis for ethical responsibility toward the poor in poor countries that goes beyond the traditional religious and moral obligations of charity. It also provides a basis for making judgments about the degree of inequality that is ethically acceptable.

But the contractarian argument cannot stand by itself. Its credibility is weakened if, in fact, the transfers do not achieve the desired consequences. Failures of analysis or design can produce worse consequences than if no assistance had been undertaken (Nye 1984, Ruttan 1996). There is no obligation to transfer resources that do not generate either immediate welfare gains or growth in the capacity of poor states to meet the needs of their citizens. It becomes important, therefore, to evaluate the consequences of development assistance and to consider the policy interventions that can lead to more effective development assistance programs.

Since the 1950s our understanding of the development process has made major advances. But we can never fully anticipate the consequences of any assistance activity or of intervention into complex and interdependent social systems. Our limited knowledge about how to give and use aid to contribute most effectively to development does not, however, protect us from an obligation to assess the consequences of our strategic or development assistance and to advance our capacity to understand the role of external assistance in the development process. This constitutes a major challenge to the social sciences to advance knowledge for the design of sustainable development assistance programs in a world characterized by social and political disorder.

PART IV

CHAPTER 10

# Postscript

I see no need to repeat here the arguments or the conclusions from earlier chapters. I would, however, like to reemphasize the perspective on the role of social science knowledge in economic development that runs through all of the chapters.

A consistent theme is that advances in social science knowledge represent a powerful source of economic growth and, more broadly, of economic development. Advances in social science knowledge contribute to the formulation of economic and social policy and to institutional design. Advances in social science knowledge represent a high payoff input into economic development. This position falls squarely within the tradition of Enlightenment political philosophy. A magnificent example of this design perspective is the U.S. Constitution.

The design perspective stands in sharp contrast to the organic or evolutionary perspective. Hayek, for example, has argued that improvements in institutional performance are the result of a process of collective learning that has passed the slow test of time and is embodied in a people's language, culture, and institutions. This accumulated knowledge is built into ways of learning and has a powerful impact on the present and on the future. Since collective learning occurs at the level of the community rather than the individual, there are severe constraints on the rational design of policies and institutions. But there can be no presumption that the institutions that emerge out of the process of social evolution will result in favorable paths of cultural, social, or economic development (Hayek 1978b; North 1994).

The induced institutional innovation model employed in this book embraces and challenges both the idealist and the evolutionary concepts of institutional innovation. Successful and productive institutional innovations are not the result of simply taking thought independent of historical and contemporary context. Nor are they determined by changes in resource endowments, cultural endowments, or technology.

The pattern model outlined in chapter 1 is built on recursive relationships among changes in resource endowments, technology, institutions, and culture. Successful institutional innovation will almost always be culture specific. It involves more than simply institutional (or technology) transfer. Advances in social science knowledge open up new and productive opportunities for institutional innovations that enhance development. In the induced innovation model there is no role for simple resource, technological, institutional, or cultural determinism. The dialectical relationships between changes in resource and cultural endowments and technical and institutional change influence the rate and direction of social, political, and economic development. And the feedback from these changes are the sources of change in resource and cultural endowments.

Finally, intellectual history conducted apart from technological and institutional history is arid, as is theoretical inquiry carried on apart from a continuing dialogue with data. I hope this book is not viewed merely as a work in intellectual history but as an inspiration for cross-disciplinary research.

# Appendix: Definitions of Culture

The definitions of culture in this appendix were selected to illustrate the progressive narrowing of the concept of culture in anthropology.

1. "Culture . . . is that complex whole which includes knowledge, belief, art, morals, law, custom, and any other capabilities and habits acquired by man as a member of Society" (Taylor 1958, 1). This definition does not distinguish social organization and social institutions from a general concept of culture.
2. "Culture consists of patterns, explicit and implicit, of and for behavior acquired and transmitted by symbols, constituting the distinctive achievement of human groups, including their embodiments in artifacts; the essential core of culture consists of traditional (i.e., historically derived and selected) ideas and especially their attached values; culture systems may, on the one hand, be considered as products of action, on the other as conditioning elements of further action" (Kroeber and Kluckholm 1952, 18).
3. "If . . . society is taken to be an organized set of individuals with a given way of life, culture is that way of life. If society is taken to be an aggregate of social relations, then culture is the content of those relations. Society emphasizes the human component, the aggregate of people and the relations between them. Culture emphasizes the component of accumulated resources, immediate as well as material, which the people inherit, employ, transmute, add to and transmit" (Firth 1951, 27).
4. "Culture is not a material phenomenon; it does not consist of things, people, behavior, or emotions. It is rather an organization of these things. It is the forms of things that

people have in mind, their models for perceiving, relating, and otherwise interpreting them" (Goodenough 1964, 36).
5. "It is useful to define the concept culture for most usages more narrowly than has been generally the case in the American anthropological tradition, restricting its reference to transmitted and created content and patterns of values, ideas, and other symbolic-meaningful systems as factors in the shaping of human behavior and the artifacts produced through behavior. On the other hand, we suggest that the term society—or more generally, social system—be used to designate the specifically relational system of interaction among individuals and collectivities" (Kroeber and Parsons 1958, 582). A purpose of the Kroeber and Parsons article was to distinguish the proper subject matter of anthropology (culture) and sociology (social system).
6. "Radcliffe-Brown and other adherents of the theory of social structure tended to avoid using the term 'culture' after the early 1930s. This avoidance is based on the claim that social anthropology studies social structure, not culture" (Singer 1968, 531). "The theory of social structure can dispense with the word 'culture' [because] it has incorporated the culture concept into the core of the theory, for the theory of social structure deals with social relations not simply as concrete actually existing objects of observations but as institutionalized and standardized modes of behavior and thought whose normal forms are socially recognized in the explicit or implicit rules to which the members of a given society tend to conform" (532).
7. "Culture . . . refers to the learned repertory of thoughts and actions exhibited independently of genetic heredity from one generation to the next" (Harris 1968, 47).
8. "The concept of culture that I espouse . . . is essentially a semiotic (symbolic) one. Believing that man is an animal suspended in webs of significance he himself has spun. I take culture to be those webs, and the analysis of it (culture) to be therefore not an experimental science in search of law but an interpretive one in search of meaning" (Geertz 1973, 5).
9. "Culture in the narrow sense is the most problematic element in modern society" (Gans 1985, 52). "The independent variable of historical evolution is not the economic, nor the technical, nor indeed the aesthetic, but the ethical" (58). The

"means of preserving order . . . are at the very heart of culture and more immediately relevant to its specific creations than the more general need to increase appetite satisfaction" (74). "The abandonment of order in a mad rush to satisfy appetites is a true breakdown of culture" (75).

10. "It may be . . . that the culture concept has served its time. Perhaps, following Foucault, it should be replaced by a vision of powerful discursive formations globally and strategically deployed. Such entities would at least no longer be closely tied to notions of organic unity, traditional continuity, and the enduring grounds of language and locale. But however the culture concept is finally transcended, it should, I think be replaced by some set of relations that preserves the concepts of differential and relativist functions and that avoids the positioning of cosmopolitan essences and human common denominators" (Clifford 1988, 274–75).

# Bibliography

Abramovitz, Moses. 1952. "Economics of Growth." In B. F. Haley, ed., *A Survey of Contemporary Economics.* Homewood, IL: Richard D. Irwin.

Abu-Lughod, Lila. 1991. "Writing Against Culture." In Richard G. Fox, ed., *Recapturing Anthropology: Working on the Past,* 137–62. Santa Fe, NM: School for America Research Press.

Acemoglu, Daron. 2002. "Technical Change, Inequality, and the Labor Market." *Journal of Economic Literature* 40:7–72.

Acemoglu, Daron, Simon Johnson, and James A. Robinson. 2001. "The Colonial Origins of Comparative Development: An Empirical Investigation." *American Economic Review* 91:1369–1401.

Adams, Adrian. 1981. "The Senegal River Valley." In J. Heyer, P. Roberts, and Gavin Williams, eds., *Rural Development in Tropical Africa.* New York: St. Martin's.

Adams, John. 2001. "Culture and Economic Development in South Asia." *Annals* 573:152–75.

Adams, Julia. 1999. "Culture in Rational Choice Theories of State Formation." In George Steinmetz, ed., *State/Culture: State Formation after the Cultural Turn,* 98–122. Ithaca: Cornell University Press.

Adelman, Irma, and Cynthia Taft Morris. 1965. "A Factor Analysis of the Interrelationship between Social and Political Variables and Per Capita Gross National Product." *Quarterly Journal of Economics* 79:555–78.

———. 1967. *Society, Politics and Economic Development: A Quantitative Approach.* Baltimore: Johns Hopkins University Press.

———. 1973. *Economic Growth and Social Equity in Developing Countries.* Stanford: Stanford University Press.

Adesina, A. A., and Jojo Baidu-Forson. 1995. "Farmers' Perception and Adoption of New Agricultural Technology: Evidence from Analyses in Burkina Faso and Guinea, West Africa." *Agricultural Economics* 13:1–9.

Aghion, P., and Peter W. Howitt. 1992. "A Model of Growth through Creative Destruction." *Econometrica* 60:323–51.

———. 1998. *Endogenous Growth Theory.* Cambridge: MIT Press.

Aghion, Philippe, Eve Caroli, and Cecilia Garcia-Peñalosa. 1999. "Inequality and Economics Growth: The Perspective of the New Growth Theories." *Journal of Economic Literature* 37:1615–60.

Ahmad, Syed. 1966. "On the Theory of Induced Innovation." *Economic Journal* 76:344–57.

Ajami, Foud. 1993. "The Summoning." *Foreign Affairs* (September/October): 2–9.

Akerlof, George A. 1970. "The Market for Lemons: Quality Uncertainty and the Market Mechanism." *Quarterly Journal of Economics* 84:488–500.

———. 1976. "The Economics of Caste and of the Rat Race and Other Woeful Tales." *Quarterly Journal of Economics* 40:599–617.

Alchian, Armen, and Harold Demsetz. 1973. "The Property Right Paradigm." *Journal of Economic History* 33:16–27.

Alesina, Alberto, and Enrico Spolaore. 1997. "On the Number and Size of Nations." *Quarterly Journal of Economics* 113:1027–56.

Alexander, Jeffrey C. 1990. "Analytical Debates: Understanding the Relative Autonomy of Culture." In Jeffrey C. Alexander and Steven Seidman, eds., *Culture and Society: Contemporary Debates*, 1–27. Cambridge: Cambridge University Press.

Allen, Beth. 2000. "The Future of Macroeconomic Theory." *Journal of Economic Perspectives* 14:143–50.

Allen, Robert C. 1982. "The Efficiency and Distributional Consequences of Eighteenth-Century Enclosures." *Economic Journal* 92:937–53.

Almond, Gabriel A. 1966. "Political Theory and Political Science." *American Political Science Review* 60:869–79.

———. 1987. "The Development of Political Development." In Myron Weiner and Samuel P. Huntington, eds., *Understanding Political Development*, 437–90. Boston: Little, Brown.

———. 1993. "The Early Impact of Downs' *An Economic Theory of Democracy* on American Political Science." In Bernard Grofman, ed., *Information, Participation, and Choice*. Ann Arbor: University of Michigan Press.

Almond, Gabriel A., Emmanuel Sivan, and R. Scott Appleby. 1995. "Fundamentalism: Genus and Species." In Martin E. Marty and R. Scott Appleby, eds., *Fundamentalisms Comprehended*, vol. 5, 399–424. Chicago: University of Chicago Press.

Alston, Julian M., Connie Chau-Kang, Michele C. Marra, Philip G. Pardey, and T. J. Wyatt. 2000. *A Meta-Analysis of Rates of Return to Agricultural R&D: Ex Pede Herculem?* Washington, DC: International Food Policy Research Institute, Research Report 113.

Alston, Julian M., Michele C. Marra, Philip G. Pardey, and T. J. Wyatt. 2001. "Research Returns Redux: A Meta-Analysis of the Returns to Agricultural R&D." *Australian Journal of Agricultural Economics* 44:185–210.

Alston, Lee J., Thráiin Eggertson, and Douglass C. North. 1996. *Empirical Studies in Institutional Change*. Cambridge: Cambridge University Press.

Amable, B. 1994. "Endogenous Growth Theory, Convergence and Divergence." In G. Silverberg and L. Soete, eds., *The Economics of Growth and Technical Change*. Aldershot, UK: Elgar.

Amin, Samir. 1976. *Unequal Development: An Essay on the Social Formation of Peripheral Capitalism*. New York: Monthly Review Press.

Ammerman, Nancy T. 1991. "North American Protestant Fundamentalism." In Martin E. Marty and R. Scott Appleby, eds., *Fundamentalisms Observed*, vol. 1, 1–65. Chicago: University of Chicago Press.

Anderson, Kym, and Yujiro Hayami. 1986. *The Political Economy of Agricultural Protection: East Asia in International Perspective*. Sydney: Allen and Unwin.

Aoki, Masahiko. 2001. *Toward Comparative Institutional Analysis*. Cambridge: MIT Press.

Arce, Alberto, and Norman Long. 2000. "Reconfiguring Modernity and Development from an Anthropological Perspective." In Alberto Arce and Norman Long, eds., *Anthropology Development and Modernities: Exploring Discourses, Counter-Tendencies and Violence*, 1–13. London: Routledge.

Arendt, Hannah. 1986. "Communicative Power." In Stephen Lukes, ed., *Power*, 58–74. New York: New York University Press. Reprinted from Hannah Arendt, *On Violence*, 41. New York: Harcourt Brace and World, 1969.

Aron, Janine. 2000. "Growth and Institutions: A Review of the Evidence." *World Bank Research Observer* 15:99–135.

Arrighi, Giovanni. 1999. "Globalization and Historical Macrosociology." In Janet L. Abu-Lughod, ed., *Sociology for the Twenty-First Century: Continuities and Cutting Edges*, 117–33. Chicago: University of Chicago Press.

Arrow, Kenneth J. 1962. "The Economic Implications of Learning by Doing." *Review of Economic Studies* 29:155–73.

———. 1969. "Classification Notes on the Production and Transmission of Technological Knowledge." *American Economic Review* 59:29–35.

———. 1983. *Social Choice and Justice*. Cambridge: Harvard University Press.

———. 1990. In Richard Swedeberg, ed., *Economics and Sociology. Redefining Their Boundaries: Conversations with Economists and Sociologists*, 133–51. Princeton: Princeton University Press.

———. 1994. "Methodological Individualism and Social Knowledge." *American Economic Review* 84:1–9.

———. 1999. "Observation in Social Capital." In Partha Dasgupta and Ismail Serageldin, eds., *Social Capital: A Multifaceted Perspective*, 3–5. Washington, DC: World Bank.

Arrow, Kenneth J., and Frank H. Hahn. 1971. *General Competitive Analysis*. Edinburgh, UK: Oliver and Boyd.

Arthur, W. Bryan, and David A. Lane. 1994. "Information Contigation." In W. Bryan Arthur, ed., *Increasing Returns and Path Dependence in the Economy*, 69–97. Ann Arbor: University of Michigan Press.

Asad, Talal, ed. 1973. *Anthropology and the Colonial Encounter*. London: Ithaca Press.

Aslund, Anders. 1995. *How Russia Became a Market Economy*. Washington, DC: Brookings.

Atkinson, Anthony B. 1969. "The Timescale of Economic Models: How Long Is the Long Run?" *Review of Economic Studies* 36:137–52.

Audrelsch, David B. 1998. "Agglomeration and the Localization of Innovative Activity." CEPR Discussion Paper 1974.

Autum, Suzanne. 1996. "Anthropologists, Development and Situated Truth." *Human Organization* 55:480–84.

Babcock, Jarvis M. 1962. "Adoption of Hybrid Corn: A Comment." *Rural Sociology* 27:332–38.

Bacha, Edmar L. 1977. "The Kuznets Curve and Beyond: Growth and Change in Inequalities." In E. Malinvaud, ed., *Economic Growth and Resources*, vol. 1, *The Major Issues*. New York: St. Martin's.

Bachrach, Peter, and Morton S. Baratz. 1962. "The Two Faces of Power." *American Political Science Review* 56:942–52.

Backus, D. K., Patrick J. Kehoe, and Timothy J. Kehoe. 1992. "In Search of Scale Effects in Trade and Growth." *Journal of Economic Theory* 58:377–409.

Ball, Richard. 2001. "Individualism, Collectivism and Economic Development." *Annals of the American Academy of Political and Social Science* 273:57–84.
Ball, Terrence. 1975. "Power, Causation and Explanation." *Polity* (1975): 189–214.
Banfield, Edward C. 1963. "American Foreign Aid Doctrine." In Robert A. Goldwin, ed., *Why Foreign Aid?* Chicago: Rand McNally.
Baran, Paul A. 1952. "The Political Economy of Backwardness." *Manchester School of Economics and Social Studies* 20:66–84.
———. 1957. *The Political Economy of Growth.* New York: Monthly Review Press.
Bardhan, Ashok. 1993. "Economics of Development and the Development of Economics." *Journal of Economic Perspectives* 7:129–42.
———. 1995. "The Contribution of Endogenous Growth Theory to the Analysis of Development Problems: An Assessment." In J. Behrman and T. N. Srinivasan, eds., *Handbook of Development Economics,* vol. 3B. New York: Elsevier Science.
Bardhan, Pranab, ed. 1989. *Conversations between Economists and Anthropologists: Methodological Issues in Measuring Change in Rural India.* Oxford: Oxford University Press.
Barnes, Barry. 1982a. "The Science-Technology Relationship: A Model and a Query." *Social Studies of Science* 12:166–72.
———. 1982b. *T. S. Kuhn and Social Science.* New York: Columbia University Press.
Baron, James N., and Michael T. Hannan. 1994. "The Impact of Economics on Contemporary Sociology." *Journal of Economic Literature* 32:1111–46.
Barro, Robert J. 1997. *Determinants of Economic Growth: A Cross Country Empirical Study.* Cambridge: MIT Press.
Barro, Robert J., and Xavier X. Sala-i-Martin. 1995. *Economic Growth.* New York: McGraw-Hill.
Bartlett, Peggy F., ed. 1980. *Agricultural Decision Making: Anthropological Contributions to Rural Development.* New York: Academic Press.
Bass, Frank M. 1969. "A New Product Growth Model of Consumer Durables." *Management Science* 5:215–27.
———. 1980. "The Relationship between Diffusion Rates, Experience Curves and Demand Elasticities for Consumer Durable Technological Innovations." *Journal of Business* 53:51–67.
Bates, Robert H. 1981. *Markets and States in Tropical Africa: The Political Basis of Agricultural Policies.* Berkeley: University of California Press.
———. 1983. *Essays on the Political Economy of Rural Africa.* Berkeley: University of California Press.
Bauer, Martin. 1995. *Resistance to New Technology: Nuclear Power, Information and Biotechnology.* Cambridge: Cambridge University Press.
Bauer, Peter T. 1954. *West African Trade.* Cambridge: Cambridge University Press.
———. 1972. *Dissent on Development.* Cambridge: Harvard University Press.
———. 1981. *Equality, the Third World, and Economic Delusion.* Cambridge: Harvard University Press.
———. 1984a. *Reality and Rhetoric: Studies in the Economics of Development.* Cambridge: Harvard University Press.
———. 1984b. "Remembrance of Studies Past: Retracing First Steps." In G. M. Meier and Dudley Seers, eds., *Pioneers in Development,* 27–43. New York: Oxford University Press.

———. 2000. *From Subsistence to Exchange and Other Essays*. Princeton: Princeton University Press.

Bauer, Peter T., and B. S Yamey. 1957. *The Economics of Underdeveloped Countries*. Chicago: University of Chicago Press.

Baumol, William J. 1986. "Productivity Growth, Convergence, and Welfare." *American Economic Review* 76:1072–85.

Baumol, William J., and Edward N. Wolff. 1988. "Productivity Growth, Convergence, and Welfare: Reply." *American Economic Review* 78:1155–59.

Beal, George M., and Joe M. Bohlen. 1957. *The Diffusion Process*. Ames: Iowa State Agricultural Experiment Station, Special Report.

Beaudoux, Etienne, and Marc Nieuwkerk. 1985. *Groupements paysans d'Afrique: Dossier pour l'action*. Paris: L'Harmattan.

Becker, Gary S. 1976. *The Economic Approach to Human Behavior*. Chicago: University of Chicago Press.

———. 1991 [1981]. *A Treatise on the Family*. Cambridge: Harvard University Press.

———. 1993. "The Economic Way of Looking at Behavior." *Journal of Political Economy* 101:385–409.

———. 1996. *Accounting for Tastes*. Cambridge: Harvard University Press.

Becker, Gary S., and Kevin M. Murphey. 2000. *Social Economy: Market Behavior in a Social Environment*. Cambridge: Harvard University Press.

Behrman, Jere R., and Raj Kumar Sah. 1984. "What Role Does Equity Play in the International Distribution of Development Aid?" In L. Taylor, M. Dyrquin, and L. E. Westphal, eds., *Economic Structure and Performance*. New York: Academic Press.

Behrman, Jere R., and T. N. Srinivasan, eds. 1995. *Handbook of Development Economics*. Vols. 13A and 313. Amsterdam: North-Holland.

Beitz, Charles. 1979. *Political Theory and International Relations*. Princeton: Princeton University Press.

Bell, David E., William C. Clark, and Vernon W. Ruttan. 1994. "Global Research Systems for Sustainable Development: Agriculture, Health and Environment." In Vernon W. Ruttan, ed., *Agriculture, Environment and Health: Sustainable Development in the Twenty-first Century*, 358–79. Minneapolis: University of Minnesota Press.

Bender, Leonard. 1986. "The Natural History of Development Theory." *Comparative Studies in Society and History* 28:3–33.

Bendroth, Margaret L. 1996. Review of *Fundamentalism Comprehended*. *Christian Century* 113 (May 22): 575–79.

Benedict, Ruth. 1946. *The Chrysanthemum and the Sword*. Boston: Houghton Mifflin.

Bennett, John W. 1976. "Anticipation, Adaptation and the Concept of Culture in Anthropology." *Science* 192:874–53.

———. 1988. "Introductory Essay." In John Bennett and John Bowen, eds., *Production and Autonomy: Anthropological Critiques of Development*. Lanham, MD: University Press of America/Society for Economic Anthropology.

Bennett, John W., and John R. Bowen, eds. 1988. *Production and Autonomy: Anthropological Studies and Critique of Development*. Lanham, MD: University Press of America.

Benton, Ted. 1978. "How Many Sociologies?" *Sociological Review* 26:217–36.

Berger, Peter L., and Michael H. Hsin-Huang, eds. 1998. *In Search of an East Asian Development Model*. New Brunswick, NJ: Transaction Books.

Berk, Richard A. 1981. "On the Compatibility of Applied and Basic Sociological Research: An Effort in Marriage Counseling." *American Sociologist* 16:204–11.

Berlin, Isaiah. 1980. "Nationalism: Past Neglect and Present Power." In *Against the Current: Essays in the History of Ideas*, 333–55. New York: Viking Press.

———. 2000. *Three Critics of the Enlightenment: Vico, Hamann, Herder.* Ed. Henry Hardy. Princeton: Princeton University Press.

Bernstein, Henry. 1971. "Modernization and the Sociological Study of Development." *Journal of Development Studies* 7 (January): 141.

Bhagwati, Jagdish. 1966. *The Economics of Underdeveloped Countries.* New York: McGraw-Hill.

———. 1970. *Amount and Sharing of Aid.* Washington, DC: Overseas Development Council.

Bijker, Wiebe E. 1995. "Sociohistorical Technology Studies." In Sheila Jasanoff, Gerald E. Markle, James C. Petersen, and Trevor Pinch, eds., *Handbook of Science and Technology Studies*, 229–56. Thousand Oaks, CA: Sage.

Bijker, Wiebe E., Thomas P. Hughes, and Trevor Pinch. 1987. *The Social Construction of Technological Systems: New Directions in the Sociology and History of Technology.* Cambridge: MIT Press.

Bikhchandani, Sushil, David Hirshleifer, and Ivo Welch. 1998. "Learning from the Behavior of Others: Conformity, Fads, and Informational Cascades." *Journal of Economics Perspectives* 12:151–70.

Binder, Leonard. 1986. "The Natural History of Development Theory." *Comparative Studies in Society and History* 28:3–33.

Binswanger, Hans P. 1974. "A Microeconomic Approach to Induced Innovation." *Economic Journal* 84:940–58.

Binswanger, Hans P., and Klaus Deininger. 1997. "Explaining Agricultural and Agrarian Policies in Developing Countries." *Journal of Economic Literature* 35:1958–2005.

Binswanger, Hans P., Robert E. Evenson, Cecilia A. Florencio, and Benjamin N. W. White, eds. 1980. *Rural Household Studies in Asia.* Singapore: University of Singapore Press.

Binswanger, Hans P., and John McIntyre. 1987. "Behavioral and Material Determinants of Production Relations in Land-Abundant Tropical Agriculture." *Economic Development and Cultural Change* 36:73–100.

Binswanger, Hans P., and Vernon W. Ruttan, eds. 1978. *Induced Innovation: Technology, Institutions and Development.* Baltimore: Johns Hopkins University Press.

Birdsal, Nancy, Allen C. Kelley, and Steven W. Sinding. 2001. *Population Matters: Demographic Change, Economic Growth and Poverty in the Developing World.* Oxford: Oxford University Press.

Bishop, Charles E. 1967. "The Urbanization of Rural America: Implications for Agricultural Economics." *Journal of Farm Economics* 49 (December): 999–1008.

Blaikie, Piers. 1978. "The Theory of Spatial Diffusion of Innovations: A Spacious Cul-De-Sac." *Progress in Human Geography* 2:268–95.

Blaug, Mark. 1961. "Economic Imperialism Revisited." *Yale Review* 50 (spring): 335–49.

———. 1974. *The Cambridge Revolution: Success or Failure?* London: Institute of Economic Affairs.

Blaut, James M. 1987. "Diffusionism: A Uniformitarian Critique." *Annals of the Association of American Geographers* 77:30–47.

Bliss, Christopher J., and Nicholas H. Stern. 1982. *Palanpur: The Economy of an Indian Village.* Oxford: Clarendon.
Blumer, Herbert. 1966. "The Idea of Social Development." *Studies in Comparative International Development* 2 (1): 3–11.
Bohm, Peter. 1985. "Comparative Analysis of Alternative Policy Instruments." In A. V. Kneese and J. Sweeney, eds., *Handbook of Natural Resource and Energy Economics,* vol. 1, 395–460. Amsterdam: North-Holland.
Booth, David. 1975. "Andre Gunder Frank: An Introduction and Appreciation." In Ivan Oxaal, Tony Barnett, and David Booth, eds., *Beyond the Sociology of Development: Economy and Society in Latin America and Africa,* 50–85. London: Routledge and Kegan Paul.
Boudon, Raymond, and Francois Bourricaud. 1989. *A Critical Dictionary of Sociology.* Chicago: University of Chicago Press. French ed., 1982, 1986.
Boulding, Kenneth E., and Tapan Mukerjee. 1970. "Unprofitable Empire: Britain in India, 1800–1967: A Critique of the Hobson-Lenin Thesis on Imperialism." *Peace Research Society Papers 14,* Rome Conference.
Boulton, Patrick, and Gerard Roland. 1997. "The Breakup of Nations: A Political Economy Analysis." *Quarterly Journal of Economics* 113:1057–90.
Bourdieu, Pierre. 1990. *The Logic of Practice.* Stanford, PA: Stanford University Press.
Bowles, Samuel. 1998. "Endogenous Preferences: The Cultural Consequences of Markets and Other Economic Institutions." *Journal of Economic Literature* 36:75–111.
Boycko, Maxim, Andrei Shleifer, and Robert Vishny. 1995. *Privatizing Russia.* Cambridge: MIT Press.
Boyer, Pascal. 2001. *Religion Explained: The Evolutionary Origins of Religious Thought.* New York: Basic Books.
Bradner, Lowell, and Murray A. Straus. 1959. "Congruence versus Profitability in the Diffusion of Hybrid Sorghum." *Rural Sociology* 24:381–83.
Braibanti, Ralph. 1969. "External Inducement of Political-Administrative Development." In Ralph Braibanti, ed., *Political and Administrative Development,* 3–106. Durham: Duke University Press.
Breton, Albert. 1964. "The Economics of Nationalism." *Journal of Political Economy* 72:376–86.
Britto, R. 1973. "Some Recent Developments in the Theory of Economic Growth: An Interpretation." *Journal of Economic Literature* 11:1343–66.
Brock, William A., and Steven Durlauf. 2001. "Growth Empirics and Reality." *World Bank Economic Review* 15:229–72.
Brown, Lawrence A. 1981. *Innovation Diffusion: A New Perspective.* New York: Methuen.
Brunner, Edmund. 1957. *The Growth Age of a Science: A Half-Century of Rural Sociological Research in the United States.* New York: Harper and Brothers.
Buchanan, James M., and Gordon Tullock. 1962. *The Calculus of Consent: Logical Foundations of Constitutional Democracy.* Ann Arbor: University of Michigan Press.
Buijsrogge, Piet. 1989. *Initiatives Paysannes en Afrique de l'Ouest.* Paris: L' Harmattan.
Burkhart, Ross E., and Michael S. Lewis-Beck. 1994. "Comparative Democracy: The Economic Development Thesis." *American Political Science Review* 88:903–10.
Burnside, A. Craig. 1996. "Production Function Regressions, Returns to Scale, and Externalities." *Journal of Monetary Economics* 37:177–201.

Busch, Lawrence. 1978. "On Understanding Understanding: Two Views of Communication." *Rural Sociology* 43:450–73.

———. 1992. "Metatheories and Better Theories: A Reply to Ruttan." *International Journal of Sociology of Agriculture and Food/Revista International de Sociologia sobre Agricultura y Alimentos* 2:44–49.

Busch, Lawrence, and William B. Lacy. 1983. *Science, Agriculture, and the Politics of Research.* Boulder: Westview.

Bush, Vannevar. 1945. *Science: The Endless Frontier.* Washington, DC: U.S. Office of Scientific Research and Development. Reprint editions, National Science Foundation, 1960, 1980.

Buttel, Frederick H., Olaf F. Larson, and Gilbert W. Gillespie Jr. 1990. *The Sociology of Agriculture.* New York: Greenwood Press.

Caldwell, Bruce. 1997. "Hayek and Socialism." *Journal of Economic Literature* 35:1856–90.

Cancian, Francesca M. 1968. "Varieties of Functional Analysis." In David L. Sills, ed., *International Encyclopedia of the Social Sciences,* vol. 6, 29–43. New York: Crowell Collier and Macmillan.

Cappell, Charles L., and Thomas M. Guterback. 1992. "Visible Colleges: The Social and Conceptual Structure of Sociological Specialties." *American Sociological Review* 57:266–73.

Cardoso, Fernando Henrique, and Enzo Faletto. 1979. *Dependence and Development in Latin America.* Berkeley: University of California Press. [Spanish ed. 1969.]

Cavallo, Domingo, and Yair Mundlak. 1982. *Agriculture and Economic Growth in an Open Economy: The Case of Argentina.* International Food Policy Research Institute, Research Report No. 36, Washington, DC.

Cernea, Michael M. 1985. *Putting People First: Sociological Variables in Rural Development.* New York: Oxford University Press.

———. 1990. "From Unused Social Knowledge to Policy Creation: The Case of Population Resettlement." Cambridge: Harvard University Institute for International Development. Development Discussion Paper 342.

———, ed. 1991a. *Putting People First: Sociological Variables in Rural Development.* 2d ed. New York: Oxford University Press.

———. 1991b. "Using Knowledge from Social Science in Development Projects." Washington, DC: World Bank, Discussion Paper 114.

———. 1996. *Social Organization and Development Anthropology: The 1995 Malinowski Award Lecture.* Washington, DC: World Bank Environmentally Sustainable Development Studies. Studies and Monographs No. 6.

Chambers, J. D., and George E. Mingay. 1966. *The Agricultural Revolution, 1750–1880.* London: B. T. Batsford.

Chambers, Robert. 1997. *Whose Reality Counts? Putting the First Last.* London: Intermediate Technology Publications.

Chari, Varadarajan V., and Hugo Hopenhayn. 1991. "Vintage Human Capital, Growth and the Diffusion of New Technology." *Journal of Political Economy* 99:1142–65.

Chari, Varadarajan V., Patrick J. Kehoe, and Ellen R. McGrattan. 1996. *The Poverty of Nations: A Quantitative Exploration.* Minneapolis Federal Reserve Bank of Minneapolis Research Department Staff Report 204.

Chen, Xianming. 2001. "Both Glue and Lubricant: Transnational Ethnic Social Capital as a Source of Asia-Pacific Subregionalism." In John D. Montgomery and Alex

Inkeles, eds., *Social Capital as a Policy Resource,* 43–61. Boston: Kluwer Academic Publishers.
Chenery, Hollis B., and T. N. Srinivasan, eds. 1988, 1989. *Handbook of Development Economics.* Vols. 1 and 2. Amsterdam: North-Holland.
Chenery, Hollis B., and Alan M. Strout. 1966. "Foreign Assistance and Economic Development." *American Economic Review* 56:679–733.
Chenery, Hollis, and Moshe Syrquin. 1975. *Patterns of Development, 1950–1970.* London: Oxford University Press.
Cheung, Steven N. S. 1969a. *The Theory of Share Tenancy.* Chicago: University of Chicago Press.
———. 1969b. "Transaction Costs, Risk Aversion and the Choice of Contractual Arrangements." *Journal of Law and Economics* 12:23–42.
Chipman, John S. 1992a. "Intra-Industry Trade, Factor Proportions and Aggregation." In Wilhelm Neuefeind and Raymond Riezman, eds., *Economic Theory and International Trade,* 67–92. Berlin: Springer-Verlag.
———. 1992b. *Intra-Industry Trade in a Loglinear Model.* Department of Economics, University of Minnesota, Minneapolis, March 10. Mimeo.
Christianson, James A., and Lorraine E. Gorkovich. 1985. "Fifty Years of Rural Sociology: States, Trends and Impressions." *Rural Sociology* 55:395–410.
Clague, Christopher. 1997. "The New Institutional Economics and Economic Development." In Christopher Clague, ed., *Institutions and Economic Development: Growth and Governance in Less-Developed and Post-Socialist Countries.* Baltimore: Johns Hopkins University Press.
———. 1999. "The Political Economy of Policy Reform: Analytical Approaches from Economics and Political Science." Department of Economics, San Diego State University, May. Mimeo.
Clague, Christopher, P. Keefer, S. Knack, and Mancur Olson. 1996. "Property and Contract Rights in Autocracies and Democracies." *Journal of Economic Growth* 1:243–76.
Clague, Christopher, and Gorden Rausser, eds. 1992. *The Emergence of Market Economies in Eastern Europe.* Oxford: Basil Blackwell.
Clark, Colin. 1940. *The Conditions of Economic Progress.* London: Macmillan. 3d ed., 1957.
Clark, Gregory. 1987. "Why Isn't the Whole World Developed? Lessons from the Cotton Mills." *Journal of Economic History* 47:141–74.
Clastres, Pierre. 1998 [1972]. *Chronicle of the Guayaki Indians.* Trans. from the French by Paul Auster. New York: Zone Books.
———. 1977. *Society against the State: The Leader as Servant and the Human Use of Power among the Indians of the America.* Trans. from the French by Robert Hurley. New York: Urizen Books.
Cleaver, Harry M., Jr. 1972. "The Contradictions of the Green Revolution." *American Economic Review* 62:177–86.
Cleveland, Harlan. 1993. *Birth of a New World: An Open Moment for International Leadership.* San Francisco: Jossey-Bass.
Clifford, James. 1988. *The Predicament of Culture: Twentieth Century Ethnography, Literature and Art.* Cambridge: Harvard University Press.
———. 1997. *Routes: Travel and Translation in the Late Twentieth Century.* Cambridge: Harvard University Press.

Coase, Ronald H. 1937. "The Nature of the Firm." *Economics*, n.s., 4:386–405.
———. 1960. "The Problem of Social Cost." *Journal of Law and Economics* 3:1–44.
Cochrane, Willard W. 1979. *The Development of American Agriculture: A Historical Analysis*, 41–47, 179–88. Minneapolis: University of Minnesota Press.
Coggins, Jay, and Vernon W. Ruttan. 1999. "U.S. Emissions Permit System." *Science* 284:263–64.
Coleman, James S. 1964. *Introduction to Mathematical Sociology*. New York: Free Press of Glencoe.
———. 1986. "Social Theory, Social Research, and a Theory of Action." *American Journal of Sociology* 91:1309–35.
———. 1988. "Social Capital in the Creation of Human Capital." *American Journal of Sociology* 94:S95–S120.
———. 1990a. "Commentary: Social Institutions and Social Theory." *American Sociological Review* 55:333–38.
———. 1990b. *Foundations of Social Theory*. Cambridge: Harvard University Press.
———. 1993. "The Rational Reconstruction of Society." *American Sociological Review* 58:1–15.
Coleman, James S., E. Q. Campbell, C. Hobson, J. McPartland, A. Mood, F. Weinfield, and R. L. York. 1966. *Equality of Educational Opportunity*. Washington, DC: U.S. Government Printing Office.
Coleman, James S., and Thomas B. Hoffer. 1987. *Public and Private High Schools: The Impact of Communities*. New York: Basic Books.
Collier, Paul, and Jon W. Gunning. 1999. "Explaining African Economic Performance." *Journal of Economic Literature* 37:64–111.
Collins, Randall. 1986. "Is Sociology in the Doldrums?" *American Journal of Sociology* 91 (May): 133–55.
———. 1990. "The Organizational Politics of the ASA." *American Sociologist* 21:311–15.
Commission of Security and Economic Assistance. 1983. *A Report to the Secretary of State*. Washington, DC: Commission (Carlucci Commission).
Commons, John R. 1950. *The Economics of Collective Action*. New York: Macmillan.
Conceição, Pedro, and James K. Galbraith. 2001. "Toward a New Kuznets Hypothesis: Theory and Evidence on Growth Inequality." In James K. Galbraith and Maureen Berner, eds., *Inequality and Industrial Change: A Global View*, 139–60. Cambridge: Cambridge University. Press.
Conklin, Howard C. 1957. *Hanusoo Agriculture in the Philippines*. Forestry Development Paper No. 12. Rome, Italy: UN Food and Agricultural Organization (UNFAO).
Connor, Linda. 1984. "Comments." *Current Anthropology* (1984): 25.
Connor, Walker. 1972. "Nation-Building or Nation-Destroying?" *World Politics* 24:319–55.
———. 1987. "Ethnonationalism." In M. Weiner and S. P. Huntington, *Understanding Political Development*, 196–220. Boston: Little, Brown.
Conway, Gordon. 1997. *The Doubly Green Revolution: Food for All in the Twenty-first Century*. Ithaca: Cornell University Press.
Cook, Joanne, Jennifer Roberts, and Georgina Waylen, eds. 2000. *Toward a Gendered Political Economy*. New York: St. Martin's.
Cooper, Richard N. 1977. "The Oceans as a Source of Revenue." In Jagdigh N. Bhag-

wati, ed., *The New International Economic Order: The North-South Debate.* Cambridge: MIT Press.

———. 1986. *Economic Policy in an Interdependent World: Essays in World Economics.* Cambridge: MIT Press.

Coser, Lewis A., and Robert Nisbet. 1975. "Merton and the Contemporary Mind: An Affectionate Dialogue." In Lewis A. Coser, ed., *The Idea of Social Structure: Papers in Honor of Robert K. Merton,* 3–10. New York: Harcourt Brace Jovanovich.

Cottril, Charlotte C., Everett M. Rogers, and Tamsy Mills. 1989. "Co-citation Analysis of the Scientific Literature of Innovation Research Traditions." *Knowledge: Creation, Diffusion, Utilization* 11:181–208.

Cox, Harvey. 1969. *The Feast of Fools: A Theological Essay on Festivity and Fantasy.* New York: Harper and Row.

Crafts, N. F. R. 1995. "Exogenous or Endogenous Growth? The Industrial Revolution Reconsidered." *Journal of Economic History* 55:745–72.

Crane, Diane. 1972. *Invisible Colleges: Diffusion of Knowledge in Scientific Communities.* Chicago: University of Chicago Press.

Crews, Frederick. 1986. "In the Big House of Theory." *New York Review of Books* 33 (May 29): 36–42.

Critchley, Will. 1990. "Catch the Rain." *CERES* 125:41–45.

———. 1991. *Looking After Our Land: Soil and Water Conservation in Dryland Africa.* Oxford: Oxfam.

Crocker, Thomas D. 1966. "The Structure of Atmospheric Pollution Control Systems." In H. Wolozin, ed., *The Economics of Air Pollution,* 61–86. New York: W. W. Norton.

Dahl, Robert A. 1957. "The Concept of Power." *Behavioral Science* 2:201–15.

Dahlman, Carl J. 1980 *The Open Field System and Beyond: A Property Rights Analysis of an Economic Institution.* Cambridge: Cambridge University Press.

Dales, James H. 1968a. "Land, Water and Ownership." *Canadian Journal of Economics* 1:791–804.

———. 1968b. *Pollution, Property and Prices.* Toronto: University of Toronto Press.

Dalrymple, Dana G. 2001. "International Agricultural Research as a Global Public Good." Washington, DC: U.S. Agency for International Development, Office of Agriculture and Food Security.

Dasgupta, Partha. 2000. "Economic Progress and the Idea of Social Capital." In Partha Dasgupta and Ismail Serageldin, eds., *Social Capital: A Multifaceted Perspective,* 325–424. Washington, DC: World Bank.

Dasgupta, Partha, and Ismail Serageldin, eds. 2000. *Social Capital: A Multifaceted Perspective.* Washington, DC: World Bank.

Dasgupta, Susmita. 1989. *Diffusion of Innovations in Village India.* New Delhi: Wiley Eastern.

Datta, Anusa, and Hamid Mohtadi. 2001. "On the Mechanism of Technology Transfer: Can Imitation Lead to Innovation?" University of Wisconsin, Milwaukee, Working Paper.

Davis, Lance E., and Douglass C. North. 1971. *Institutional Change and American Economic Growth.* Cambridge: Cambridge University Press.

Davis, Stephen. 1979. *The Diffusion of Process Innovations.* Cambridge: Cambridge University Press.

De George, Richard T., and Fernande M. De George. 1972. *The Structuralists: From Marx to Lévi-Strauss.* Garden City, NY: Doubleday.

De Long, J. Bradford. 1988. "Productivity Growth, Convergence and Welfare: Comment." *American Economic Review* 78:1138–54.

de Janvry, Alain. 1973. "A Socioeconomic Model of Induced Innovations for Argentine Agricultural Development." *Quarterly Journal of Economics* 87:410–35.

de Schweinitz, Karl. 1964. *Industrialization and Democracy: Economic Necessities and Political Possibilities.* Glencoe, IL: Free Press.

De Tocqueville, Alexis. 1966 [1835]. *Democracy in America.* New York: Knopf.

De Waal, Alex. 2002. "Anthropology and the Aid Encounter." In Jeremy MacClancy, ed., *Exotic No More: Anthropology on the Front Lines,* 251–69. Chicago: University of Chicago Press.

Deiros, Pablo A. 1991. "Protestant Fundamentalism in Latin America." In Martin E. Marty and R. Scott Appleby, eds. *Fundamentalisms Observed,* 132–96. Chicago: University of Chicago Press.

Demsetz, Harold. 1964. "The Exchange and Enforcement of Property Rights." *Journal of Law and Economics* 7:11–26.

Dennison, Edward F. 1962. "The Sources of Economic Growth in the United States and the Alternatives before Us." Supplementary Paper 13. New York: Committee for Economic Development.

Deutsch, Karl. 1953. *Nationalism and Social Communication: An Inquiry into the Foundations of Nationality.* Cambridge: Harvard University Press.

Deyo, Frederic C., ed. 1987. *The Political Economy of New Asian Industrialism.* Ithaca: Cornell University Press.

Dixit, Avinash K. 1973. "Models of Dual Economies." In J. H. Wimlees and Nicholas H. Stem, eds., *Models of Economic Growth,* 325–52. New York: Wiley.

———. 1996. *The Making of Economic Policy: A Transaction-Cost Politics Perspective.* Cambridge: MIT Press.

Dollar, David, and Edward N. Wolff. 1993. *Competitiveness, Convergence and International Specialization.* Cambridge: MIT Press.

Domar, E. 1946. "Capital Expansion, Rate of Growth and Employment." *Econometrica* 14:137–47.

———. 1947. "Expansion and Employment." *American Economic Review* 37 (1): 343–55.

Dosi, Giovanni. 1984. *Technical Change and Industrial Transformation.* New York: St. Martin's.

Dosi, Giovanni, L. Orsenigo, and Gerald Silverberg. 1986. *Innovation Diversity and Diffusion: A Self-Organization Model.* Sussex: University of Sussex Science Policy Research Unit.

Douglas, Mary. 1986. *How Institutions Think.* Syracuse, NY: Syracuse University Press.

Dovring, Folke. 1959. "The Share of Agriculture in a Growing Population." *Monthly Bulletin of Agricultural Economics and Statistics* (FAO) 8, Aug., Sept., 1–11.

Downs, Anthony. 1957. *An Economic Theory of Democracy.* New York: Harper and Row.

Drucker, Peter. 1994. "The New Superpower: The Overseas Chinese." *Wall Street Journal,* December 20, A14.

Duesenberry, James. 1960. "Comment on 'An Economic Analysis of Fertility.'" In National Bureau Committee for Economic Research, ed., *Demographic and Eco-*

*nomic Change in Developed Countries,* 231–34. Princeton: Princeton University Press.

Duvall, Raymond D. 1978. "Dependence and Dependence Theory: Notes toward Precision of Concept and Argument." *International Organization* 32:51–78.

Easterly, William P. 1995. "Explaining Miracles: Growth Regressions Meet the Gang of Four." In T. Ito and A. O. Krueger, eds., *Growth Theories in Light of East Asian Experience,* 267–84. Chicago: University of Chicago Press.

———. 2001a. "Can Institutions Resolve Ethnic Conflict?" *Economic Development and Cultural Change* 49:687–706.

———. 2001b. *The Elusive Quest for Growth: Economists' Adventures and Misadventures in the Tropics.* Cambridge: MIT Press.

Easterly, William P., and Ross E. Levine. 1997. "Africa's Growth Tragedy." *Quarterly Journal of Economics* 62:1203–50.

———. 2001. "It's Not Factor Accumulation: Stylized Facts and Growth Models." *World Bank Economic Review* 15:177–219.

Echevarria, Christina E. 1995. "Agricultural Development vs. Industrialization: Effects of Trade." *Canadian Journal of Economics* 28:631–47.

———. 1997. "Changes in Sectoral Composition Associated with Economic Growth." *International Economic Review* 38:431–52.

———. 2000. "Non-Homothetic Preferences and Growth." *Journal of International Trade and Economic Development* 9:151–71.

Echeverria, Ruben. 1990. "Assessing the Impact of Agricultural Research." In Ruben Echeverria, ed., *Methods for Diagnosing Research System Constraints and Assessing the Impact of Agricultural Research,* vol. 2. The Hague: International Service for National Agricultural Research.

Eckhaus, Richard S. 1977. *Appropriate Technology for Developing Countries.* Washington, DC: National Academy of Sciences Press.

———. 1987. "Appropriate Technology: The Movement Has Only a Few Clothes On." *Issues in Science and Technology* 4:62–71.

Eckstein, Harry. 1982. "The Idea of Political Development: From Dignity to Efficiency." *World Politics* 34:451–86.

Eckstein, Susan. 2002. "Globalization and Mobilization: Resistance to Neoliberalism in Latin America." In Mario F. Guillén, Randall Collins, Paula England, and Marshall Meyer, eds., *The New Economic Sociology: Developments in an Emerging Field.* New York: Russell Sage Foundation.

Eggertsson, Thráinn. 1992. "Analyzing Institutional Success and Failure: A Millennium of Common Mountain Pastures in Iceland." *International Review of Law and Economics* 12:423–37.

———. 1994. "The Economics of Institutions in Transition Economies." In Salvatore Schiavo-Campo, ed., *Institutional Change and the Public Sector in Transitional Economies.* Washington, DC: World Bank.

———. 1997. "No Experiments, Monumental Disasters: Why It Took a Thousand Years to Develop a Specialized Fishing Industry in Iceland." *Journal of Economic Behavior and Organization* 30:1–23.

———. 2003. *Imperfect Institutions: Poverty of Nations and Institutional Policy.* Ann Arbor: University of Michigan Press.

Ehrlich, Isaac. 1990. "The Problem of Development: Introduction." *Journal of Political Economy* 98:S1–S11.

Eichelberger, James. 1969. "Power Problems of a Revolutionary Government." In Miles Copeland, ed., *The Game of Nations*, 71–75. London: Weidenfeld and Nicholson.

Eisenstadt, Samuel N. 1969. "Social Changes: Differentiation and Evolution." *American Sociological Review* 29 (June): 375–86.

———. 1973. *Tradition, Change, and Modernity*. New York: John Wiley.

———. 1987. "Introduction: Historical Traditions, Modernization and Development." In Samuel N. Eisenstadt, ed., *Patterns of Modernity*, vol. 1, *The West*. New York: New York University Press.

Elbasha, E., and Terry L. Roe. 1995. "Environment in Three Classes of Endogenous Growth Models." St. Paul: University of Minnesota Economic Development Center Bulletin 95-6, Aug.

Ellerman, A. Denny, Paul L. Joskow, Richard Schmalensee, Juan-Pablo Montero, and Elizabeth M Bailey. 2000. *The Market for Clean Air: The U.S. Acid Rain Program*. Cambridge: Cambridge University Press.

Ellsworth, Lynn. 1988. "Mutual Insurance and Non-Market Transactions among Farmers in Burkina-Faso." Ph.D. diss., Department of Agricultural Economics, University of Wisconsin, Madison.

Elster, Jon. 1989. *Nuts and Bolts for the Social Sciences*. Cambridge: Cambridge University Press.

Embree, John F. 1950. "Thailand—A Loosely Structured Social System." *American Anthropologist* 52:181–93.

Emmanuel, Arghiri. 1972. *Unequal Exchange: A Study of the Imperialism of Trade*. New York: Monthly Review Press.

Ennis, James G. 1992. "The Social Organization of Sociological Knowledge: Modeling the Intersection of Specialties." *American Sociological Review* 57:259–65.

Ensminger, Jean. 1992. *Making a Market: The Institutional Transformation of an African Society*. New York: Cambridge University Press.

———. 1997. "Changing Property Rights: Reconciling Formal and Informal Rights to Land in Africa." In John N. Drobak and John V. C. Nye, eds., *The Frontiers of the New Institutional Economics*. San Diego: Academic Press.

———. 2002. "Experimental Economics: A Powerful New Method for Theory Testing in Anthropology." In Jean Ensminger, ed., *Theory in Economic Anthropology*. Walnut Creek, CA: Altamira Press.

Escobar, Arturo. 1984. "Discourse and Power in Development: Michel Foucault and the Relevance of His Work to the Third World." *Alternatives* 10:377–400.

———. 1988. "Power and Visibility: Development and Management of the Third World." *Cultural Anthropology* 3:428–33.

———. 1991. "Anthropology and the Development Encounter: The Making and Marketing of Development Anthropology." *American Anthropologist* 18:658–82.

———. 1995. *Encountering Development: The Making and Unmaking of the Third World*. Princeton: Princeton University Press.

———. 1999. "After Nature: Steps to an Antiessentialist Political Ecology." *Current Anthropology* 40:1–16.

Etzioni, Amitai. 1968. *The Active Society: A Theory of Societal and Political Processes*. New York: Free Press.

Fafchamps, Marcel. 1992. "Solidarity Networks in Preindustrial Society: Rational

Peasants with a Moral Economy." *Economic Development and Cultural Change* 41:148–74.
Faia, Michael A. 1993. *What's Wrong with the Social Sciences? The Perils of the Postmodern.* Lanham, MD: University Press of America.
Fan, Shengen. 1991. "Effects of Technological Change and Institutional Reform Production Growth in Chinese Agriculture." *American Journal of Agricultural Economics* 73:266–75.
Fan, Shengen, Peter Hazel, and Sukhadeo Thorat. 1999. *Linkages between Government Spending, Growth and Poverty in Rural India.* Washington, DC: International Food Policy Research Institute.
Feder, Gershan, Richard E. Just, and David Zilberman. 1985. "Adoption of Agricultural Innovations in Developing Countries: A Survey." *Economic Development and Cultural Change* 33:255–98.
Feeney, David. 1982. *The Political Economy of Productivity: Thai Agricultural Development, 1880–1975.* Vancouver, Canada: University of British Columbia Press.
———. 1988. "The Demand and Supply of Institutional Arrangements." In Vincent Ostrum, David Feeney, and Hartmunt Pichl, eds., *Rethinking Institutional Analysis and Development.* San Francisco: International Center for Economic Growth.
———. 2002. "The Coevolution of Property Rights Regimes for Land, Man and Forests in Thailand, 1790–1990." In John F. Richards, ed., *Land, Property and the Environment,* 179–220. Oakland, CA: ICS Press.
Fei, John C. H. 1989. "Chinese Cultural Values and Industrial Capitalism." *Conference Proceedings on Confucianism and Economic Development in East Asia,* 257–78. Taipei, Taiwan, Republic of China.
Fei, John C. H., and Gustav Ranis. 1964. *Development of the Labor Surplus Economy: Theory and Policy.* Homewood, IL: Irwin.
Fellner, William. 1961. "Two Propositions in the Theory of Induced Innovations." *Economic Journal* 71:305–8.
Fetini, Habib. 1993. "Institutional Deficiencies and Indigenous Responses—A Case of Institutional Renovation and Diffusion: The *Groupements Naam* and the 6-S NGO in the Sahel." Washington, DC: World Bank. Mimeo.
Field, Alexander J. 1981. "The Problem with Neoclassical Institutional Economics: A Critique with Special Reference to the North/Thomas Model of Pre-1500 Europe." *Explorations in Economic History* 18:174–98.
Findlay, Ronald. 1978. "Relative Backwardness, Direct Foreign Investment, and the Transfer of Technology: A Simple Dynamic Model." *Quarterly Journal of Economics* 42:1–16.
Findlay, Ronald, and Ronald W. Jones. 2001. "Economic Development from an Open Economy Perspective." In Deepak Lal and Richard H. Snape, eds., *Trade Development and Political Economy: Essays in Honor of Anne O. Krueger,* 159–73. New York: Palgrave.
Fine, Ben. 2001. *Social Capital versus Social Theory: Political Economy and Social Science at the Turn of the Millennium.* London: Routledge.
Finkler, Kaja. 1979. "Employing Econometric Techniques in Economic Anthropology." *American Ethnologist* 6:675–81.
Firth, Raymond. 1951. *Elements of Social Organization.* London: Watts.
———. 1981. "Engagement and Detachment: Reflections on Applying Social Anthropology to Social Affairs." *Human Organization* 40:193–201.

Fisher, J. C., and R. H. Pry. 1971. "A Sample Substitution Model of Technological Change." *Technological Forecasting and Social Change* 3:75–88.

Fliegel, Frederick. 1993. *Diffusion Research in Rural Sociology: The Record and Prospects for the Future.* Westport, CT: Greenwood Press.

Flora, Peter, and Jens Alber. 1981. "Modernization and Development of the Welfare State in Western Europe." In Peter Flora and Arnold Heidenheimer, eds., *The Development of the Welfare State in Europe and America*, 37–80. New Brunswick, NJ: Transaction.

Flyvberg, Bent. 2001. *Making Social Science Matter: Why Social Inquiry Fails and How It Can Succeed Again.* Cambridge: Cambridge University Press.

Fogel, Robert W. 1992. "Douglass C. North and Economic Theory." In John N. Drobak and John V. C. Nye, eds., *The Frontiers of the New Institutional Economics*, 13–28. San Diego: Academic Press.

———. 2000. *The Fourth Great Awakening and the Future of Egalitarianism.* Chicago: University of Chicago Press.

Foster, George M. 1965. "Peasant Society and the Image of Limited Good." *American Anthropologist* 67 (pt. 1): 293–315.

Foucault, Michel. 1972. *The Archaeology of Knowledge.* New York: Harper and Row.

Frank, Andre Gunder. 1966. *Capitalism and Underdevelopment in Latin America.* New York: Monthly Review Press.

———. 1967. "The Development of Underdevelopment." *Monthly Review* 18:17–31.

———. 1969. *Capitalism and Underdevelopment in Latin America.* New York: Monthly Review Press.

Frank, Robert H. 1992. "Melding Sociology and Economics: James Coleman's *Foundations of Social Theory.*" *Journal of Economic Literature* 30:147–70.

———. 1996. "The Political Economy of Preference Falsification: Timur Kuran's *Private Truths and Public Lies.*" *Journal of Economic Literature* 34:115–23.

Freeman, John R. 1985. *The Politics of Indebted Economic Growth.* Denver: University of Denver Graduate School of International Studies.

———. 1989. *Democracy and Markets: The Politics of Mixed Economies.* Ithaca: Cornell University Press.

Friedland, William H., and Amy Benton. 1975. *Destalking the Wily Tomato: A Case Study in Social Consequences in California Agricultural Research.* Davis: University of California Department of Applied Behavioral Sciences, Mimeograph 15.

Friedland, William H., Amy E. Benton, and Robert P. Thomas. 1978. *Manufacturing Green Gold: The Conditions and Consequences of Lettuce Harvest Mechanization.* Davis: University of California Department of Applied Behavioral Sciences.

Frolich, Norman, Joe A. Oppenheimer, and Oran R. Young. 1971. *Political Leadership and Collective Goods.* Princeton: Princeton University Press.

Fuglie, Keith O., and Catherine A. Kascak. 2001. "Adoption and Diffusion of Natural Resource Conserving Agricultural Technology." *Review of Agricultural Economics* 23:386–403.

Fukuyama, Francis. 1989. "The End of History." *National Interest* 16:3–18.

———. 1995. "Social Capital and the Global Economy." *Foreign Affairs* 74:89–103.

———. 1996. "Still a Dangerous Place." *Wall Street Journal* (Nov. 7).

Furubotn, Eirik G., and Svetozar Pejovich. 1972. "Property Rights and Economic Theory: A Survey of Recent Literature." *Journal of Economic Literature* 10:1137–61.

Fusfeld, Daniel R. 1980. "The Conceptual Framework of Modern Economics." *Journal of Economic Issues* 14:1–52.

Galor, Oded, and Daniel Tsiddon. 1996. "Income Distribution and Growth: The Kuznets Hypothesis Revisited." *Economica* 63:103–17.

Gans, Eric. 1985. *The End of Culture: Toward a Generative Anthropology*. Berkeley: University of California Press.

Gartell, C. David, and John W. Gartell. 1985. "Social Status and Agricultural Innovations: A Meta-Analysis." *Rural Sociology* 50:38–50.

Geertz, Clifford. 1963. *Agricultural Involution in Indonesia: The Process of Ecological Change*. Berkeley: University of California Press.

———. 1964. "Ideology as a Cultural System." In D. E. Apter, ed., *Ideology and Discontent*. Glencoe, IL: Free Press; London: Collier Macmillan.

———. 1972. "Deep Play: Notes on the Balinese Cockfight." *Daedalus* 101:1–37.

———. 1973. *The Interpretation of Cultures*. New York: Basic Books.

———. 2000. *Available Light: An Anthropological Reflection on Philosophical Topics*. Princeton: Princeton University Press.

Gellner, David. 1982. "Max Weber: Capitalism and the Religion of India." *Sociology* 16:527–43.

Gellner, Ernest. 1983. *Nations and Nationalism*. Ithaca: Cornell University Press.

———. 1998. *Language and Solitude: Wittgenstein, Malinowski, and the Hapsburg Dilemma*. New York: Cambridge University Press.

———. 1999. "The Coming of Nationalism and Its Interpretation: The Myths of Nation and Class." In Samuel Bowles, Mario Franzini, and Ugo Pagano, eds, *The Politics and Economics of Power*, 179–224. London: Routledge.

Gentil, Dominique. 1986. *Les Mouvements Coopératifs en Afrique de l'Ouest: Interventions de l'État ou Organisations Paysannes?* Paris: L'Harmattan.

Gerschenkron, Alexander. 1962. *Economic Backwardness in Historical Perspective*. Cambridge: Harvard University Press.

———. 1968. *Continuity in History and Other Essays*. Cambridge: Harvard University Press.

———. 1972. *An Economic Spurt That Failed: Four Lectures on Austrian History*. Princeton: Princeton University Press.

Ghai, Yash. 2000. "Autonomy as a Strategy for Diffusing Conflict." In Paule Stern and Daniel Druckman, eds., *International Conflict Resolution after the Cold War*, 483–530. Washington, DC: National Academy Press.

Giddens, Anthony. 1977. *Studies in Social and Political Theory*. New York: Basic Books.

Gilbert, James. 1997. *Redeeming Culture: American Religion in an Age of Science*. Chicago: University of Chicago Press.

Gladwin, Christina H. 1979. "Production Functions and Decision Models: Complementary Models." *American Ethnologist* 6:653–74.

———. 1989. "On the Division of Labor between Economics and Economic Anthropology." In Stuart Plattner, ed., *Economic Anthropology*. Stanford: Stanford University Press.

Gold, Bella. 1981. "Technological Diffusion in Industry: Research Needs and Shortcomings." *Journal of Industrial Economics* 29:247–69.

Goodenough, W. H. 1964. "Cultural Anthropology and Linguistics." In D. H. Hymes, ed., *Language in Culture and Society*. New York: Harper.

Goodheart, Eugene. 1980. "The Romantic Critique of Industrial Civilization." *Alternative Futures* 3:126–38.

Goodin, Robert E., and Hans-Dieter Klingeman. 1996. *A New Handbook of Political Science*. New York: Oxford University Press.

Goulder, A. 1970. *The Coming Crisis in Western Sociology*. New York: Basic Books.

Gow, David D. 2002. "Anthropology and Development: Evil Twin or Moral Narrative?" *Human Organization* 61:299–313.

Granovetter, Mark. 1985. "Economic Action and Social Structure: A Theory of Embeddedness." *American Journal of Sociology* 91:481–510.

———. 1990. "Granovetter (Interview)." In Richard Swedberg, ed., *Economics and Sociology, Redefining Their Boundaries: Conversations with Economists and Sociologists*. Princeton: Princeton University Press.

Gray, Virginia. 1973. "Innovation in the States: A Diffusion Study." *American Political Science Review* 67:1174–85.

Greenberg, Joel. 1997. "Did This Arab Die for Selling Land to Jews?" *New York Times* (May 12): A3.

Gregory, Chris A. 1982. *Gifts and Commodities*. New York: Academic Press.

Greif, Avner. 1994. "Cultural Beliefs and the Organization of Society: A Historical and Theoretical Reflection on Collectivist and Individualist Societies." *Journal of Political Economy* 102:912–50.

Griliches, Zvi. 1957. "Hybrid Corn: An Exploration of the Economics of Technical Change." *Econometrica* 25:501–22.

———. 1958. "Research Costs and Social Reform: Hybrid Corn and Related Inventions." *Journal of Political Economy* 66:419–31.

———. 1960. "Congruence versus Profitability: A False Dichotomy." *Rural Sociology* 25:354–56.

———. 1963. "The Sources of Measured Productivity Growth: United States Agriculture, 1940–60." *Journal of Political Economy* 71:331–46.

———. 1992. "The Search for R&D Spillovers." *Scandinavian Journal of Economics* 94:29–47.

———. 1994. "Productivity, R&D and the Data Constraint." *American Economic Review* 84:1–23.

———. 1996. "The Discovery of the Residual: A Historical Note." *Journal of Economic Literature* 34:1324–30.

Gross, Paul R., and Norman Levitt. 1994. *Higher Superstition: The Academic Left and Its Quarrels with Science*. Baltimore: Johns Hopkins University Press.

Gross, Paul R., Norman Levitt, and Martin W. Lewis, eds. 1996. *The Flight from Science and Reason*. Baltimore: Johns Hopkins University Press (for New York Academy of Sciences).

Grossman, Gene M., and Elhanan Helpman. 1990a. "Comparative Advantage and Long-Run Growth." *American Economic Review* 80:796–815.

———. 1990b. "Trade, Innovation and Growth." *American Economic Review* 80:86–91.

———. 1991a. *Innovation and Growth in the Global Economy*. Cambridge: MIT Press.

———. 1991b. "Trade, Knowledge Spillovers, and Growth." *European Economic Review* 35:517–26.

———. 1994. "Endogenous Innovation in the Theory of Growth." *Journal of Economic Perspectives* 8:23–44.

Grossman, Gene M., and Alan B. Krueger. 1995. "Economic Growth and the Environment." *Quarterly Journal of Economics* 110:353–77.
Groves, T., Roy Radner, and Stanley Reiter, eds. 1987. *Information, Incentives and Economic Mechanisms.* Minneapolis: University of Minnesota Press.
Gruber, W. H., and Donald G. Marquis, eds. 1969. *Factors in the Transfer of Technology.* Cambridge: MIT Press.
Grübler, Arnulf. 1991. "Diffusion: Long-Term Patterns and Discontinuities." *Technical Forecasting and Social Change* 39:159–80.
———. 1992. "Introduction to Diffusion Theory." In Robert Ayres, William Haywood, and Ideri Tohijou, eds., *Computer Integrated Manufacturing,* vol. 3, *Models, Case Studies and Forecasts of Diffusion,* 3–53. London: Chapman and Hall.
———. 1998. *Technology and Global Change.* Cambridge: Cambridge University Press.
Gudeman, Stephen. 1986. *Economics as Culture: Models and Metaphors of Livelihood.* London: Routledge and Kegan Paul.
———, ed. 1998. *Economic Anthropology.* Cheltenham, UK: Elgar.
———. 2001. *The Anthropology of Economy: Community, Market, and Culture.* London: Blackwell.
Guillén, Mario F., Randall Collins, Paula England, and Marshall Meyer, eds. 2002. *The New Economic Sociology: Developments in an Emerging Field.* New York: Russell Sage Foundation.
Gunatillke, Godfrey. 1994. "Health Policy for Rural Areas: Sri Lanka." In Vernon W. Ruttan, ed., *Agriculture, Environment and Health: Sustainable Development in the Twenty-first Century,* 208–34. Minneapolis: University of Minnesota Press.
Gunnell, John G. 1993. *The Descent of Political Theory: The Genealogy of an American Vocation.* Chicago: University of Chicago Press.
Gupta, Dipak K. 2002. "Economics and Collective Identity: Explaining Collective Action." In Shoshana Grossbard-Shechtman and Christopher Clague, eds., *The Expansion of Economics: Toward a More Inclusive Social Science,* 239–65. Armonk, NY: M. E. Sharp.
Guttman, Joel M. 1982. "Can Political Entrepreneurs Solve the Free-Rider Problem?" *Journal of Economic Behavior and Organization* 3:1–10.
Habermas, Jürgen. 1990. *The Philosophical Discourse of Modernity: Twelve Lectures.* Trans. Fredric Lawrence. Cambridge: MIT Press.
Hackenberg, Robert A. and Beverly H. Hackenberg. 1999. "You Can Do Something! Forming Policy from Applied Projects, then and Now." *Human Organization* 58:1–15 (Molinowski Award Lecture).
Hagen, Everett E. 1962. *On the Theory of Social Change: How Economic Growth Begins.* Homewood, IL: Dorsey Press.
———, ed. 1980. *The Economics of Development.* Homewood, IL: Richard D. Irwin.
Hägerstrand, Torsten. 1952. *The Propagation of Innovation Waves.* Lund, Sweden: Lund Studies in Geography.
———. 1953. "Innovations for Luppet Ur Korolgisk Synpunk." Lund, Sweden: University of Lund, Department of Geography Bulletin 15.
———. 1969. *Innovation Diffusion as a Spatial Process.* Chicago: University of Chicago Press.
Hahn, Robert W., and Robert N. Stavins. 1991. "Incentive Based Environmental Regulation: A New Era from an Old Idea?" *Ecological Law Quarterly* 18:1–42.

Hammond, Peter B. 1966. *Yatenga: Technology in the Culture of a West African Kingdom.* New York: Free Press.
Hammond, Peter J., and Andres Rodriguez-Clare. 1993. "On Endogenizing Long-Run Growth." *Scandinavian Journal of Economics* 95:391–425.
Harberger, Arnold. 1990. "Reflections on the Growth Process." Department of Economics, University of California at Los Angeles, June. Mimeo.
———. 1993. "The Search for Relevance in Economics." *American Economic Review* 83:1–22 (Richard T. Ely Lecture).
Harris, Marvin E. 1966. "The Cultural Ecology of India's Sacred Cattle." *Current Anthropology* 12:57–59.
———. 1968. *The Rise of Anthropological Theory: A History of the Theory of Culture.* New York: Thomas Y. Crowell.
———. 1971. "Comment on Alan Heston's 'An Approach to the Sacred Cow of India.'" *Current Anthropology* 12:199–201.
———. 1977. *Cannibals and Kings: The Origins of Culture.* New York: Random House.
———. 1979. *Cultural Materialism: The Struggle for a Science of Culture.* New York: Random House.
———. 1999. *Theories of Culture in Modern Times.* Walnut Creek, CA: Sage.
Harrison, David. 1988. *The Sociology of Modernization and Development.* London: Unwin Hyman.
Harrison, Lawrence E. 1988. *The Sociology of Modernization and Development.* London: Unwin and Hyman.
Harrison, Lawrence E., and Samuel P. Huntington, eds. 2000. *Culture Matters: How Values Shape Human Progress.* New York: Basic Books.
Harrison, Paul. 1987. *The Greening of Africa: Breaking through in the Battle for Land and Food.* London: Paladin Grafton.
Harrod, R. F. 1939. "An Essay in Dynamic Theory." *Economic Journal* 49:14–33.
———. 1948. *Toward a Dynamic Economics.* London: Macmillan.
Harsanyi, John C. 1960. "Explanation and Comparative Dynamics in Social Science." *Behavioral Science* 5:136–45.
———. 1962. "Measurement of Social Power, Opportunity Costs, and the Theory of Two-Person Bargaining Games." *Behavioral Science* 7:67–80. Reprinted in K. W. Rothschild, *Power in Economics.* Baltimore: Penguin, 1971.
Hart, Michael, and Antonio Negri. 2000. *Empire.* Cambridge: Harvard University Press.
Hatch, Elvin. 1997. "A Humanistic Theory of Theory." *Cultural Dynamics* 9:301–24.
Hayami, Yujiro. 1997. *Development Economics: From the Poverty to the Wealth of Nations.* New York: Oxford University Press.
Hayami, Yujiro, and Toshihiko Kawagoe. 1993. *The Agrarian Origins of Commerce and Industry: A Study of Peasant Marketing in Indonesia.* New York: St. Martin's.
Hayami, Yujiro, and Masao Kikuchi. 1981. *Asian Village Economy at the Crossroads: An Economic Approach to Institutional Change.* Tokyo: University of Tokyo Press; Baltimore: Johns Hopkins University Press.
———. 2000. *A Rice Village Saga: Three Decades of Green Revolution in the Philippines.* London: Macmillan.
Hayami, Yujiro, Masao Kikuchi, Luisa Barnbo, and Ester B. Marciano. 1989. "Transformation of a Laguna Village in the Two Decades of Green Revolution." Los Baños, Laguna, Philippines: International Rice Research Institute, Agricultural Economics Paper 89-117.

Hayami, Yujiro, and Keijira Otsaka. 1993. *The Economics of Contract Choice.* Oxford: Oxford University Press.
Hayami, Yujiro, and Vernon W. Ruttan. 1970. "Agricultural Productivity Differences among Countries." *American Economic Review* 60:895–911.
———. 1985 [1971]. *Agricultural Development: An International Perspective.* Baltimore: Johns Hopkins University Press.
Hayek, Friedrich A. 1935. *Collectivist Economic Planning: Critical Studies in the Possibilities of Socialism.* London: Routledge.
———. 1967. "The Results of Human Action but Not of Human Design." In Friedrich A. Hayek, *Studies in Philosophy, Politics and Economics,* 96–105. Chicago: University of Chicago Press.
———. 1978a. "The Atavism of Social Justice." In *New Studies in Philosophy, Politics, Economics, and the History of Ideas.* London: Routledge and Kegan Paul.
———. 1978b. *New Studies in Philosophy, Politics, Economics and the History of Ideas.* London: Routledge and Kegan Paul.
Hays, Sharon. 2000. "Constructing the Centrality of Culture—and Deconstructing Sociology." *Contemporary Sociology* 29:594–602.
Hayter, Teresa. 1971. *Aid as Imperialism.* London: Penguin.
Hedstrom, Peter, and Richard Swedberg. 1998. "Social Mechanisms: An Introductory Essay." In Peter Hedstrom and Richard Swedberg, eds., *Social Mechanisms: An Analytical Approach to Social Theory,* 1–31. Cambridge: Cambridge University Press.
Heertje, A., and Mark Perlman, eds. 1990. *Evolving Technology and Market Structure: Studies in Schumpeterian Economics.* Ann Arbor: University of Michigan Press.
Helliwell, John F. 1992. *Empirical Linkages between Democracy and Economic Growth.* Cambridge, MA: National Bureau of Economic Research Working Paper 4066.
Henrich, Joseph. 2001. "Cultural Transmission and the Diffusion of Innovations: Adoption Dynamics Indicate that Biased Cultural Transmission Is the Predominant Force in Behavioral Change." *American Anthropologist* 103:992–1013.
Herbst, Jeffrey L. 2000. *States and Power in Africa: Comparative Lessons in Authority and Control.* Princeton: Princeton University Press.
Heston, Alan. 1971. "An Approach to the Sacred Cow of India." *Current Anthropology* 12:191–200.
Hicks, John R. 1969. *A Theory of Economic History.* Oxford: Oxford University Press.
———. 1985. *Methods of Dynamic Economics.* Oxford: Clarendon.
Hirschman, Albert O. 1958. *The Strategy of Economic Development.* New Haven: Yale University Press.
———. 1965. "Obstacles to Development: A Classification and a Quasi-Vanishing Act." *Economic Development and Cultural Change* 1:385–93.
———. 1982. "Rival Interpretations of Market Society: Civilizing, Destructive, or Feeble?" *Journal of Economic Literature* 20:1463–84.
Hoff, Karla, and Joseph E. Stiglitz. 1993. "Imperfect Information and Rural Credit Markets: Puzzles and Policy Perspectives." In Karla Hoff, Avishay Braverman, and Joseph E. Stiglitz, *The Economics of Rural Organization: Theory, Practice and Policy,* 33–52. New York: Oxford University Press.
Hollingsworth, J. Robert, and Robert Boyer, eds. 1997. *Contemporary Capitalism: The Embeddedness of Institutions.* Cambridge: Cambridge University Press.
Holmberg, Allan R. 1955. "Participant Intervention in the Field." *Human Organization* 14:23–26.

Holt, Robert T., and John M. Richardson Jr. 1970. "Competing Paradigms in Comparative Politics." In Robert T. Holt and John E. Turner, eds., *The Methodology of Comparative Research*, 21–24. New York: Free Press.

Holt, Robert T., and John E. Turner. 1966. *The Political Basis of Economic Development: An Exploration in Comparative Political Analysis*. Princeton, NJ: Van Nostrand.

———. 1970. "The Methodology of Comparative Research." In Robert T. Holt and John E. Turner, eds., *The Methodology of Comparative Research*, 1–20. New York: Free Press.

———. 1975. "Crisis and Sequence in Collective Theory Development." *American Political Science Review* 69:979–94.

Homans, George C. 1961. *Social Behavior: Its Elementary Forms*. London: Routledge and Kegan Paul.

Hopper, W. David. 1965. "Allocative Efficiency in a Traditional Indian Agriculture." *Journal of Farm Economics* 47:611–24.

Horowitz, Irving Louis. 1972. *Three Worlds of Development: The Theory and Practice of Stratification*. New York: Oxford University Press.

———. 1982. *Beyond Empire and Revolution: Militarization and Consolidation in the Third World*. New York: Oxford University Press.

———. 1993. *The Decomposition of Sociology*. New York: Oxford University Press.

Hoselitz, Bert F. 1952. "Non-economic Barriers to Economic Development." *Economic Development and Cultural Change* 1:8–21.

———. 1960. *Sociological Aspects of Economic Growth*. Glencoe, IL: Free Press.

Hough, Jerry F. 2001. *The Logic of Economic Reform in Russia*. Washington, DC: Brookings Institutions Press.

Hunt, Robert C. 1988. "Size and Structure of Authority in Canal Irrigation Systems." *Journal of Anthropological Research* 44:335–55.

———. 1989. "Appropriate Social Organization? Water Users Associations in Bureaucratic Canal Irrigation Systems." *Human Organization* 48:79–90.

———. 2000. "Labor Productivity and Agricultural Development: Boserup Revisited." *Human Ecology* 28:251–76.

Huntington, Samuel P. 1965. "Political Development and Political Decay." *World Politics* 17:386–430.

———. 1968. *Political Order in Changing Societies*. New Haven: Yale University Press.

———. 1971a. "Foreign Aid for What and for Whom, Part 1." *Foreign Policy* 2 (1): 161–89.

———. 1971b. "Foreign Aid for What and for Whom, Part 2." *Foreign Policy* 2 (2): 114–34.

———. 1987. "The Goals of Development." In Myron Weiner and Samuel P. Huntington, eds., *Understanding Political Development*, 437–90. Boston: Little, Brown.

———. 1991a. "Religion and the Third Wave." *National Interest* (summer): 29–42.

———. 1991b. *The Third Wave: Democratization in the Late Twentieth Century*. Norman: University of Oklahoma Press.

———. 1993. "The Clash of Civilizations." *Foreign Affairs* 72:22–49.

———. 1996. *The Clash of Civilizations and the Remaking of World Order*. New York: Simon and Schuster.

Huntington, Samuel P., and Jorge I. Dominiquez. 1973. "Political Development." In

Fred I. Greenstein and Nelson W. Polsby, eds., *Macropolitical Theory*, 1–113. Reading, MA: Addison-Wesley.
Hurwicz, Leonid. 1972a. "On Informationally Decentralized Systems." In C. B. McGuire and Roy Radner, eds., *Decision and Organization*, 297–336. Amsterdam: North-Holland.
———. 1972b. "Organizational Structures for Joint Decision Making: A Designer's Point of View." In Matthew Tuite, Roger Chisholm, and Michael Radnor, eds., *Interorganizational Decision Making*, 37–44. Chicago: Aldine.
———. 1996. "Institutions as Families of Game Forms." *Japanese Economic Review* 47:113–32.
———. 1998. "Issues in the Design of Mechanisms and Institutions." In Edna T. Loehman and D. Marc Kilgour, eds., *Designing Institutions for Environmental and Resource Management*, 29–56. Cheltenham, UK: Elgar.
Iannaccone, Laurence R. 1993. "Heirs to the Protestant Ethic? The Economics of American Fundamentalists." In Martin E. Marty and R. Scott Appleby, eds., *Fundamentalisms and the State: Remaking Politics, Economics, and Militance*, 342–66. Chicago: University of Chicago Press.
———. 1998. "Introduction to the Economics of Religion." *Journal of Economic Literature* 36:1465–95.
———. 2002. "A Marriage Made in Heaven? Economic Theory and Religious Studies." In Shoshana Grossbard-Shechtman and Christopher Clague, eds., *The Expansion of Economics: Toward a More Inclusive Social Science*, 203–23. Armonk, NY: M. E. Sharpe.
Ingham, Geoffrey. 1996. "Some Recent Changes in the Relationship between Economics and Sociology." *Cambridge Journal of Economics* 20:235–43.
Inglehart, Ronald. 1977. *The Silent Revolution: Changing Values and Political Styles among Western Publics*. Princeton: Princeton University Press.
———. 1990. *Culture Shift in Advanced Industrial Society*. Princeton: Princeton University Press.
———. 1997. *Modernization and Post Modernism: Cultural, Economic and Political Changes in Forty-three Societies*. Princeton: Princeton University Press.
Inkeles, Alex. 2000. "Measuring Social Capital and Its Consequences." In John D. Montgomery and Alex Inkeles, eds., *Social Capital as a Policy Resource*, 245–68. Boston: Kluwer Academic Publishers.
Inkeles, Alex, and Masamichi Sasaki, eds. 1996. *Comparing Nations and Cultures: Readings in a Cross-Disciplinary Perspective*. New York: Prentice-Hall.
International Crops Research Institute for the Semi-Arid Tropics (ICRISAT). 1985. *Annual Report of ICRISAT/Burkina Economics Program*. Ouagadougou: ICRISAT.
Isaac, Jeffrey C. 1987a. "Beyond the Three Faces of Power: A Realist Critique." *Polity* 20:4–31.
———. 1987b. *Power and Marxist Theory: A Realist View*. Ithaca: Cornell University Press.
Isaacman, Allan. 1995. *Cotton Is the Mother of Poverty: Peasants, Work and Rural Struggle in Colonial Mozambique, 1930–61*. Portsmouth, NY: Heinemann.
Isenman, Paul, and Hans W. Singer. 1977. "Food Aid: Disincentive Effects and Their Policy Implications." *Economic Development and Cultural Change* 25:205–38.
Isham, Jonathan, Sunder Ramaswamy, and Thomas Kelly, eds. 2003. *Social Capital and Well-Being in Developing Countries*. Aldershot, UK: Elgar.

Ishikawa, Shigeru. 1981. *Essays on Technology, Employment and Institutions in Economic Development.* Tokyo: Kinokuniya.

Islam, Nazuri. 1995. "Growth Empirics: A Panel Data Approach." *Quarterly Journal of Economics* 60:1128–70.

Ito, Takatoshi, and Anne O. Krueger, eds. 1995. *Growth Theories in Light of East Asian Experience.* Chicago: University of Chicago Press.

Izard, Michel. 1985a. *Yatenga Précolonial: L'Ancien Royaume du Burkina.* Paris: Karthala.

———. 1985b. *Gens du Pouvoir, Gens de la Terre.* Cambridge: Cambridge University Press.

Jacobs, Jane. 1984. *Cities and the Wealth of Nations: Principles of Economic Life.* New York: Random House.

Jacobson, David. 1991. *Reading Ethnography.* Albany: State University of New York Press.

Jaffe, Adam B., M. Trajtenberg, and R. Henderson. 1993. "Geographic Localization of Knowledge Spillovers as Evidenced by Patent Citations." *Quarterly Journal of Economics* 108:577–98.

Jasanoff, Sheila, Gerald E. Markle, James C. Petersen, and Trevor Pinch, eds. 1995. *Handbook of Science and Technology Studies.* Thousand Oaks, CA: Sage.

Jasso, Guillermina. 2000. "The Tripartite Structure of Social Science Analysis." Department of Sociology, New York University. Mimeo.

Jha, Prabhat, Anne Mills, Kara Hanson, Lilani Kumaranayke, Lesong Conteh, Christoph Kurowski, Son Nam Nguyan, Valeria Oliveria Curz, Kent Ransom, Laura M. E. Vaz, Shengchao Yu, Oliver Morton, and Jeffrey Sachs. 2002. "Improving the Health of the Global Poor." *Science* 395 (March 15): 2036–39.

Johnson, D. Gale. 1950. "Resource Allocation under Share Contracts." *Journal of Political Economy* 58:111–23.

Johnson, Harry. 1963. *The Canadian Quandary: Economic Policies.* Toronto: McGraw-Hill.

Jones, Charles I. 1995a. "Time Series Tests of Endogenous Growth Models." *Quarterly Journal of Economics* 110:495–526.

———. 1995b. "R&D-Based Models of Economic Growth." *Journal of Political Economy* 103:758–84.

Jones, E. L. 1988. *Growth Recurring: Economic Change in World History.* Oxford: Clarendon.

Jones, Ronald W., and Sugata Marjit. 2001. "The Role of International Fragmentation in the Development Process." *American Economic Review* 91:363–66.

Jorgenson, Dale W. 1961. "The Development of the Dual Economy." *Economic Journal* 71:309–34.

Jorgenson, Dale W., and Zvi Griliches. 1969. "The Explanation of Productivity Change." *Review of Economic Studies* 34:249–83.

Jorgenson, Dale W., and Eric Yip. 2002. *Whatever Happened to Productivity Growth?* In Dale W. Jorgenson, ed., *Econometrics,* vol. 3, 181–212. Cambridge: MIT Press.

Joskow, Paul L., Richard Schmalensee, and Elizabeth M. Bailey. 1998. "The Market for Sulfur Dioxide Emissions." *American Economic Review* 88:669–85.

Kaldor, Nicholas. 1957. "A Model of Economic Growth." *Economic Journal* 67:591–624.

Kanbur, Ravi. 1997. "Income Distribution and Development." Washington, DC: World Bank. Mimeo.
Kapur, Devesh, John P. Lewis, and Richard Webb. 1997. *The World Bank: The First Half Century.* Washington, DC: Brookings Institution Press.
Karshenas, M., and Paul Stoneman. 1995. "Technological Diffusion." In Paul Stoneman, ed., *Handbook of the Economics of Innovation and Technological Change.* Oxford: Basil Blackwell.
Kaseje, Dan C. 1994. "Health Systems for Rural Areas: Kenya." In Vernon W. Ruttan, ed., *Agriculture, Environment and Health: Sustainable Development in the Twenty-first Century.* Minneapolis: University of Minnesota Press.
Katz, Elihu. 1960. "Communication Research and the Image of Society: Convergence of Two Traditions." *American Journal of Sociology* 65:435–40.
Katz, Elihu, Martin L. Levin, and Herbert Hamilton. 1963. "Traditions of Research on the Diffusion of Innovation." *American Sociological Review* 28:237–52.
Katz, M. L., and C. Shapiro. 1986. "Technology Adoption in the Presence of Network Externalities." *Journal of Political Economy* 94:722–841.
Kawagoe, Toshihiko, Yujiro Hayami, and Vernon W. Ruttan. 1985. "The Intercountry Agricultural Production Function and Productivity Differences among Countries." *Journal of Development Economics* 19:113–32.
Keene, Donald. 1969. *The Japanese Discovery of Europe, 1720–1830.* Rev. ed. Stanford: Stanford University Press.
Keller, Wolfgang. 2001. "Knowledge Spillovers at the World Technology Frontier." Cambridge, MA: National Bureau of Economic Research WP# 8150 (May).
———. 2002. "Geographic Localization of International Technology Diffusion." *American Economic Review* 92:120–42.
Kendrick, John W. 1961. *Productivity Trends in the United States.* Princeton: Princeton University Press.
Kenlow, Peter, and Andres Rodriguez-Clare. 1997. "The Neoclassical Revival in Growth Economics: Has It Gone too Far?" *Macroeconomics Annual.* Cambridge, MA: National Bureau of Economic Research.
Kennedy, Charles. 1964. "Induced Bias in Innovation and the Theory of Distribution." *Economic Journal* 74:541–47.
Keohane, Robert O. 1984. *After Hegemony: Cooperation and Discord in the World Political Economy.* Princeton: Princeton University Press.
———. 1988. "The Rhetoric of Economics as Viewed by a Student of Politics." In Arjo Klamer, Donald N. McCloskey, and Robert M. Solow, eds., *The Consequences of Economic Rhetoric.* New York: Cambridge University Press.
Kepel, Gilles. 2000. *Jihad: The Trail of Political Islam.* Cambridge: Harvard University Press.
Keyes, Charles F. 1993. "Buddhist Economics and Buddhist Fundamentalism in Burma and Thailand." In Martin E. Marty and R. Scott Appleby, eds., *Fundamentalisms and the State: Remaking Politics, Economics, and Militance,* 367–409. Chicago: University of Chicago Press.
Khan, Azizur Rahman. 1977. "Growth and Inequality in the Rural Philippines." In *Poverty and Landlessness in Rural Asia,* 243–49. Geneva: International Labor Office.
Kikuchi, Masao, and Yujiro Hayami. 1980. "Inducements to Institutional Innovations in an Agrarian Community." *Economic Development and Cultural Change* 29:21–36.
Kim, Jong-Il, and Lawrence J. Lau. 1994. "The Sources of Economic Growth of the

East Asian Newly Industrialized Countries." *Journal of the Japanese and International Economics* 8:235–71.
Kimball, Miles S. 1988. "Farmers' Cooperatives as Behavior toward Risk." *American Economic Review* 78:224–32.
Kindleberger, Charles P. 1952. "Review of The Economy of Turkey, The Economic Development of Guatemala, and Report on Cuba." *Review of Economics and Statistics* 34 (November): 391–92.
King, Robert G., and Sergio T. Rebelo. 1990. "Public Policy and Economic Growth: Developing Neoclassical Implications." *Journal of Political Economy* 98:S126–S150.
———. 1993. "Transitional Dynamics and Economic Growth in the Neoclassical Model." *American Economic Review* 83:908–31.
Kitschelt, Herbert. 1993. "Social Movements, Political Parties and Democratic Theory." *Annals of the American Academy of Political and Social Science* 528 (July): 13–29.
Klitgaard, Robert. 1991. *In Search of Culture: A Progress Report on Culture and Development*. Pietermaritzburg, South Africa, University of Natal. Mimeo.
Klonglan, Gerald E., and E. Walter Coward Jr. 1970. "The Concept of Symbolic Adoption: A Suggested Interpretation." *Rural Sociology* 35:77–83.
Knack, Stephen, and Philip Keefer. 1997. "Does Social Capital Have an Economic Payoff? A Cross-Country Investigation." *Quarterly Journal of Economics* 112:1251–88.
Knight, Frank H. 1952. "Institutionalism and Empiricism in Economics." *American Economic Review* 42:45–55.
Knorr-Cetina, Karin. 1981. *The Manufacture of Knowledge: An Essay on the Constructivist and Contextual Nature of Science*. New York: Pergamon Press.
Knudson, Mary K., and Vernon W. Ruttan. 1988. "Research and Development of a Biological Innovation: Commercial Hybrid Wheat." *Food Research Institute Studies* 21:45–68.
Kolodko, Grzegorz W. 2000. *From Shock to Therapy: The Political Economy of Postsocialist Transformation*. New York: Oxford University Press.
Koppel, Bruce M., ed. 1995. *Induced Innovation Theory and international Agricultural Development: A Reassessment*. Baltimore: Johns Hopkins University Press.
Kornai, Janos. 1990. *The Road to a Free Economy*. New York: Norton.
———. 2000. "What the Change of System from Socialism to Capitalism Does and Does Not Mean." *Journal of Economic Perspectives* 14:27–42.
Krasner, Stephen D. 1974. "Trade in Raw Materials: The Benefits of Capitalist Alliances." In Stephen Rosen and James R. Kurth, eds., *Testing Theories of Economic Imperialism*. Lexington, MA: D. C. Heath.
Krasner, Stephen D., and Daniel T. Froats. 1998. "Minority Rights and the Westphalian Model." In David A. Lake and Donald Rothchild, eds., *The International Spread of Ethnic Conflict: Fear, Diffusion, and Escalation*, 227–74. Princeton: Princeton University Press.
Krishna, Raj. 1980. "The Economic Development of India." *Scientific American* (September): 168–78.
Kroeber, A. L., and Clyde Kluckholm. 1952. *Culture: A Critical Review of Concepts and Definitions*. Peabody Museum of American Archeology and Ethnology Papers, vol. 47, no. 1:18. Cambridge: Harvard University Peabody Museum.

Kroeber, A. L., and Talcott Parsons. 1958. "The Concept of Culture and of Social Systems." *American Sociological Review* 23:582–83.
Krueger, Anne O. 1968. "Factor Endowments and Per Capita Income Differences among Countries." *Economic Journal* 78:641–59.
———. 1974. "The Political Economy of the Rent-Seeking Society." *American Economic Review* 64:291–303.
———. 1983. *Trade and Employment in Developing Countries*. Vol. 3, *Synthesis and Conclusions*. Chicago: University of Chicago Press.
———. 1986. "Aid in the Development Process." *World Bank Research Observer* 1:57–78.
———. 1997. "Trade Policy and Economic Development: How We Learn." *American Economic Review* 87:1–22.
Krueger, Anne O., Constantine Michalopoulos, and Vernon W. Ruttan. 1989. *Aid and Development*. Baltimore: Johns Hopkins University Press.
Krugman, Paul R. 1995. *Development Geography and Economic Theory*. Cambridge: MIT Press.
Kuhn, Thomas S. 1970 [1962]. *The Structure of Scientific Revolutions*. Chicago: University of Chicago Press.
Kuper, Adam. 1999. *Culture: The Anthropologists' Account*. Cambridge: Harvard University Press.
Kuran, Timur. 1993a. "Fundamentalisms and the Economy." In Martin E. Marty and R. Scott Appleby, eds., *Fundamentalisms and the State: Remaking Politics, Economics, and Militance*, 298–301. Chicago: University of Chicago Press.
———. 1993b. "The Economic Impact of Islamic Fundamentalism." In Martin E. Marty and R. Scott Appleby, eds., *Fundamentalisms and the State: Remaking Politics, Economics, and Militance*, 302–41. Chicago: University of Chicago Press.
———. 1995a. "Islamic Economics and the Islamic Subeconomy." *Journal of Economic Perspectives* 9:155–73.
———. 1995b. *Private Truths, Public Lies: The Social Consequences of Preference Falsification*. Cambridge: Harvard University Press.
———. 1997. "Islam and Underdevelopment: An Old Puzzle Revisited." *Journal of Institutional and Theoretical Economics* 153:40–71.
———. 1998. "Moral Overload and Its Alleviation." In Avner Ben-Nor and Louis Putterman, eds., *Economics, Values, and Organization*, 231–66. Cambridge: Cambridge University Press.
———. 2001a. "The Provision of Public Goods under Islamic Law: Origins, Impact, and Limitations of the Waqf System." *Law and Society Review* 35:841–97.
———. 2001b. "The Islamic Commercial Crisis: Institutional Roots of Economic Underdevelopment in the Middle East." Department of Economics, University of Southern California, Los Angeles. Mimeo.
Kuznets, Simon S. 1955. "Economic Growth and Income Inequality." *American Economic Review* 45:1–28.
———. 1966a. *Economic Growth of Nations: Total Output and Production Structure*. New Haven: Yale University Press.
———. 1966b. *Modern Economic Growth: Rate Structure and Spread*. New Haven: Yale University Press.
———. 1971. *Economic Growth of Nations: Total Output and Production Structure*. Cambridge: Harvard University Press.

Lal, Deepak. 1988. *Cultural Stability and Economic Stagnation: India c1500 BC–AD 1980.* Oxford: Oxford University Press.

———. 1993. "The Economic Impact of Hindu Revivalism." In Martin E. Marty and R. Scott Appleby, eds., *Fundamentalisms and the State: Remaking Politics, Economics, and Militance,* 410–26. Chicago: University of Chicago Press.

———. 1998. *Unintended Consequences: The Impact of Factor Endowments, Culture, and Politics on Long-Run Economic Performance.* Cambridge: MIT Press.

———. 2000. *The Poverty of Development Economics.* 2d ed. Cambridge: MIT Press.

Lam, D. Kin-Kong, and L. Lee. 1982. "Guerrilla Capitalism and the Limits of Statistical Theory: Comparing the Chinese NICs." In Cal Clark and Steve Chan, eds., *The Evolving Pacific Basin in the Global Political Economy—Domestic and International Linkages,* 107–24. London: Lynne Rienner.

Lange, Oscar. 1938. "On the Economic Theory of Socialism." In Oscar Lange and F. Taylor, *On the Economic Theory of Socialism,* ed. Benjamin E. Lippincott. Minneapolis: University of Minnesota Press.

Lanjouw, Peter, and Nicholas Stern, eds. 1998. *The Economic Development of Palanpur over Five Decades.* Oxford: Clarendon.

Lansing, J. Stephen. 1991. *Priests and Programmers: Technologies of Power in the Engineered Landscape of Bali.* Princeton: Princeton University Press.

Lappe, Francis M. 1980. *Aid as an Obstacle: Twenty Questions about Our Foreign Aid and the Hungry.* San Francisco: Institute for Food and Development Policy.

Lasswell, Harald D., and Abraham Kaplan. 1950. *Power and Society: A Framework for Political Inquiry.* New Haven: Yale University Press.

Latour, Bruno, and Steve Woolgar. 1979. *Laboratory Life: The Social Construction of Scientific Facts.* Beverly Hills, CA: Sage.

Lau, Lawrence J., and Pan A. Yotopoulos. 1989. "The Meta-Production Function Approach to Technological Change in World Agriculture." *Journal of Development Economics* 31:241–69.

Leach, Edmund R. 1954. *Political Systems of Highland Burma.* Boston: Beacon Press.

Lenin, Vladimir I. 1964. *The Development of Capitalism in Russia.* Moscow: Progress Publishers.

Lerner, Daniel. 1958. *The Passing of Traditional Society.* New York: Free Press.

Lévi-Strauss, Claude. 1966a. "Anthropology: Its Achievements and Its Future." *Current Anthropology* 7:124–27.

———. 1966b. *The Savage Mind.* Chicago: University of Chicago Press.

Levine, Daniel H. 1995. "Protestants and Catholics in Latin America: A Family Portrait." In Martin E. Marty and R. Scott Appleby, eds., *Fundamentalisms Comprehended,* 155–78. Chicago: University of Chicago Press.

Levine, Ross E., and D. Renelt. 1992. "A Sensitivity Analysis of Cross Country Regressions." *American Economic Review* 82:942–63.

Levy, Jacob T. 2000. *The Multiculturalism of Fear.* Oxford: Oxford University Press.

Levy, Marion J., Jr. 1968. "Structural-Functional Analysis." In David L. Sills, ed., *International Encyclopedia of the Social Sciences,* vol. 6, 21–43. New York: Crowell Collier and Macmillan.

———. 1989. "Confucianism and Modernization." In *Conference Proceedings on Confucianism and Economic Development in East Asia,* 555–64. Taipei, Taiwan, Republic of China.

Lewis, Bernard. 2002. *What Went Wrong? The Clash between Islam and Modernity in the Middle East.* New York: Oxford University Press.
Lewis, W. Arthur. 1954. "Economic Development with Unlimited Supplies of Labor." *Manchester School of Economic and Social Studies* 22 (May): 139–91.
Lin, Justin Yifu. 1988. "The Household Responsibility System in China's Agricultural Reform: A Theoretical and Empirical Study." *Economic Development and Cultural Change* 36:S199–224.
———. 1994. "Chinese Agriculture: Institutional Change and Performance." In T. N. Srinivasan, ed., *Agriculture and Trade in China and India.* San Francisco: ICS Press.
———. 1995. "The Needham Puzzle: Why the Industrial Revolution Did Not Originate in China." *Economic Development and Cultural Change* 43:269–92.
Lin, Justin Yifu, Cai Fang, and Zhou Li. 1996. *The China Miracle: Development Strategy and Economic Reform.* Hong Kong: Chinese University Press.
Lin, Nan. 2001. *Social Capital: A Theory of Social Structure and Action.* Cambridge: Cambridge University Press.
Lionberger, Herbert L. 1960. *Adoption of New Ideas and Practices: A Summary of Research Dealing with the Acceptance of Technological Change in Agriculture with Implications in Facilitating Social Change.* Ames: Iowa State University Press.
Lipset, Seymour Martin. 1959. "Some Social Requisites of Democracy: Economic Development and Political Legitimacy." *American Political Science Review* 53:69–105.
———. 1994. "The Social Requisites of Democracy Revisited." *American Sociological Review* 59:1–22.
Lissoni, F., and James S. Metcalf. 1994. "Diffusion of Innovation: Ancient and Modern: A Review of the Moen Themes." In M. Dodgeson and R. Rothwell, eds., *Industrial Innovation.* Aldershot, UK: Elgar.
Little, Peter D., and Michael Painter. 1995. "Discourse, Politics, and the Development Process: Reflections on Escobar's Anthropology and the Development Encounter." *American Ethnologist* 22:602–16.
Loury, Glen. 1977. "A Dynamic Model of Racial Income Differences." In P. A. Wallace and A. M. LaMond, eds., *Women, Minorities and Employment Discrimination,* 153–86. Lexington, MA: D. C. Heath.
———. 1981. "Intergenerational Transfers and the Distribution of Earnings." *Econometrica* 49:843–67.
Lucas, Robert E., Jr. 1988. "On the Mechanics of Economic Development." *Journal of Monetary Economics* 22 (1): 3–42.
———. 1990. "Why Doesn't Capital Flow from Rich to Poor Countries?" *American Economic Review* 80:92–96.
———. 1993. "Making a Miracle." *Econometrica* 61:251–72.
Lukes, Steven, ed. 1986. *Power.* New York: New York University Press.
Lumsdaine, David H. 1993. *Moral Vision in International Politics: The Foreign Aid Regime, 1949–1989.* Princeton: Princeton University Press.
Lyotard, Jean-François. 1984. *The Postmodern Condition: A Report on Knowledge.* Minneapolis: University of Minnesota Press. French ed., 1979.
———. 1990. "The Postmodern Condition." In Jeffrey C. Alexander and Steven Seidman, eds., *Culture and Society: Contemporary Debate.* Cambridge: Cambridge University Press.

MacPherson, C. B. 1985. *The Rise and Fall of Economic Justice, and Other Papers.* Oxford: Oxford University Press.
Macrae, Duncan, Jr. 1977. "Review Essay: The Sociological Economics of Gary S. Becker." *American Journal of Sociology* 83 (5): 1244–58.
Maddison, Angus. 1979. "Per Capita Output in the Long Run." *Kyklos* 32 (1/2): 412–29.
———. 1982. *Phases of Capitalist Development.* New York: Oxford University Press.
———. 1995. *Monitoring the World Economy, 1820–1992,* 41–42. Paris: OECD.
———. 2001. *The World Economy: A Millennial Perspective.* Paris: OECD.
Mahajan, V. E. Muller, and Frank M. Bass. 1990. "New Product Diffusion Models in Marketing: A Review and Directions for Research." *Journal of Marketing* 54:1–26.
Mahalanobis, P. C. 1953. "Some Observations on the Process of Growth of National Income." *Sankhy* 12:307–12.
———. 1955. "The Approach of Operational Research to Planning in India." *Sankhy* 16:3–62.
Maizels, Alfred, and Machiko Nissanke. 1984. "Motivations for Aid to Developing Countries." *World Development* 12 (9): 879–900.
Maldonado, Jorge E. 1993. "Building 'Fundamentalism' from the Family in Latin America." In Martin E. Marty and R. Scott Appleby, eds., *Fundamentalisms and Society: Reclaiming the Sciences, the Family, and Education,* 214–39. Chicago: University of Chicago Press.
Malinowski, Bronislaw. 1930. "The Rationalization of Anthropology and Administration." *Africa* 3:405–30.
Mankiw, N. Gregory. 1995. "The Growth of Nations." *Brookings Papers on Economic Activity* 1:275–376.
Mankiw, N. Gregory, D. Romer, and D. N. Weil. 1992. "A Contribution to the Empirics of Economic Growth." *Quarterly Journal of Economics* 107:407–38.
Mansfield, Edward D., and Jack Snyder. 1995. "Democratization and War." *Foreign Affairs* 74:79–97.
Mansfield, Edwin. 1961. "Technological Change and the Rate of Imitation." *Econometrica* 29:741–66.
———. 1963a. "Intrafirm Rates of Diffusion of an Innovation." *Review of Economics and Statistics* 45:348–54.
———. 1963b. "The Speed of Response of Firms to New Techniques." *Quarterly Journal of Economics* 77:290–311.
Manski, Charles F. 2000. "Economic Analysis of Social Interactions." *Journal of Economic Perspectives* 14:115–36.
Marchette, Ceasre. 1991. "Modeling Innovation Diffusion." In B. Henry, ed., *Forecasting Technology and Innovation.* Brussels and Luxembourg: ECSC, EEC, and EAEC.
Marchette, Ceasre, and Nebojsa Nakićenović. 1979. *The Dynamics of Energy Systems and the Logistical Substitution Model.* Laxenburg, Austria: International Institute for Applied Systems Analyses, RR 79–13.
Marcus, G. E., and M. M. J. Fischer. 1986. *Anthropology as Cultural Critique: An Experimental Moment in the Human Sciences.* Chicago: University of Chicago Press.
Marsh, Robert M. 1979. "Does Democracy Hinder Economic Development in Latecomer Developing Nations?" *Comparative Social Research* 2:215–48.
Marty, Martin E. 1996. "Too Bad We're So Relevant: The Fundamentalism Project Projected." *Bulletin of the American Academy of Arts and Sciences* 59 (March): 22–37.

Marty, Martin E., and R. Scott Appleby, eds. 1991a. "Conclusion: An Interim Report on a Hypothetical Family." In Martin E. Marty and R. Scott Appleby, eds., *Fundamentalisms Observed*, 814–42. Chicago: University of Chicago Press.
———. 1991b. *Fundamentalisms Observed.* Chicago: University of Chicago Press.
———, eds. 1993a. *Fundamentalisms and Society: Reclaiming the Sciences, the Family, and Education.* Chicago: University of Chicago Press.
———. 1993b. "Introduction: A Sacred Cosmos, Scandalous Code, and Defiant Society." In Martin E. Marty and R. Scott Appleby, eds., *Fundamentalisms and Society: Reclaiming the Sciences, the Family, and Education,* 1–19. Chicago: University of Chicago Press.
———, eds. 1993c. *Fundamentalisms and the State: Remaking Politics, Economics, and Militance.* Chicago: University of Chicago Press.
———, eds. 1994. *Accounting for Fundamentalisms: The Dynamic Character of Movements.* Chicago: University of Chicago Press.
———, eds. 1995. *Fundamentalisms Comprehended.* Chicago: University of Chicago Press.
Marx, Karl. 1867. *Capital.* Vol. 1. New York: International Publishers.
———. 1913. *A Contribution to the Critique of Political Economy,* 11–12. Chicago: Kerr.
———. 1936 [1906]. *Capital, A Critique of Political Economy.* Ed. Friedrich Engels. New York: Modern Library; Charles H. Kerr.
———. 1963. *The Eighteenth Brumaire of Louis Bonaparte.* New York: International Publishers.
———. 1967. *Capital.* Vol. 3. New York: International Publishers.
———. 1968. *Theories of Surplus Value.* Moscow: Progress Publishers.
Marx, Leo. 1964. *The Machine in the Garden: Technology and the Pastoral Ideal in America.* New York: Oxford University Press.
Matlon, Peter J. 1987. "The West African Semi-Arid Tropics." In J. W. Mellor, C. L. Delgado, and M. J. Blackie, eds., *Accelerating Food Production in Sub-Saharan Africa.* Baltimore: Johns Hopkins University Press.
Matsuyama, Kiminori. 1992. "Innovation and Growth in the Global Economy." *Journal of International Economics* 33:383–95.
Mauss, Marcel. 1954. *The Gift.* Glencoe, IL: Free Press.
Mazur, R. E., and S. Tuinji Titilola. 1992. "Social and Economic Dimensions of Local Knowledge Systems in African Agriculture." *Sociologia Ruralis* 32:264–86.
McGrattan, Ellen R. 1998. "In Defense of AK Growth Models." *Federal Reserve Bank of Minneapolis Quarterly Review* 22:13–27.
McKinlay, R. D., and Anthony Mughan. 1984. *Aid and Arms to the Third World: An Analysis of the Distribution and Impact of U.S. Official Transfers.* New York: St. Martin's.
McKinnon, Ronald I. 1993. *The Economic Order of Economic Liberalization: Financial Control in the Transition Economy.* Baltimore: Johns Hopkins University Press.
McNeill, William H. 1982. *The Pursuit of Power.* Chicago: University of Chicago Press.
Medema, Stephen G., ed. 1995. *The Legacy of Ronald Coase in Economic Analysis.* Vols. 1 and 2. Aldershot, UK: Elgar.
Meier, Gerald M., and Joseph E. Stiglitz, eds. 2001. *Frontiers of Development Economics: The Future in Perspective.* New York: Oxford University Press.
Mellor, John W., and Bruce F. Johnston. 1984. "The World Food Equation: Interrela-

tions among Development, Employment, and Food." *Journal of Economic Literature* 22 (June): 531–74.
Merton, Robert. 1968. *Social Theory and Social Structure.* New York: Free Press.
Messer, Ellen. 2002. "Anthropologists in a World With and Without Human Rights." In Jeremy MacClancy, ed., *Exotic No More: Anthropology in the Front Lines,* 319–37. Chicago: University of Chicago Press.
Metcalf, James S. 1982. "The Diffusion of Invention: An Interpretive Survey." In Giovanni Dosi, Christopher Freeman, Richard Nelson, Gerald S. Silverberg, and Luc Soete, eds., *Technical Change and Economic Theory.* New York: Pinter.
Meyers, Ramon H. 1989. "Confucianism and Economic Development: Mainland China, Hong Kong and Taiwan." *Conference Proceedings on Confucianism and Economic Development in East Asia,* 281–304. Taipei, Taiwan, Republic of China.
Miller, Gary J. 1997. "The Impact of Economics on Contemporary Political Science." *Journal of Economic Literature* 35:1173–1204.
Mokyr, Joel. 1992. "Technological Inertia in Economic History." *Journal of Economic History* 52:325–38.
———. 1998. "Innovation and Its Enemies: The Economic and Political Roots of Technological Inertia." In Mancur Olson and and Satu Kahkonen, eds., *A Not So Dismal Science,* 67–104. Oxford: Oxford University Press.
Mokyr, Joel. 2002. *The Gifts of Athena: Historical Origins of the Knowledge Economy.* Princeton, NJ: Princeton University Press.
Montgomery, John D. 1957. *Forced to Be Free: The Artificial Revolutions in Germany and Japan.* Chicago: University of Chicago Press.
———. 1969. "The Quest for Political Development." *Comparative Politics* 1:285–95.
———. 1990. "How Facts Replace Fads: Social Science and Social Development." *Comparative Politics* 22:237–48.
———. 1998. "The Next Thousand Years." *World Policy Journal* 15:77–81.
Montgomery, John D., and Alex Inkeles. 2001. *Social Capital as a Policy Resource.* Boston: Kluwer Academic Publishers.
Moore, Barrington, Jr. 1966. *The Social Origins of Dictatorship and Democracy: Lord and Peasant in the Making of the Modern World.* Boston: Beacon Press.
Moore, Mick. 1994. "How Difficult Is It to Construct Market Relations? A Commentary on Platteau." *Journal of Development Studies* 30:818–30.
Moore, Wilbert E. 1964a. "Predicting Discontinuities in Social Change." *American Sociological Review* 29 (June): 331–38.
———. 1964b. "Social Aspects of Development." In Robert E. L. Farris, ed., *Handbook of Modern Sociology,* 882–911. Chicago: Rand McNally.
Moran, Emilio. 1996. *Transforming Societies, Transforming Anthropology.* Ann Arbor: University of Michigan Press.
Morgenthau, Hans. 1962. "A Political Theory of Foreign Aid." *American Political Science Review* 56 (2): 301–9.
Morris, Cynthia Taft. 1992. "Politics, Development, and Equity in Five Land-Rich Countries in the Latter Nineteenth Century." *Research in Economic History* 14:1–68.
Morris, Cynthia Taft, and Irma Adelman. 1988. *Comparative Patterns of Economic Development, 1850–1914.* Baltimore: Johns Hopkins University Press.
Mosley, Paul. 1985. "The Political Economy of Foreign Aid: A Model of the Market for a Public Good." *Economic Development and Cultural Change* 33:373–93.

Mumford, Lewis. 1934. *Techniques and Civilization.* New York: Harcourt Brace and World.
———. 1964. *The Pentagon of Power.* New York: Harcourt Brace Jovanovich.
Murphy, Kevin M., Andrei Shleifer, and Robert W. Vishny. 1989. "Industrialization and the Big Push." *Journal of Political Economy* 97:1003–26.
———. 1992. "The Transition to a Market Economy: Pitfall of Partial Reform." *Quarterly Journal of Economics* 107:889–906.
Murrell, Peter. 1995. "The Transaction According to Cambridge Mass." *Journal of Economic Literature* 33:164–78.
Murrell, Peter, and Yijang Wang. 1993. "When Privatization Should Be Delayed: The Effect of Communist Legacies on Organizational and Institutional Reform." *Journal of Comparative Economics* 17:385–406.
Murtha, Thomas P. 2002. Letter. March 28.
Murtha, Thomas P., Stefanie A. Lenway, and Jeffrey A. Hart. 2001. *Managing New Industry Creation: Global Knowledge Formation and Entrepreneurship in High Technology.* Stanford: Stanford University Press.
Musmann, Klaus, and William H. Kennedy. 1989. *Diffusion of Innovations: A Select Bibliography.* New York: Westpoint Press.
Musu, I., and M. Lines. 1995. "Endogenous Growth and Environmental Preservation." In G. Boero and A. Silberston, *Environmental Economics.* New York: St. Martin's.
Myrdal, Gunnar. 1957. *Rich Lands and Poor.* New York: Harper.
———. 1968. *Asian Drama: An Inquiry into the Poverty of Nations.* New York: Pantheon.
———. 1984. "International Inequality and Foreign Aid in Retrospect." In G. M. Meier and D. Seers, eds., *Pioneers in Development,* 151–65. New York: Oxford University Press.
Nair, Kusum. 1979. *In Defense of the Irrational Peasant.* Chicago: University of Chicago Press.
Nakićenović, Nebojsa, and Arnulf Grübler, eds. 1991. *Diffusion of Technologies and Social Behavior.* Berlin: Springer-Verlag.
Narayan, Deepa. 2000. "Bonds and Bridges: Social Capital and Poverty." World Bank Poverty Group, Washington, DC. Mimeo.
Narayan, Deepa, and Lant Pritchett. 1999. "Cents and Sociability: Household Income and Social Capital in Rural Tanzania." *Economic Development and Cultural Change* 47:871–989.
Nasbeth, L., and G. F. Ray. 1974. *The Diffusion of New Industrial Processes: An International Study.* Cambridge: Cambridge University Press.
National Commission on Philanthropy and Civic Renewal. 1998. *National Index of Civic Engagement.* Storrs: University of Connecticut.
Needham, Joseph. 1954. *Science and Civilization in China.* Cambridge: Cambridge University Press.
Nelson, Richard R. 1998. "The Agenda for Growth Theory: A Different Point of View." *Cambridge Journal of Economics* 22:497–520.
Nelson, Richard R., and Howard Pack. 1997. "The Asian Miracle and Modern Growth Theory." School of International and Public Affairs, Columbia University and School of Public Planning and Government, University of Pennsylvania. Mimeo.
Nelson, Richard R., and Sidney G. Winter. 1982. *An Evolutionary Theory of Economic Change.* Cambridge: Harvard University Press.

———. 2002. "Evolutionary Theorizing in Economics." *Journal of Economic Perspectives* 16:23–46.
Nerlove, Marc L. 1974. "Household and Economy: Toward a New Theory of Population and Economic Growth." *Journal of Political Economy* 82, Part 2: S200–S218.
Nerlove, Marc L., and Lakishmi K. Raut. 1997. "Growth Models with Endogenous Population: General Framework." In M. R. Rosenzweig and O. Stark, eds., *The Handbook of Population and Family Economics.* Amsterdam: Elsevier.
Netting, Robert. 1974. "Agrarian Ecology." *Annual Review of Anthropology* 3:21–56.
———. 1982. "The Ecological Perspective: Holism and Scholasticism in Anthropology." In E. Adamson Hoebel, Richard Currier, and Susan Kaiser, *Crisis in Anthropology: View from Spring Hill, 1980,* 271–319. New York: Garland Publishing.
———. 1993. *Smallholders, Households: Farm Families and the Ecology of Intensive Sustainable Agriculture.* Stanford: Stanford University Press.
Nisbet, Robert A. 1969. *Social Change and History: Aspects of Western Ideas of Development.* New York: Oxford University Press.
———. 1970. "Developmentalism: A Critical Analysis." In John C. McKinney and Edward Tiryakian, eds., *Theoretical Sociology: Perspectives and Developments,* 168–204. New York: Appleton-Century-Crofts.
Nordhaus, William D. 1973. "Some Skeptical Thoughts on the Theory of Induced Innovation." *Quarterly Journal of Economics* 87:209–19.
North, Douglass C. 1981. *Structure and Change in Economic History.* New York: Norton.
———. 1983. "A Theory of Economic Change." *Science* 219:163–64.
———. 1990a. *Institutions, Institutional Change and Economic Performance.* Cambridge: Cambridge University Press.
———. 1990b. "A Transaction Cost Theory of Politics." *Journal of Theoretical Politics* 2:355–67.
———. 1991. "Institutions." *Journal of Economic Perspectives* 5:97–112.
———. 1994. "Economic Performance through Time." *American Economic Review* 84:359–68. 1993 Nobel Award Lecture.
North, Douglass C., and Robert Paul Thomas. 1970. "An Economic Theory of the Growth of the Western World." *Economic History Review* 23 (1): 1–17.
———. 1973. *The Rise of the Western World.* London: Cambridge University Press.
North, Douglass C., and Barry R. Weingast. 1989. "Constitutions and Commitment: Evaluation of the Institutions of Public Choice in Seventeenth Century England." *Journal of Economic History* 59:803–32.
Nowak, P. 1992. "Why Farmers Adopt Production Technology." *Journal of Soil and Water Conservation* 47:14–16.
Nugent, Jeffrey B., and Nicolas Sanchez. 1993. "Tribes, Chiefs, and Transhumance: A Comparative Institutional Analysis." *Economic Development and Cultural Change* 42:87–113.
Nye, Joseph S., Jr. 1984. "Ethical Dimensions in International Involvement in Land Reform." In John D. Montgomery, ed., *International Dimensions of Land Reform.* Boulder: Westview.
Oberoi, Harjot. 1995. "Mapping India Fundamentalisms through Nationalism and Modernity." In Martin E. Marty and R. Scott Appleby, eds., *Fundamentalisms Comprehended,* 96–114. Chicago: University of Chicago Press.

Olson, Mancur. 1965. *The Logic of Collective Action: Public Goods and the Theory of Groups.* Cambridge: Harvard University Press.
———. 1968. *The Logic of Collective Action: Public Goods and the Theory of Groups.* New York: Schocken Books.
———. 1982. *The Rise and Decline of Nations: Economic Growth, Stagflation, and Social Rigidities.* New Haven: Yale University Press.
———. 1996. "Big Bills Left on the Sidewalk: Why Some Nations Are Rich and Others Poor." *Journal of Economic Perspectives* 10:3–24.
———. 2000. *Power and Prosperity: Outgrowing Communist and Capitalist Dictatorships.* New York: Basic Books.
Ortner, Sherry B. 1994. "Theory in Anthropology since the Sixties." In Nicholas B. Dirks, Geoff Eley, and Sherry B. Ortner, eds., *Culture/ Power/ History,* 372–411. Princeton: Princeton University Press.
Osborne, Evan. 2001. "Culture, Development and Government: Reservations in India." *Economic Development and Cultural Change* 49:659–85.
Ostrom, Elinor. 1990. *Governing the Commons: The Evolution of Institutions for Collective Action.* Cambridge: Cambridge University Press.
———. 1992. *Crafting Institutions for Self-Governing Irrigation Systems.* San Francisco: Institute for Contemporary Studies Press.
———. 1997. "Investing in Capital: Institutions and Incentives." In Christopher Clague, ed., *Institutions and Economic Development: Growth and Governance in Less Developed and Post-Socialist Countries.* Baltimore: Johns Hopkins University Press.
———. 1998. "A Behavioral Approach to the Rational Choice Theory of Conflict Resolution." *American Political Science Review* 92:1–22.
———. 2000. "Collective Action and the Evolution of Social Norms." *Journal of Economic Perspectives* 14:137–58.
Ostrom, Elinor, and James Walker. 1997. "Neither Markets nor States: Linking Transformation Processes in Collective Action Arenas." In Dennis C. Mueller, ed., *Perspectives on Public Choice: A Handbook,* 35–72. Cambridge: Cambridge University Press.
Ostrom, Elinor, Roy Gardner, and James Walker. 1994. *Rules, Games, and Common Pool Resources.* Ann Arbor: University of Michigan Press.
Ostrom, Elinor, Clark Gibson, Sujai Shivakumar, and Krister Andersson. 2002. *Aid, Incentives and Sustainability: An Institutional Analysis of Development Cooperation.* Stockholm, Sweden: Swedish International Development Cooperation Agency (SIDA).
Ostrom, Vincent. 1976a. "The American Experiment in Constitutional Choice." *Public Choice* 27:1–12.
———. 1976b. "Response to William Riker's Comments." *Public Choice* 27:16–19.
Otsuka, Keijiro, Hiroyuki Chuma, and Yujiro Hayami. 1992. "Land and Labor Contracts in Agrarian Economies." *Journal of Economic Literature* 30:1965–2018.
Otsuka, Keijiro, and Yujiro Hayami. 1998. "Theories of Share Tenancy: A Critical Survey." *Economic Development and Cultural Change* 37:31–68.
Ouedraogo, Bernard L. 1990. *Entraide Villageoise et Développement: Groupements Paysans au Burkina Faso.* Paris: L'Harmattan.
Pack, Howard. 1994. "Endogenous Growth Theory: Intellectual Appeal and Empirical Shortcomings." *Journal of Economic Perspectives* 8:55–72.
———. 2001. "Technological Change and Growth in East Asia: Macro versus Micro

Perspectives." In Joseph E. Stiglitz and Shahid Yusuf, eds., *Rethinking the East Asia Miracle,* 95–142. New York: Oxford University Press.

Packingham, Robert A. 1973. *Liberal America and the Third World: Political Development Ideas in Foreign Aid and Social Science.* Princeton: Princeton University Press.

Palley, Thomas L. 1996. "Growth Theory in a Keynesian Mode: Some Keynesian Foundations for New Endogenous Growth Theory." *Journal of Post Keynesian Economics* 19:113–36.

Pardey, Philip G., and Nienke M. Beintema. 2001. *Slow Magic: Agricultural R&D a Century after Mendel.* Washington, DC: International Food Policy Research Institute.

Parente, Stephen L., and Edward C. Prescott. 1991. "Technology Adoption and Growth." Minneapolis: Federal Reserve Bank of Minneapolis Staff Report 136.

———. 1993. "Changes in the Wealth of Nations." *Federal Reserve Bank of Minneapolis Quarterly Review* 17 (spring): 3–16.

———. 1994. "Barriers to Technology Adoption and Development." *Journal of Political Economy* 102:298–321.

———. 2000. *Barriers to Riches.* Cambridge: MIT Press.

Park, Albert F., and Bruce Johnston. 1995. "Rural Development and Dynamic Externalities in Taiwan's Structural Transformation." *Economic Development and Cultural Change* 44:181–208.

Parsons, Talcott. 1934. "Some Reflections on the Nature and Significance of Economics." *Quarterly Journal of Economics* 48 (May): 511–45.

———. 1948. *The Structure of Social Action.* New York: McGraw-Hill.

———. 1963. "On the Concept of Political Power." *Proceedings of the American Philosophical Society* 3:232–62.

———. 1964. "Evolutionary Universals in Society." *American Sociological Review* 29 (June): 339–57.

———. 1969. *Politics and Social Structure.* New York: Free Press.

Parsons, Talcott, and Neil Smelser. 1956. *Economy and Society.* London: Routledge and Kegan Paul.

Pavlowitch, Steven K. 1994. "Who Is Balkanizing Whom? The Misunderstanding between the Debris of Yugoslavia and an Unprepared West." *Daedalus* 123:203–33.

Pearse, Andrew. 1980. *Seeds of Plenty, Seeds of Want.* Oxford: Oxford University Press.

Peet, Richard, and Elaine Hartwick. 1999. *Theories of Development.* New York: Guilford Press.

Persson, Tonsten, and Guido Tabellini. 2001. *Political Economics: Explaining Economic Policy.* Cambridge: MIT Press.

Peterson, Willis. 1989. "Rates of Return on Development Assistance Capital: An International Comparison." *Kyklos* 42:203–17.

Pincus, John A. 1963. "The Cost of Foreign Aid." *Review of Economics and Statistics* 45 (November): 360–67.

Piore, Michael J. 1996. "Review of 'The Handbook of Economic Sociology.'" *Journal of Economic Literature* 34:741–54.

Platteau, Jean-Philippe. 1991. "Traditional Systems of Social Security and Hunger Insurance: Past Achievements and Modern Challenges." In E. Ahmad, J. Droze, J. Hills, and A. Sen, eds., *Social Security in Developing Countries.* Oxford: Clarendon.

———. 1994a. "Behind the Market Stage Where Real Societies Exist—Part I: The Role of Public and Private Order Institutions." *Journal of Development Studies* 30:533–77.

———. 1994b. "Behind the Market Stage Where Real Societies Exist—Part II: The Role of Moral Norms." *Journal of Development Studies* 30:578–817.
———. 2000. *Institutions, Social Norms, and Economic Development*. Amsterdam, Netherlands: Harwood Academic Publishers.
Platteau, Jean-Philippe, and Yujiro Hayami. 1998. "Resource Endowments and Agricultural Development: Africa versus Asia." In Yujiro Hayami and Masahiko Aoki, eds., *The Institutional Foundations of East Asian Economic Development*. London: Macmillan.
Plattner, Stewart, ed. 1989a. *Economic Anthropology*. Stanford: Stanford University Press.
———. 1989b. "Marxism." In Stuart Plattner, ed., *Economic Anthropology*. Stanford: Stanford University Press.
Pogge, Thomas W. 1986. "Liberalism and Global Justice: Hoffman and Norden on Morality in International Affairs." *Philosophy and Public Affairs* 15 (winter): 67–81.
Polsby, Nelson W. 1960. "How to Study Community Power: The Pluralist Alternative." *Journal of Politics* 22:478–84.
Popkin, Samuel L. 1979. *The Rational Peasant: The Political Economy of Rural Society in Vietnam*. Berkeley: University of California Press.
Popper, Karl R. 1968. *The Logic of Scientific Discovery*. New York: Harper. German ed., Vienna: Julius Springer, 1935.
Portes, Alejandro. 1997. "Neoliberalism and the Sociology of Development: Emerging Trends and Unanticipated Facts." *Population and Development Review* 23:229–59.
———. 1998. "Social Capital: Its Origins and Applications in Modern Sociology." *Annual Review of Sociology* 22:39–55.
Posner, Richard A. 1980. "A Theory of Primitive Society, with Special Reference to Law." *Journal of Law and Economics* 23:1–53.
Pourgerami, Abbas. 1988. "The Political Economy of Development: A Cross-National Causality Test of the Development-Democracy-Growth Hypothesis." *Public Choice* 58:123–41.
Pradervand, Pierre. 1989. *Listening to Africa: Developing Africa from the Grassroots*. New York: Praeger.
Pred, Alan. 1969. "Postscript." In T. Hagerstrand, *Innovation Diffusion as a Spatial Process*, 299–324. Chicago: University of Chicago Press.
Prescott, Edward C. 1988. "Robert M. Solow's Neoclassical Growth Model: An Influential Contribution to Economics." *Scandinavian Journal of Economics* 90 (1): 7–12.
———. 1997. "Needed: A Theory of Total Factor Productivity." Minneapolis: Federal Reserve Bank of Minneapolis, Research Department Staff Report 242.
———. 2002. "Prosperity and Depression: 2002 Richard T. Ely Lecture." *American Economic Review* 92:1–15.
Prewitt, Kenneth. 1982. "The Impact of the Developing World on U.S. Social-Science Theory and Methodology." In Lawrence D. Stifel, Ralph K. Davidson, and James S. Coleman, eds., *Social Sciences and Public Policy in the Developing World*. Lexington, MA: D. C. Heath.
Pritchett, Lant. 1995. *Divergence, Big Time*. Washington, DC: World Bank Policy Research Working Paper 1522, Oct.
———. 1997. "Our Friend 'Recent Growth Research.'" Paper presented at American Economic Association Annual Meetings, Jan. 5.

———. 2000. "Understanding of Economic Growth: Searching for Hills among Plateaus, Mountains and Plains." *World Bank Economic Review* 14:221–50.
———. 2001. "Comment on 'Growth Empirics and Reality' by William A. Brock and Steven N. Durlauf." *World Bank Economic Review* 15:273–75.
Przeworski, Adam. 1997. "Democratization Revisited." *Items (Social Science Research Council)* 51:6–11.
Putnam, Robert D. 2000. *Bowling Alone: The Collapse and Revival of American Community.* New York: Simon and Schuster.
Putnam, Robert D., Robert Leonardi, and Rafella Y. Nanette. 1993. *Making Democracy Work: Civic Traditions in Modern Italy.* Princeton: Princeton University Press.
Putnam, Robert T. 1993. *Making Democracy Work: Civic Traditions in Modern Italy.* Princeton: Princeton University Press.
Pye, Lucian W. 1962. *Politics, Personality and Nation Building: Burma's Search for Identity.* New Haven: Yale University Press.
———. 1966. *Aspects of Political Development.* Boston: Little, Brown.
Pye, Lucian W., and Mary W. Pye. 1985. *Asian Power and Politics: The Cultural Dimensions of Authority.* Cambridge: Harvard University Press.
Pye, Lucian W., and Sidney Verba, eds. 1965. *Political Culture and Political Development.* Princeton: Princeton University Press.
Rabinow, Paul, and William M. Sullivan, eds. 1987. *Interpretive Social Science: A Second Look.* Berkeley: University of California Press.
Randall, Adrian. 1991. *Before the Luddites.* Cambridge: Cambridge University Press.
Ranis, Gustav. 1961. "Analytics of Development: Dualism." In Hollis B. Chenery and T. N. Srinivasan, eds., *Handbook of Development Economics,* vol. 1, 73–92. Amsterdam: Elsevier Science Publishers.
Ranis, Gustav, and John C. H. Fei. 1961. "A Theory of Economic Development." *American Economic Review* 51:533–65.
Rausser, Gordon C., Erik Lichtenberg, and Ralph Lattimore. 1982. "Developments in the Theory and Empirical Application of Endogenous Government Behavior." In Gordon C. Rausser, ed., *New Directions in Econometric Modeling and Forecasting in U.S. Agriculture,* 547–614. New York: Elsevier.
Rawls, John. 1971. *A Theory of Justice.* Cambridge: Harvard University Press.
Reddy, N. Mohan, and Liming Zhao. 1990. "International Technology Transfer: A Review." *Research Policy* 19:285–307.
Redfield, Robert, and W. Lloyd Warner. 1940. "Cultural Anthropology in Modern Agriculture." In *Yearbook of Agriculture, Farmers in a Changing World,* 983–93. Washington, DC: U.S. Government Printing Office.
Reij, Chris. 1989. *Indigenous Soil and Water Conservation in Africa.* Sustainable Agriculture Programme, Gatekeeper Series 27. London: International Institute for Environment and Development.
Reij, Chris, Paul Mulder, and Louis Begemann. 1988. *Water Harvesting for Plant Production.* Technical Paper 91. Washington, DC: World Bank.
Resnick, Stephen, and Richard Wolff. 1987. *Knowledge and Class.* Chicago: University of Chicago Press.
Restivo, Sol. 1995. "The Theory Landscape in Science Studies: Sociological Traditions." In Sheila Jasanoff, Gerald E. Markle, James C. Petersen, and Trevor Pinch, eds., *Handbook of Science and Technology Studies.* Thousand Oaks, CA: Sage.

Rhoades, Robert E. 1984. *Breaking New Ground: Agricultural Anthropology.* Lima, Peru: International Potato Center.
Ricardo, David. 1911. *The Principles of Political Economy and Taxation.* London: J. M. Dent and Sons.
Richerson, Peter J., Robert Boyd, and Brian Paciotti. 2002. "An Evolutionary Theory of Commons Management." In Elinor Ostrom, Thomas Dietz, Nives Dolšak, Paul C. Stein, Susan Stonich, and Elke U. Weber, eds., *The Drama of the Commons.* Washington, DC: National Academy Press.
Riddell, Roger C. 1986. "The Ethics of Foreign Aid?" *Development Policy Review* 4 (March): 24–43.
———. 1987. *Development Assistance Reconsidered.* London: James Currey.
Riker, William. 1976. "Comments on Vincent Ostrom's Paper." *Public Choice* 27:13–17.
Rivera-Batiz, Luis A., and Pacel M. Romer. 1991. "Economic Integration and Endogenous Growth." *Quarterly Journal of Economics* 106:531–55.
Robbins, Lionel. 1932. *An Essay on the Nature and Significance of Economic Science.* London: Macmillan.
Robelo, S. 1991. "Long Run Policy Analysis and Long Run Growth." *Journal of Political Economy* 99:500–521.
Robinson, James A. 2002. "States and Power in Africa by Jeffrey Herbst: A Review Essay." *Journal of Economic Literature* 40:510–19.
Rodriguez, Francisco, and Dani Rodrik. 1999. *Trade Policy and Economic Growth: A Skeptic's Guide to the Cross-National Evidence.* Cambridge, MA: National Bureau of Economic Research Working Paper 7081.
Roe, Terry, and Hamid Mohtadi. 2001. "International Trade and Growth: An Overview Using the New Growth Theory." *Review of Agricultural Economics* 23:423–40.
Rogers, Everett M. 1962. *Diffusion of Innovations.* New York: Free Press of Glencoe.
———. 1969. "Motivations, Values and Attitudes of Subsistence Farmers: Toward a Subculture of Peasantry." In Clifton R. Wharton Jr., ed., *Subsistence Agriculture and Economic Development,* 111–35. Chicago: Aldine.
———. 1983. *Diffusion of Innovations.* 3d ed. New York: Free Press-Collier Macmillan.
———. 1995. *Diffusion of Innovations.* 4th ed. New York: Free Press.
Rogers, Everett M., and A. Eugene Havens. 1962. "Adoption of Hybrid Corn: A Comment." *Rural Sociology* 27:327–30.
Rogers, Everett M., and Floyd F. Shoemaker. 1971. *Communication of Innovations: A Cross Cultural Approach.* New York: Free Press.
Rogers, Everett M., and Lynne Svenning. 1969. *Modernization Among Peasants: The Impact of Communication.* New York: Holt, Rinehart and Winston.
Rokken, Stein. 1974. "Dimensions of State Formation and Nation Building." In Charles Tilly, ed., *The Formation of National States in Western Europe.* Princeton: Princeton University Press.
Roling, N. 1985. "Extension Science: Increasingly Preoccupied with Knowledge Systems." *Sociologia Ruralis* 24:269–90.
———. 1988. *Information Systems in Agricultural Development.* Cambridge: Cambridge University Press.
Romer, Paul M. 1983. "Dynamic Competitive Equilibria with Externalities, Increasing Returns and Unbounded Growth." Ph.D. diss., Department of Economics, University of Chicago.

———. 1986. "Increasing Returns and Long-Run Growth." *Journal of Political Economy* 94:1002–37.
———. 1987. "Crazy Explanations for the Productivity Slowdown." In S. Fisher, ed., *NBER Macroeconomic Annual.* Cambridge: MIT Press.
———. 1990. "Endogenous Technological Change." *Journal of Political Economy* 98:S71–S102.
———. 1993. "Idea Gaps and Object Gaps on Economic Development." *Journal of Monetary Economics* 32:543–73.
———. 1994. "The Origins of Endogenous Growth." *Journal of Economic Perspectives* 8:3–22.
———. 1996. "Why, Indeed, in America? Theory, History and the Origins of Modern Economic Growth." Bureau of Economic Research Working Paper 5442, Jan.
Rosecrance, Richard. 1998. "Review of *The Clash of Civilizations and the Remaking of World Order.*" *American Political Science Review* 92:978–990.
Rosen, Steven J. 1974. "The Open Door Imperative of U.S. Foreign Policy." In Stephen Rosen and James R. Kurth, eds., *Testing Theories of Economic Imperialism.* Lexington, MA: D. C. Heath.
Rosenberg, Nathan. 1972. "Factors Affecting the Diffusion of Technology." *Explorations in Economic History* 10:3–33.
———. 1976. *Perspectives on Technology.* Cambridge: Cambridge University Press.
———. 1982. "Marx as a Student of Technology." *Inside the Black Box: Technology and Economics,* 34–54. Cambridge: Cambridge University Press.
———. 1994. "Joseph Schumpeter: Radical Economist." In Nathan Rosenberg, ed., *Exploring the Block Box: Technology, Economics and History,* 47–61.
———. 2000. *Schumpeter and the Endogeneity of Technology: Some American Perspectives.* London: Routledge.
Rosenstein-Rodan, P. N. 1943. "Problems of Industrialization in Eastern and Southeastern Europe." *Economic Journal* 53:202–11.
Rossi, Peter H. 1980. "The Challenge and Opportunities of Applied Social Research." *American Sociological Review* 45:889–904.
Rostow, Walter W. 1956. "The Take-Off into Self-Sustained Growth." *Economic Journal* 66:25–48.
———. 1960. *The Stages of Economic Growth: A Non-Communist Manifesto.* Cambridge: Cambridge University Press.
———. 1971. *Politics and the Stages of Growth.* London: Cambridge University Press.
———. 1978. *The World Economy: History and Prospect.* Austin: University of Texas Press.
———. 1985. *Eisenhower, Kennedy, and Foreign Aid.* Austin: University of Texas Press.
Rothschild, K. W. 1971. *Power in Economics.* Baltimore: Penguin.
Rucht, D. 1995. "The Impact of Anti-Nuclear Power Movements in International Comparison." In Martin Bauer, ed., *Resistance to New Technology: Nuclear Power, Information Technology and Biotechnology,* 277–91. Cambridge: Cambridge University Press.
Rueschemeyer, Dietrich, Evelyne Huber Stephens, and John D. Stephens. 1992. *Capitalist Development and Democracy.* Chicago: University of Chicago Press.
Runge, C. Ford. 1977. "American Agricultural Assistance and the New International Economic Order." *World Development* 5:225–46.
———. 1981a. "Common Property Externalities: Isolation, Assurance, and Resource

Depletion in a Traditional Grazing Context." *American Journal of Agricultural Economics* 63:595–606.

———. 1981b. "Institutions and Common Property Externalities: The Assurance Problem in Economic Development." Ph.D. diss., University of Wisconsin, Madison.

———. 1999. *Stream, River, Delta: Induced Innovation and Environmental Values in Economics and Policy.* St. Paul: University of Minnesota, Center for International Food and Agricultural Policy WP 99–2.

Ruttan, Lore M., and Monique Borgerhoff-Mulder. 1999. "Are East African Pastoralists Truly Conservationists?" *Current Anthropology* 40:621–37.

Ruttan, Vernon W. 1959. "Usher and Schumpeter on Invention, Innovation, and Technological Change." *Quarterly Journal of Economics* 73:596–606.

———. 1969. "Equity and Productivity Issues in Modern Agrarian Reform Legislation." In Ugo Papi and Charles Nunn, eds., *Economic Problems of Agriculture in Industrial Societies,* 581–600. London: Macmillan; New York: St. Martin's Press.

———. 1982. *Agricultural Research Policy.* Minneapolis: University of Minnesota Press.

———. 1984. "Social Science Knowledge and Institutional Change." *American Journal of Agricultural Economics* 66:549–59. American Agricultural Economics Association Fellows Lecture, 1984.

———. 1988. "Cultural Endowments and Economic Development: What Can We Learn From Anthropology?" *Economic Development and Cultural Change* 36:S247–72.

———. 1989. "Why Foreign Economic Assistance?" *Economic Development and Cultural Change* 37:411–24.

———. 1991."What Happened to Political Development?" *Economic Development and Cultural Change* 39 (2): 265–92.

———. 1992. "The Sociology of Development and Underdevelopment: Are There Lessons for Economists?" *International Journal of Sociology of Agriculture and Food/Revista Internacional de Sociologia sobre Agricultura y Alimentos* 2:17–38.

———. 1994a. *Agriculture, Environment and Health: Sustainable Development in the Twenty-first Century.* Minneapolis: University of Minnesota Press.

———. 1994b. "Social Science Knowledge and Institutional Change." *American Journal of Agricultural Economics* 66 (December): 203–23.

———. 1995. "Cultural Endowments and Economic Development: Implications for the Chinese Economies." *China Economic Review* 6 (1): 91–104.

———. 1996a. *United States Development Assistance Policy: The Domestic Politics of Foreign Economic Aid.* Baltimore: Johns Hopkins University Press.

———. 1996b. "What Happened to Technology Adoption-Diffusion Research?" *Sociologia Ruralis* 36:51–73.

———. 1997. "Induced Innovation, Evolutionary Theory and Path Dependence: Sources of Technical Change." *Economic Journal* 107:1520–29.

———. 1998. "The New Growth Economics and Development Economics." *Journal of Development Studies* 35:1–26.

———. 1999. "The Transition to Agricultural Sustainability." *Proceedings of the National Academy of Sciences* 96:5960–67.

———. 2001a. "Does the U.S. Foreign Economic Assistance Program Have a Future?" In Deepak Lal and Richard Snape, eds., *Trade Development and Political Economy: Essays in Honor of Anne O. Krueger,* 285–303. New York: Palgrave.

———. 2001b. *Technology, Growth and Development: An Induced Innovation Perspective.* New York: Oxford University Press.

———. 2002. "Imperialism and Competition in Anthropology, Sociology, Political Science and Economics: A Perspective from Development Economics." In Shoshana Grossbard-Shechtman and Christopher Clague, eds., *The Expansion of Economics: Toward a More Inclusive Social Science,* 49–67. Armonk, NY: M. E. Sharp.

Ruttan, Vernon W., and Yujiro Hayami. 1984. "Toward a Theory of Induced Institutional Innovation." *Journal of Development Studies* 20:203–22.

———. 1995a. "Induced Innovation Theory and Agricultural Development: A Personal Account." In B. M. Koppel, ed., *Induced Innovation Theory and International Agricultural Development,* 22–38. Baltimore: Johns Hopkins University Press.

———. 1995b. "Induced Innovation Theory and Agricultural Development: A Reassessment." In B. M. Koppel, ed., *Induced Innovation Theory and International Agricultural Development,* 169–88. Baltimore: Johns Hopkins University Press.

Ryan, Bruce, and Neal C. Gross. 1943. "The Diffusion of Hybrid Seed Corn in Two Iowa Communities." *Rural Sociology* 8:14–24.

———. 1950. *Acceptance and Diffusion of Hybrid Seed Corn in Two Iowa Communities.* Ames: Iowa State Agricultural Experiment Station Bulletin 272.

Sachs, Jeffrey D., and Katharina Pistor. 1997. "Introduction: Progress, Pitfalls, Scenarios and Lost Opportunities." In Jeffrey D. Sachs and Katharina Pistor, eds., *The Rule of Law and Economic Reform in Russia.* Boulder: Westview.

Sahlins, Marshall D. 1960. "Evolution: Specific and General." In Marshall D. Sahlins and Elman R. Service, eds., *Evolution and Culture.* Ann Arbor: University of Michigan Press.

———. 1976. *Culture and Practical Reason.* Chicago: University of Chicago Press.

———. 1985. *Islands of History.* Chicago: University of Chicago Press.

———. 1993. "Goodbye to *Tristes Tropes:* Ethnography in the Context of Modern History." *Journal of Modern History* 65:1–25.

———. 1995. *How "Natives" Think: About Captain Cook, for Example.* Chicago: University of Chicago Press.

Saideman, Stephen M. 1998. "Is Pandora's Box Half Empty or Half Full? The Limited Virulence of Secessionism and the Domestic Sources of Disintegration." In David A. Lake and Donald Rothchild, eds., *The International Spread of Ethnic Conflict: Fear, Diffusion, and Escalation,*127–50. Princeton: Princeton University Press.

Sala-i-Martin, Xavier X. 1997. "I Just Ran Four Million Regressions." Cambridge, MA: National Bureau of Economic Research, Working Paper 6252.

Saleth, B. Maria, and Ariel Dinar. 2002. *Water Institutions and Sector Performance: A Quantitative Analysis with Cross Country Data.* Washington, DC: World Bank (draft).

Samuelson, Paul A. 1948. *Foundations of Economic Analysis.* Harvard Economic Studies, vol. 80. Cambridge: Harvard University Press.

Samuelsson, Kurt. 1961. *Religion and Economic Action.* New York: Basic Books.

Sanders, John H., Joseph G. Nagy, and Sunder Ramaswamy. 1990. "Developing New Agricultural Technologies for the Sahelian Countries: The Burkina Faso Case." *Economic Development and Cultural Change* 39:1–22.

Sanders, John H., and Sunder Ramaswamy. 1992. "Impacts of New Technologies in Burkina Faso and the Sudan: Implications for Future Technology Design." In *Pro-*

*ceedings of a Workshop on Social Science Research and the CRSPs,* INTSORMIL Publication 93-3. Lexington: University of Kentucky.

Sandler, Todd. 2001. *Economic Concepts for the Social Sciences.* Cambridge, UK: Cambridge University Press.

Sartorius, Rolf. 1984. "Persons and Property." In R. G. Frey, ed., *Utility and Rights.* Minneapolis: University of Minnesota Press.

Satz, Debra, and John Ferejohn. 1994. "Rational Choice and Social Theory." *Journal of Philosophy* 51:71-87.

Savonnet, G. 1958. "Méthodes employées par certaines populations de Haute-Volta pour lutter contre l'érosion." *Notes Africaines* 78:38-40.

Sawyer, Darwin O. 1977. "Review Essay: Social Rules and Economic Firms: The Sociology of Human Capital." *American Journal of Sociology* 83 (5): 1259-69.

Schmitz, John A., Jr. 1993. "Early Progress on the 'Problem of Economic Development.'" *Federal Reserve Bank of Minneapolis Quarterly Review* 17 (spring): 17-33.

Schneider, Ben Ross. 1995. "Democratic Consolidations: Some Broad Comparisons and Sweeping Arguments." *Latin American Research Review* 30:215-34.

Schotter, Andrew. 1981. *The Economic Theory of Social Institutions.* Cambridge: Cambridge University Press.

Schran, Peter. 1975. "On the Yenan Origins of Current Economic Policies." In Dwight H. Perkins, ed., *China's Modern Economy in Historical Perspective,* 279-302. Stanford: Stanford University Press.

Schuh, G. Edward. 1986. *The United States and the Developing Countries: An Economic Perspective.* Washington, DC: National Planning Association.

Schultz, T. Paul. 1988. "Education Investment and Returns." In Hollis Chenery and T. N. Srinivasan, eds., *Handbook for Development Economics,* vol. 1, 543-630. Amsterdam: Elsevier Science Publishers.

Schultz, Theodore W. 1960. "Value of U.S. Farm Surpluses to Underdeveloped Countries." *Journal of Farm Economics* 62:1019-30.

———. 1961. "Investment in Human Capital." *American Economic Review* 51:1-17.

———. 1964. *Transforming Traditional Agriculture.* New Haven: Yale University Press.

———. 1975. "The Value of the Ability to Deal with Disequilibria." *Journal of Economic Literature* 13:827-46.

Schumacher, E. F. 1973. *Small Is Beautiful: A Study of Economics As If People Mattered.* London: Bland and Briggs.

Schumpeter, Joseph. 1934 [1912]. *The Theory of Economic Development.* Cambridge: Harvard University Press. Translated by Redvers Opie from the revised 1926 German edition.

———. 1939. *Business Cycles.* 2 vols. New York: McGraw-Hill.

———. 1942. *Capitalism, Socialism and Democracy.* New York: Harper and Brothers.

Scott, James C. 1976. *The Moral Economy of the Peasant: Rebellion and Subsistence in South East Asia.* New Haven: Yale University Press.

———. 1985. *Weapons of the Weak: Everyday Forms of Peasant Resistance.* New Haven: Yale University Press.

Scoville, James G. 1996. "Labor Market Underpinnings of a Caste System: Following the Coase Theorem." *American Journal of Economics and Sociology* 55:385-94.

Selznik, Phillip. 1949. *TVA and the Grass Roots.* Berkeley: University of California Press.

Sen, Amartya K. 1967. "Isolation and the Social Rates of Discount." *Quarterly Journal of Economics* 81:112–24.
———. 1981. *Poverty and Famines: An Essay on Entitlement and Deprivation.* Oxford: Clarendon.
———. 1983. "Development: Which Way Now?" *Economic Journal* 93:745–62.
———. 1999. *Development as Freedom.* New York: Knopf.
Serageldin, Ismail, and Christiaan Grootaert. 2000. "Defining Social Capital: An Integrating View." In Partha Dasgupta and Ismail Serageldin, eds., *Social Capital: A Multifaceted Perspective,* 40–58. Washington, DC: World Bank.
Sewell, John W., and John A. Mathieson. 1982. *The Third World: Exploring U.S. Interests.* Foreign Policy Association Headline Series no. 259. Washington, DC: Foreign Policy Association.
Sewell, William H., Jr. 1987. "Theory of Action, Dialectic and History: Comment on Coleman." *American Journal of Sociology* 93 (July): 166–77.
Shklar, Judith. 1989. "The Liberalism of Fear." In Nancy L. Rosenbloom, ed., *Liberalism and the Moral Life.* Cambridge: Harvard University Press.
Shue, Henry. 1980. *Basic Rights: Substance, Affluence, and U.S. Foreign Policy.* Princeton: Princeton University Press.
Shultz, George. 1984. "Foreign Aid and U.S. National Interests." In *Realism, Strength, Negotiation: Key Foreign Policy Statements of the Reagan Administration.* Washington, DC: Department of State, Bureau of Public Affairs.
Shweder, Richard A. 2000a. "Moral Maps, 'First World' Conceits, and the New Evangelists." In Lawrence E. Harrison and Samuel P. Huntington, eds., *Culture Matters: How Values Shape Human Progress,* 158–88. New York: Basic Books.
———. 2000b. "Rethinking the Object of Anthropology (and Ending Up Where Kroeber and Kluckholm Began)." *Items and Issues* 1 (summer 2000): 7–9.
Silverberg, Gerald. 1991. "Adoption and Diffusion of Technology as a Collective Evolutionary Process." *Technological Forecasting and Social Change* 39:67–80.
Silverberg, Gerald, Giovanni Dosi, and Lluc Orsenigo. 1988. "Innovation, Diversity and Diffusion: A Self Organization Model." *Economic Journal* 98:1032–54.
Simon, Herbert A. 1953. "Notes on the Observation and Measurement of Power." *Journal of Politics* 15:500–516.
Simon, Robert L. 1984. "Troubled Waters: Global Justice and Ocean Resources." In Tom Regan, ed., *Earthbound.* New York: Random House.
Singer, Milton. 1968. "The Concept of Culture." In David L. Sills, ed., *International Encyclopedia of the Social Sciences,* 527–43. New York: Macmillan.
Singer, Peter. 1977. "Reconsidering the Famine Relief Argument." In Peter Brown and Henry Shue, eds., *Food Policy.* New York: Free Press.
Skinner, Elliott Percival. n.d. "Traditional Institutions and Economic Development: The Mossi *NAAM.*" Columbia University. Mimeo.
———. 1964. *The Mossi of the Upper Volta: The Political Development of the Sudanese People.* Stanford: Stanford University Press.
Smale, Melinda, Paul Heisey, and Howard Leathers. 1995. "Maize of the Ancestors and Modern Varieties: The Microeconomics of High-Yielding Variety Adoption in Malawi." *Economic Development and Cultural Change* 43:351–68.
Smale, Melinda, and Vernon W. Ruttan. 1997. "Social Capital and Technical Change: The *Groupements Naam* of Burkina Faso." In Christopher Clague, ed., *Institutions and Economic Development: Growth and Governance in Less-Developed and Post-Socialist Countries.* Baltimore: Johns Hopkins University Press.

Smelser, Neil J. 1963. "Mechanisms of Change and Adjustment to Change." In Bert F. Hoselitz and Wilbert E. Moore, eds., *Industrialization and Society*, 32–54. Paris: United Nations Economic and Social Council.
———. 1990. "Can Individualism Yield a Sociology?" *Contemporary Sociology* 19:778–83.
Smelser, Neil J., and Richard Swedberg, eds. 1994a. *The Handbook of Economic Sociology*. Princeton: Princeton University Press.
———. 1994b. "The Sociological Perspective on the Economy." In Neil J. Smelser and Richard Swedberg, eds., *The Handbook of Economic Sociology*, 3–26. Princeton: Princeton University Press.
Smith, Adam. 1937. *Wealth of Nations*. Ed. Edward Cannon. New York: Random House.
Smith, Peter H. 1969. *Politics and Beef in Argentina: Patterns of Conflict and Change*. New York: Columbia University Press.
———. 1974. *Argentina and the Failure of Democracy: Conflict among Elites, 1880–1955*. Madison: University of Wisconsin Press.
Snedeker, George. 2000. "Defining the Enlightenment: Jürgen Habermas and the Theory of Communicative Reason." *Dialectical Anthropology* 25:239–50.
So, Alvin Y. 1990. *Social Change and Development: Modernization, Dependency, and World Systems Theories*. London: Sage.
Sobel, Dava. 1995. *Longitude: The True Story of a Lone Genius Who Solved the Greatest Scientific Problem of His Time*. New York: Walker.
Sobel, Joel. 2002. "Can We Trust Social Capital?" *Journal of Economic Literature* 40:139–54.
Soete, L. 1985. "International Diffusion of Technology, Industrial Development and Technological Leapfrogging." *World Development* 13:409–22.
Sokal, Alan D. 1996a. "Transgressing the Boundaries: An Afterword." *Dissent* 43:93–99.
———. 1996b. "Transgressing the Boundaries: Toward a Transformative Hermeneutics of Quantum Gravity." *Social Text* 14:217–52.
Solow, Robert M. 1956. "A Contribution to the Theory of Economic Growth." *Quarterly Journal of Economics* 70:65–94.
———. 1957. "Technical Change and the Aggregate Production Function." *Review of Economics and Statistics* 39:312–20.
———. 1970. *Growth Theory: An Exposition*. Oxford: Oxford University Press.
———. 1985. "Economic History and Economics." *American Economic Review* 75:327–31.
———. 1988. "Growth Theory and After." *American Economic Review* 78 (3): 307–17.
———. 1995. "Perspectives on Growth Theory." *Journal of Economic Perspectives* 8:45–54.
———. 1997. *Learning from 'Learning by Doing': Lessons for Economic Growth*. Stanford: Stanford University Press.
———. 2001. "After Technical Progress and the Aggregate Production Function." In Charles R. Hulten, Edwin Dean, and Michael J. Harper, eds., *New Development in Productivity Analysis*, 173–78. Chicago: University of Chicago Press.
Somer, Murat. 2001. "Cascades of Ethnic Polarization: Lessons from Yugoslavia." *Annals* 573:127–51.
Soto, Hernando de. 1989. *The Other Path: The Invisible Revolution in the Third World*. New York: Harper and Row.

Srinivasan, T. N. 1991. "Development Thought, Strategy and Policy: Then and Now." Background paper prepared for World Development Report. Oct. 1991. Mimeo. Washington, DC: World Bank.

———. 1995. "Long-Run Growth Theories and Empirics: Anything New." In Takatoshi Ito and Anne O. Krueger, eds., *Growth Theories in the Light of East Asian Experience,* 37–70. Chicago: University of Chicago Press.

Stavins, Robert N. 1998. "What Can We Learn from the Grand Policy Experiment? Lessons from $SO_2$ Allowances Trading." *Journal of Economic Perspectives* 12:69–88.

Steinmetz, George. 1999. "Introduction: Culture and the State." In George Steinmetz, ed., *State/Culture: State Formation after the Cultural Turn,* 1–49. Ithaca: Cornell University Press.

Stern, Nicholas. 1991. "The Determinants of Growth." *Economic Journal* 101:122–33.

Steward, Julian Hayes. 1955. "Cultural Evolution." In E. Adamson Hobel, *Readings in Anthropology,* 322–40. New York: McGraw-Hill.

Stewart, Francis, ed. 1987a. *Macro-Policies for Appropriate Technologies in Development Countries.* Boulder: Westview.

———. 1987b. "The Case for Appropriate Technology: A Reply to R. S. Eckaus." *Issues in Science and Technology* 14:101–9.

Stigler, George J., and Gary S. Becker. 1977. "De Gustibus Non Est Disputandum." *American Economic Review* 67:76–90.

Stiglitz, Joseph E. 1974. "Incentives and Risk Sharing in Sharecropping." *Review of Economic Studies* 41:219–55.

———. 1986. "The New Development Economics." *World Development* 14:257–65.

———. 1988. "Economic Organization, Information, and Development." In Hollis Chenery and T. N. Srinivasan, eds., *Handbook of Development Economics,* vol. 1, 91–160. Amsterdam: Elsevier Science Publishers.

———. 1989. "Markets, Market Failure and Development." *American Economic Review* 79:197–202.

———. 1999. "The World Bank at the Millennium." *Economic Journal* 109:F577–98.

———. 2000a. "The Contributions of the Economics of Information to Twentieth Century Economics." *Quarterly Journal of Economics* 115 (4): 1441–78.

———. 2000b. "Formal and Informal Institutions." In Partha Dasgupta and Ismail Serageldin, eds., *Social Capital: A Multifaceted Perspective,* 59–70. Washington, DC: World Bank.

———. 2002a. "Information and the Change in the Paradigm in Economics." *American Economic Review* 92:460–501. 2001 Nobel Lecture.

———. 2002b. *Globalization and Its Discontents.* New York: W. W. Norton.

Stiglitz, Joseph E., and Andrew Weiss. 1981. "Credit Rationing in Markets with Imperfect Information." *American Economic Review* 71:393–410.

Stinchcombe, Arthur L. 1975. "Merton's Theory of Social Structure." In Lewis A. Coser, ed., *The Idea of Social Structure: Papers in Honor of Robert K. Merton,* 11–33. New York: Harcourt Brace Jovanovich.

Stocking, George W., Jr. 1984. *Functionalism Historicized: Essays on British Social Anthropology. History of Anthropology,* vol. 2. Madison: University of Wisconsin Press.

Stokey, Nancey. 1991. "Human Capital, Product Quality, and Growth." *Quarterly Journal of Economics* 11:586–616.

Stoll, David. 1994. "Jesus Is Lord of Guatemala: Evangelical Reform in a Death Squad State." In Martin E. Marty and R. Scott Appleby, eds., *Accounting for Fundamen-*

talisms: The Dynamic Character of Movements, 94–123. Chicago: University of Chicago Press.
Stoneman, Paul. 2002. *The Economics of Technological Diffusion.* Oxford: Blackwell.
Sugden, Robert. 1989. "Spontaneous Order." *Journal of Economic Perspectives* 3:85–97.
Swan, T. W. 1956. "Economic Growth and Capital Accumulation." *Economic Record* 32:343–61.
Swedberg, Richard. 1990a. *Economics and Sociology, Redefining Their Boundaries: Conversations with Economists and Sociologists.* Princeton: Princeton University Press.
———. 1990b. "Socioeconomics and the New Battle of the Methods: Toward a Paradigm Shift?" *Journal of Behavioral Economics* 19 (2): 141–54.
———. 1991. *Schumpeter: A Biography.* Princeton: Princeton University Press.
———. 1998. *Max Weber and the Idea of Economic Sociology.* Princeton: Princeton University Press.
Syrquin, Moshe. 1994. "Structural Transformation and the New Growth Theory." In Luigi L. Pasinetti and Robert M. Solow, *Economic Growth and the Structure of Long-Term Development.* New York: St. Martin's.
Tai, Paul H. 1989. "Measuring the Economic Impact of Confucianism: Empirical Evidence from a Survey." *Conference Proceedings on Confucianism and Economic Development in East Asia,* 199–236. Taipei, Taiwan, Republic of China.
Tawney, Richard H. 1962. *Religion and the Rise of Capitalism.* Gloucester, MA: Peter Smith.
Tax, Sol. 1963 [1953]. *Penny Capitalism: A Guatemalan Indian Economy.* Chicago: University of Chicago Press.
Taylor, Robert E. 1989. *Ahead of the Curve: Shaping New Solutions to Environmental Problems.* New York: Environmental Defense Fund.
Teece, David J. 1977. "Technology Transfer by Multinational Firms: The Resource Cost of Transferring Technological Know-How." *Economic Journal* 87:242–61.
Temple, Jonathan. 1999. "The New Growth Evidence." *Journal of Economic Literature* 37:112–56.
Temple, Jonathan, and Paul A. Johnson. 1998. "Social Capability and Economic Growth." *Quarterly Journal of Economics* 113:965–90.
Tendler, Judith. 1997. *Good Government in the Tropics.* Baltimore: Johns Hopkins University Press.
Thirtle, Colin G., and Vernon W. Ruttan. 1987. *The Role of Demand and Supply in the Generation and Diffusion of Technical Change.* London: Harwood Academic Publishers.
Thomas, J. K., H. Ladewig, and W. A. McIntosh. 1990. "The Adoption of Integrated Pest Management Practices among Texas Cotton Growers." *Rural Sociology* 55:395–410.
Thompson, Michael, Richard Ellis, and Aaron Wildavsky. 1990. *Cultural Theory.* Boulder: Westview.
Thorsby, David. 2001. *Economics and Culture.* Cambridge: Cambridge University Press.
Tietenberg, Tom. 2002. "The Tradable Permits Approach to Protecting the Commons: What Have We Learned?" In Elinor Ostrom, Thomas Dietz, Nives Dolšak, Paul C. Stern, Susan Stonich, and Elke U. Welser, eds., *The Drama of the Commons.* Washington, DC: National Academy Press.
Tilly, Charles. 1990. *Coercion, Capital, and European States:* AD *990–1992.* Cambridge: Blackwell.

Tollison, Robert D. 1982. "Rent Seeking: A Survey." *Kyklos* 35:575–602.
Tomich, Thomas P., Peter Kilby, and Bruce F. Johnston. 1995. *Transforming Agrarian Economies: Opportunities Seized, Opportunities Missed.* Ithaca: Cornell University Press.
Tornell, Aaron. 1997. "Economic Growth and Decline with Endogenous Property Rights." *Journal of Economic Growth* 2:219–50.
Tribe, Keith. 1978. *Land, Labor and Economic Discourse.* London: Routledge and Kegan Paul.
Tucker, Robert. 1977. *The Inequality of Nations.* New York: Basic.
Tullock, Gordon. 1967. "The Welfare Costs of Tariffs, Monopolies and Theft." *Western Economic Journal* 5:224–32.
Tylor, E. B. 1958. *Primitive Culture: Researches into the Development of Mythology, Philosophy, Religion, Language, Art and Custom,* vol. 1, *Origins of Culture.* 1971 reprint. Glouster, MA: Smith.
U.S. General Accounting Office. 2000. *Foreign Assistance: International Efforts to Aid Russia's Transition Have Mixed Results.* Washington, DC: GAO-01-8.
Uzawa, Hirofami. 1961. "On a Two-Sector Model of Economics Growth, F, Studies." *Review of Economics* 29:40–47.
———. 1963. "On a Two-Sector Model of Economics Growth, IF." *Review of Economics Studies* 30:105–18.
———. 1965. "Optimum Technical Change in an Aggregative Model of Economic Growth." *International Economic Review* 6:18–31.
Valente, T. W., and Everett M. Rogers. 1995. "The Origins and Development of the Diffusion of Innovations Paradigm as an Example of Scientific Growth." *Science Communication* 16:242–73.
Van de Klundert, T., and S. Smulders. 1992. "Reconstructing Growth Theory: A Survey." *Economist* 140 (2): 177–203.
van de Ven, A., H. L. Angle, and M. S. Poole, eds. 1989. *Research on the Management of Innovation: The Minnesota Studies.* New York: Harper and Row.
Van Den Berg, Axel. 1998. "Is Sociological Theory Too Grand for Social Mechanisms." In Peter Hedstrom and Richard Swedberg, eds., *Social Mechanism: An Analytical Approach to Social Theory,* 204–37. Cambridge: Cambridge University Press.
van der Veer, Peter. 1994. "Hindu Nationalism and the Discourse of Modernity: The Vishva Hindu Parishad." In Martin E. Marty and R. Scott Appleby, eds., *Accounting for Fundamentalisms: The Dynamic Character of Movements,* 653–68. Chicago: University of Chicago Press.
Vayda, Andrew P., and Bradley B. Walters. 1999. "Against Political Ecology." *Human Ecology* 27:167–79.
Vernon, Raymond. 1966. "International Investment and International Trade in the Product Cycle." *Quarterly Journal of Economics* 80:190–207.
———. 1979. "The Product Cycle Hypothesis in a New International Environment." *Oxford Bulletin of Economics and Statistics* 41:255–67.
Verspagen, Bart. 1996. "Endogenous Innovation in Neoclassical Growth Models: A Survey." *Journal of Macroeconomics* 14:631–62.
von Weizsacker, Carl C. 1971. "Note on Endogenous Change of Tastes." *Journal of Economic Theory* 3:345–72.
Wallerstein, Immanuel. 1979. *The Capitalist World Economy.* Cambridge: Cambridge University Press.

Warren, D. Michael, L. Jan Slikkerver, and David Brokensha. 1995. *The Cultural Dimension of Development: Indigenous Knowledge Systems.* London: Intermediate Technology Publications.
Watson-Verran, Helen, and David Turnbull. 1995. "Science and Other Indigenous Knowledge Systems. In Sheila Jasanoff, Gerald E. Markle, James C. Petersen, and Trevor Pinch, eds., *Handbook of Science and Technology Studies.* Thousand Oaks, CA: Sage.
Weber, Eugene. 1976. *Peasants into Frenchmen: The Modernization of Rural France, 1870–1914.* Stanford: Stanford University Press.
Weber, Max. 1958. *The Protestant Ethic and the Spirit of Capitalism.* New York: Charles Scribner's Sons. 1st English ed. 1930.
Weinberg, Steven. 1992. *Dreams of a Final Theory.* New York: Pantheon.
Weiner, Martin J. 1981. *English Culture and the Decline of the Industrial Spirit, 1850–1980.* Cambridge: Cambridge University Press.
Weingast, Barry R. 1997. "The Political Foundations of Limited Government: Parliament and Sovereign Debt in Seventeenth and Eighteenth Century England." In John N. Drabak and John V. C. Nye, eds., *The Frontiers of the New Institutional Economics,* 213–46. San Diego: Academic Press.
West, Francis I. 1983. "The Effectiveness of Military Assistance as an Instrument of U.S. Foreign Policy." Papers prepared for the Commission of Security and Economic Assistance, Washington, DC.
White, Benjamin. 1983. "'Agricultural Involution' and Its Critics: Twenty Years After." *Bulletin of Concerned Asian Scholars* 15:18–31.
White, Harrison C. 1981. "Where Do Markets Come From?" *American Journal of Sociology* 87:517–47.
———. 1990. "Control to Deny Chance, But Thereby Muffling Identity." *Contemporary Sociology* 19:783–88.
White, Harrison C., and Robert C. Eccles. 1987. "Producers Markets." In John Eatwell, Murray Milgate, and Peter Newman, *The New Palgrave: A Dictionary of Economics,* 4 vols. 984–86. London: Macmillan.
White, Leslie. 1949. *The Science of Cultures: A Study of Man and Civilization.* New York: Farrar and Strauss.
Whitehead, Alfred North. 1925. *Science and the Modern World.* New York: Macmillan.
Wiarda, Howard J. 1985. "Toward a Nonethnocentric Theory of Development: Alternative Concepts from the Third World." In Howard J. Wiarda, ed., *New Directions in Comparative Politics,* 127–50. Boulder: Westview.
Wiener, M. J. 1981. *English Culture and the Decline of the Industrial Spirit, 1850–1980.* London: Cambridge University Press.
Williamson, J. G. 1990. "What Washington Means by Policy Reform." In John Williamson, ed., *Latin American Adjustment: How Much Has Happened?* Washington, DC: Institute for International Economics.
———. 1995. "Globalization, Convergence and History." Cambridge, MA: National Bureau of Economic Research Working Paper 5259, Sept.
———. 2000. "What Should the World Bank Think about the Washington Consensus?" *World Bank Research Observer* 15:251–64.
Williamson, Oliver E. 1979. "Transaction Cost Economics: The Governance of Contractual Economics." *Journal of Law and Economics* 22:233–61.
———. 1985. *The Economic Institutions of Capitalism.* New York: Free Press.

———. 2000. "The New Institutional Economics: Taking Stock, Looking Ahead." *Journal of Economic Literature* 38:595–613.
Wittfogel, Karl A. 1957. *Oriental Despotism: A Comparative Study of Total Power*. New Haven: Yale University Press.
Wolpert, Lewis. 1992. *The Unnatural Nature of Science*. London: Faber and Faber.
Wong, John, and Arline Wong. 1989. "Confucian Values as a Social Framework for Singapore's Economic Development." *Conference Proceedings on Confucianism and Economic Development in East Asia,* 503–40. Taipei, Taiwan, Republic of China.
Woolcock, Michael. 1998. "Social Capital and Economic Development." *Theory and Society* 27:151–99.
Woolcock, Michael, and Deepa Narayan. 2000. "Social Capital: Implications for Development Theory, Research, and Policy." *World Bank Research Observer* 15:225–49.
World Bank. 1990. *World Development Report: Poverty*. Oxford: Oxford University Press.
———. 1993. *The East Asian Miracle: Economic Growth and Public Policy*. New York: Oxford University Press.
———. 1995. *Workers in an Integrating World*. New York: Oxford University Press.
———. 1997. *World Development Report: The State in a Changing World*. Oxford: Oxford University Press.
———. 2003. *Cultivating Productivity: The Foundations of East Asia's Future Growth*. June 10, 2002 Draft.
Wright, Peter. 1985. "Water and Soil Conservation by Farmers." In H. W. Ohm and J. G. Nagy, eds., *Appropriate Technologies for Farmers in Semi-Arid West Africa*. Purdue: Office of International Programs in Agriculture, Purdue University.
Wrong, Dennis H. 1961. "The Oversocialized Conception of Man in Modern Sociology." *American Sociological Review* 26:183–93.
———. 1979. *Power: Its Forms, Bases and Uses*. New York: Harper and Row.
Yamagishi, Toshio, Karen S. Cook, and Motoki Watabe. 1998. "Uncertainty, Trust and Commitment Formation in the United States and Japan." *American Journal of Sociology* 104:165–94.
Yapa, Lakshman S. 1991. "Innovation Diffusion and Paradigms of Development." In Carville Earl and Marion Kenzer, eds., *Concepts in Human Geography,* 231–69. London: Rowman and Littlefield.
Young, Alwyn. 1995. "The Tyranny of Numbers: Confronting the Statistical Realities of East Asian Growth Experience." *Quarterly Journal of Economics* 110:641–80.
Younger, S., and E. G. Bonkoungou. 1985. "Burkina Faso: The Projet Agro-Forestier—A Case Study of Agricultural Research and Extension." In R. Bheenick, ed., *Successful Development in Africa*. EDI Development Policy Case Series, Analytical Case Studies 1. Washington, DC: World Bank.
Yusuf, Shahid. 2001. "The East Asian Miracle at the Millennium." In Joseph E. Stiglitz and Shahid Yusuf, eds., *Rethinking the East Asian Miracle,* 1–53. New York: Oxford University Press.
Zhou, Kate Xiao. 1996. *How the Farmers Changed China: Power of the People*. Boulder: Westview.
Zusman, Pinhas. 1976. "The Incorporation and Measurement of Social Power in Economic Models." *International Economic Review* 17:447–61. Reprinted in K. W. Rothschild, *Power in Economics*. Baltimore: Penguin, 1971.

# Author Index

Acemoglu, Daron, 26
Adams, Adrian, 213
Adelman, Irma, 35, 37–38, 102, 104–5, 124
Alchian, Armen, 25
Alesina, Alberto, 248
Alexander, Jeffrey, 87–88
Almond, Gabriel A., 108
Amin, Samir, 80
Appleby, R. Scott, 216n. 2, 217, 219
Arendt, Hannah, 122
Arrow, Kenneth J., 180n. 9, 186, 260

Backus, D. K., 152
Banfield, Edward C., 258
Baran, Paul, 80–81, 110
Bardhan, Ashok, 148
Barro, Robert J., 119
Bass, Frank M., 186–87
Bates, Robert, 94
Bauer, Peter T., 35, 39–40, 94
Baumol, William J., 168
Beal, George M., 175
Begemann, Louis, 208
Beitz, Charles, 259
Benedict, Ruth, 44n. 12, 52, 227–32
Bhagwati, Jagdish, 102–3
Binswanger, Hans, xvi, 3, 29, 71
Bliss, Christopher, 43
Boas, Franz, 42, 49, 52, 64
Bohlen, Joe M., 175
Bonkoungou, E. G., 209
Bourdieu, Pierrre, 87–88
Buchanan, James M., 246
Buijsrogge, Piet, 212

Busch, Lawrence, 183

Cai Fang, 231
Chari, Varadarajan, 164n. 40, 190
Chenery, Holis B., 141
Cheung, Stephen, 160
Clague, Christopher, 106
Clark, Colin, 72, 157
Clastres, Pierre, 53n. 21, 65
Cleveland, Harlan, 266
Coase, Ronald H., xvi, 22, 25, 160
Coleman, James S., 27–28, 71, 72n. 6, 88, 90–92, 94, 96, 98
Commons, John R., 5, 8n. 8
Crane, Diane, 171
Crocker, Thomas, 20

Dahl, Robert, 124
Dales, James H., 20
Dasgupta, Parthe, 92
Datta, Anusa, 192
De Long, J. Bradford, 168
Demsetz, Harold, 16n. 16, 25
Dennison, Edward F., 168
Derrida, Jacques, 54
de Sweinitz, Karl, 102, 103
Dinar, Ariel, 197
Dixit, Avinash K., 161
Domar, E., 139–41
Dominquez, Jorge I., 110n. 12, 116
Dosi, Giovanni, 189
Downs, Anthony, 125, 126
Durkheim, Émile, 69

Easterly, William P., 247

327

Eckstein, Harry, 109–11
Eisenstadt, Samuel N., 74–75
Elster, Jon, 89
Escobar, Arturo, 54, 57–58, 65, 66

Fafchamps, Marcel, 204
Feder, Gershan, 187
Fei, John C., 227, 229
Findlay, Ronald, 195–96
Firth, Raymond, 42
Fliegal, Frederick, 197–98
Flora, Peter, 124
Fogel, Robert, 159n. 33
Foster, George, 35
Foucault, Michel, 54, 58, 58n. 25, 59, 65
Frank, Andre Gunder, 80–81, 110–11
Frank, Robert, 71n. 5, 90
Fukuyama, Francis, 116, 123

Gans, Eric, 48, 49
Geertz, Clifford, 50–52, 53n. 21, 65
Gellner, Ernest, 233, 244, 245–46
Gentil, Dominique, 210–12
Gerschenkron, Alexander, 101–2, 103
Giddens, Anthony, 87–88
Greif, Avner, 26
Griliches, Zvi, 148, 152, 168, 179, 184–86, 198
Gross, Neal C., 174–75
Grossman, Gene M., 149, 191, 192n. 24
Grübler, Arnulf, 187, 189
Gudeman, Stephen, 59, 61–62, 63, 65, 91

Habermas, Jürgen, 87–88
Hagan, Everett E., 35, 36–37
Hägerstrand, Torsten, 177, 198
Hammond, Peter B., 203
Harris, Marvin, 49
Harrison, David, 73n. 8, 79, 208, 212
Harrod, R. F., 139–41
Hart, Jeffrey A., 193
Hartwick, Elaine, 87
Hayami, Yujiro, 29, 41, 43, 48, 59n. 26, 71, 191n. 23; on induced technical change, xvi, 3, 8, 10, 25

Hayek, Friedrich A., 8nn. 8–9, 10, 126, 258, 271
Hegel, Georg Wilhelm Friedrich, 72
Helpman, Elhanan, 149, 192n. 24
Herbsi, Jeffrey, 243
Hicks, Sir John R., v, 167–68
Hirschman, Albert O., 41
Hobbes, Thomas, 258
Holt, Robert T., 107–8, 112–13
Hopenhayn, Hugo, 190
Horowitz, Irving Louis, 79
Hoselitz, Bert F., 35–36, 70
Hunt, Robert C., 43n. 9, 128n. 37
Huntington, Samuel P., 109, 110n. 12, 115–16, 123–24, 242, 255
Hurwicz, Leonid, 16n. 16

Iannaccone, Laurence R., 221–22
Inglehart, Ronald, 26, 118–19
Isaac, Jeffrey C., 121
Islam, Nazuri, 151

Johnson, D. Gale, 160
Johnson, James A., 26
Jones, Charles I., 151–52
Jorgenson, Dale W., 168
Just, Richard E., 187

Katz, Elihu, 171, 174, 186
Kehoe, Patrick J., 152, 164n. 40
Kehoe, Timothy J., 153
Kendrick, John W., 168
Kikuchi, Masao, 10, 25, 43, 48
Kimball, Miles S., 204
Kindleberger, Charles P., 69
Knight, Frank H., 5
Kornai, Janos, 196
Krishna, Raj, 227
Kuhn, Thomas S., 84
Kuran, Timur, 220, 223, 224, 236–38
Kuznets, Simon S., 72, 159

Lacan, Jacques, 54
Lacy, William, 183
Lal, Deepak, 232–35
Lam, Danny Kin-Kong, 229
Lanjouw, Peter, 43
Lee, Ian, 229

Lenin, Vladimir, 80, 257
Lenway, Stefanie A., 193
Lerner, Daniel, 74
Levine, Ross E., 150n. 21, 151, 153
Lévi-Strauss, Claude, 44n. 13, 54, 65, 67
Levy, Jacob, 246–47
Levy, Marion J., 230, 235, 238
Li, Zhou, 231
Lin, Justin Yifu, 231
Lipset, Seymour Martin, 117
Lucas, Robert E., Jr., 144–50, 154–55, 163n. 38, 166
Lyotard, Jean-François, 54, 56, 57, 58n. 25

Maddison, Angus, 168
Mankiw, N. Gregory, 150
Mansfield, Edwin, 184, 186–87, 198
Marsh, Robert M., 116
Marty, Martin E., 216n. 2, 217, 219
Marx, Karl, 6, 8n. 9, 69, 101n. 3, 196
McCloskey, Donald (Deirdre), 56
Mead, Margaret, 44n. 12, 52
Merton, Robert, 83–84
Meyers, Ramon H., 229
Mills, C. Wright, 124
Mohtadi, Hamid, 192
Mokyr, Joel, 194
Morgan, Lewis Henry, 52
Morris, Cynthia Taft, 35, 37–38, 102, 104–5, 124
Mulder, Paul, 208
Murtha, Thomas P., 193
Myrdal, Gunnar, 35, 38–39, 40

Nagy, Joseph G., 207
Nair, Kusum, 35
Narayan, Deepa, 93
Netting, Robert, 43n. 10
Nkrumah, Kwame, 112
North, Douglass C., 5nn. 2–3, 25
Nozick, Robert, 258
Nugent, Jeffrey B., 205

Olson, Mancur, 5n. 4, 25–26, 126, 156, 204
Ostrom, Elinor, 22n. 19, 114, 125, 126–28, 130–31, 204–5
Ostrom, Vincent, 113
Ouedraogo, Bernard L., 202, 204–6, 209–13

Packingham, Robert A., 108
Parente, Stephen L., 164n. 40, 194–96
Parsons, Talcott, 75–77, 88, 90, 114, 121–22; on structural-functional model, 70–71, 73
Peet, Richard, 87
Piore, Michael J., 68, 98
Platteau, Jean-Philippe, 204
Popkin, Samuel L., 204
Popper, Karl R., 83–84
Posner, Richard A., 204
Pradervand, Pierre, 212
Prebish, Raul, 80
Prescott, Edward C., 142, 143, 156, 164n. 40, 194–96
Pritchett, Lant, 93
Przeworski, Adam, 119
Putnam, Robert, 91, 130–31
Pye, Lucian W., 108n. 9, 114–15, 120, 122

Radcliffe-Brown, A. R., 44n. 12, 49
Ramaswamy, Sunder, 207, 209
Rawls, John, 258–59, 268
Reij, Chris P., 208
Reilly, William, 20
Renelt, D., 150n. 21, 151, 153
Ricardo, David, 136
Riker, William, 113, 114
Robertson, Pat, 221
Robinson, James A., 26
Rogers, Everett, 171–74, 176–78, 182nn. 11–12, 184n. 15, 197–98
Romer, Paul M., 144–50, 155, 164n. 40
Rostow, Walt W., 101n. 3, 102, 103–4, 152n. 22
Rueschemeyer, Dietrich, 117
Runge, C. Ford, 259
Ruttan, Vernon W., xvi, 3, 29–30, 41, 59n. 26, 71, 191n. 23
Ryan, Bruce, 174–75

Sahlins, Marshall D., 50, 64
Saleth, B. Maria, 197
Sanchez, Nicolas, 205
Sanders, John H., 207, 209
Savonnet, G., 208
Schultz, Theodore W., xvi, 41, 155n. 28
Schumpeter, Joseph, 62n. 27, 136–39
Scott, James C., 194, 204
Simon, Herbert A., 124
Singer, Milton, 50
Smith, Adam, 136
Solow, Robert M., 141–43, 145n. 14, 150, 154, 158, 163n. 39, 167
Spencer, Herbert, 52, 69
Stalin, Josef, 101
Stephens, Evelyne Huber, 117
Stephens, John D., 117
Stern, Nicholas, 43
Stiglitz, Joseph E., 160, 162
Stokes, Donald D., 86
Stone, Richard, 72
Swan, Trevor W., 141–43

Thomas, J. K., 25

Tullock, Gordon, 246
Turner, John E., 107–8, 112–13
Tylor, Edward B., 52

Uzawa, Hirofami, 146

Van Den Berg, Axel, 87–88
van der Veer, Peter, 225
Vernon, Raymond, 190

Wallerstein, Immanuel, 80
Weber, Max, 26–28, 69, 72n. 7, 219, 224–26, 232–34
Weil, D. N., 150
White, Leslie, 44n. 12, 64
Wittfogel, Karl A., 24, 231
Wittgenstein, Ludwig, 51
Wong, Arline, 231
Wong, John, 231

Young, Alwyn, 151
Younger, S., 209

Zilbeman, David, 187

# Subject Index

Abortion, 221
Adoption, of technology, 171, 175–77, 180, 185–86, 191–92, 198
Africa, 33, 94, 105–6, 124n. 37, 164; food production in, 66, 264; nation building in, 241, 243; sub-Saharan, 200–202, 207–10, 213–14, 243
*Agricultural Development* (Hayami and Ruttan), 41
Agriculture, 3–4, 9, 157, 232n. 11, 253, 263–64; in China, 129; research of, 17, 182–84, 264; swidden, 59–60; technology of, 191n. 23, 197, 210. *See also* Food
Algeria, 123
Alliance for Progress, 254
Anglo-American model, of political development, 108
Anthropology, 30, 42, 52–53, 62n. 27, 215; interpretive, 50, 65; materialist, 44–49, 50, 64, 65; postmodernism in, 54–57, 59, 65; of technology diffusion, 172. *See also* Cultural endowments; Ethnography
Arabia, 223
Argentina, 17–18, 241n. 22
Asceticism, in North America, 230
Asia, 24; irrigation systems in, 128n. 37; leadership in, 115; nation building in, 241; potato producers in, 66; preindustrial economy of, 164. *See also specific country*

Asia, East, 24, 59, 146, 165; demographic transition in, 158; economics of, 226, 227–32, 238–40; rates of growth in, 151; schooling levels in, 163n. 38
Asia, South, 18, 33, 38–40, 222, 226, 232, 239
Asia, Southeast, 40, 79
*Asian Drama: An Inquiry into the Poverty of Nations* (Myrdal), 38–40
Assets, redistribution of, 260
Assistance, development, 132, 177, 260–62, 266, 268. *See also* Development; Diffusion of technology
Austria, 185, 242
Authoritarianism, 7, 101, 103, 105, 116–17, 132–33
Autocracy, 225

Banking: credit and savings, 138, 140, 164, 210–11, 228; Islamic, 222–23. *See also* Financial functioning; World Bank
*Barriers to Riches* (Parente and Prescott), 194–95
Base, social capital as, 91, 93
Behavioral characteristics, of Chinese economy, 227–28
Brazil, 253
Bureaucracy, 76
Burkina Faso, 92, 200, 202, 207–8, 210, 213–14
Bush, George H. W., 20
Bush Report (1945), 84

## Subject Index

Capital: assimilation in Confucianism, 232; contribution of, 146–47; cultural, 92n. 28; returns to, 150; social, 201–2, 206, 213–14
Capitalism, 26–28, 80–81, 94, 135, 222, 229, 239; authoritarian, 103, 105; costs of, 161; in equilibrium economics, 138–41; role of Protestantism in, 219, 224–25
*Capitalism, Socialism and Democracy* (Schumpeter), 138
Carlucci Commission, 254, 255
Caste system, 233, 235n. 17
Cattle, 48
Center on Economic Development and Cultural Change, 36
Central planning, 101, 132, 165–66. See also Democracy
Change: vs. stability, 5; theory of, 3–4. See also Development; Growth; Institutional innovation
Chile, 106, 193
China, 33, 102–3; agriculture in, 129; economy of, 112–13, 226–32; political development in, 19, 115, 239. See also Asia
Christianity, 220–21, 234–35. See also Religions
*Christian Perspectives: A Journal of Free Enterprise*, 221
Church, separation of state and, 216
Class, 6, 81, 117–19, 245
Classical growth economics, 136–37
Clean Air Act of 1990, 21
Coasian bargains, 10, 126
Cold War, 78
Collective, 8n. 8, 26, 129. See also Cooperation; *Naam*
Colleges, 221. See also Education
Colombia, 178
Colonialism, 27, 53, 94, 215, 243
Communalism, 236, 250
Communication, of technology diffusion, 171, 174, 175–77
*Communication of Innovations* (Rogers), 176
Communism, 19, 232. See also Marxism

Community, 63
Confucianism, 115, 228, 229–32, 240, 268
Consequences, unintended, 8, 58n. 25, 63, 92, 99, 238, 249. See also Externalities
Constitutional systems, 249–50
Constructivist perspective, 8
Consumption, 104
Contractarian theory, 258–60, 268
Convergence: of diffusion traditions, 176; in growth economics, 168; of productivity and income, 198–99; within sociology, 183
Cooperation, 203–4, 230. See also Collective; *Naam*
Corn, hybrid, 174–75, 179, 184, 185n. 18, 198
Corporations, 230–31, 237
Credit and savings, 138, 140, 164, 210–11, 222–23, 228
Cross-country analysis, 116, 150–53, 156
Cultural endowments, 4–5, 18, 29, 92n. 28, 130, 272; collectivist vs. individualistic, 26; equilibrium framework of, 22–24, 64; role in political and economic development, 33, 113–16, 226–32, 238–39; and social capital, 201–2, 206, 213–14
Culture, 42, 44–49, 51, 67, 241, 247, 273–75
Czechoslovakia, 241n. 22
Czech Republic, 250

Decision theory, 125, 154
Decolonization, 215. See also Colonialism
Democracy, 76–77, 105–8, 128n. 25; authoritarian, 101, 103; income level in, 132; qualitative, 204; role in economic development, 117–19. See also Government
Demographics, 24n. 22, 158. See also Population
Deng Xiaoping, 232
Dependency, 17, 79, 110–11

Design, of institutional innovations, 8, 105–6, 125–33, 130–32, 246–50, 271
Developed countries, 70, 132, 135, 192, 251–53, 256, 263. *See also specific country*
Development, 11n. 13, 40, 69, 85, 101n. 3, 263–64; assistance for, 99, 132, 177, 256, 260–62, 268; deconstructing, 52; of institutions, 237–38; of knowledge, 144–47; knowledge in, 69–70, 96–97; process of, 29, 58, 97; of technology, 3–4, 147–48, 184–88, 197–99. *See also* Economic development; Political development; Social development
Development economics, xv, 30, 116–17, 154–60, 131–34; research on, 148–49, 153. *See also* Growth economics
Dialectical relationship, 64–65
Difference principle, 258–59
Differentiation, 73, 76, 105
*Diffusion of Innovations* (Rogers), 174, 182n. 12
Diffusion of technology, 171–99, 210; challenges to traditions in, 175–81; communication of, 171, 174, 175–77; diffusion of research in, 184–86; divergence of traditions in, 197–99; economics of, 171–73, 182, 184–85; equilibrium models of, 77–78, 137–38, 144–45, 156, 186–88, 234–35; evolutionary models of, 75–78, 188–90, 271; geography of, 171, 173; of hybrid corn, 174–77, 179, 184, 185n. 18, 198; international, 190–93; new sociological paradigm of, 181–84; research traditions of, 171–75, 177–78; resistance to, 193–97
Dike construction, 128n. 37, 200–201, 206–9, 212, 213
Disease environment, 27
Disequilibrium, 7, 77; in agricultural production practices, 129; between economic potential and performance, 240; in productivity and income, 192; in property rights, 9–10; in resource allocation, 5–6. *See also* Equilibrium models
Distributive justice, 257–61. *See also* Ethics; Redistribution of assets

East Asia. *See* Asia, East
Ecology, political, 66n. 31
Economic development, 69, 101, 119, 266; indicators of, 37; knowledge of, 166; political basis of, 112–16; research on, 134; role of cultural endowments in, 33, 63, 113–16, 215, 226–32, 238–39; role of democracy in, 117–19; role of social science knowledge in, 271–72
*Economic Development and Cultural Change* (journal), 36
Economic growth: impact of ethnicity on, 246–48; microlevel changes on, 166; rate of, 140, 142, 163, 164n. 40; research on, 143, 148–49; sources of, 90, 147; stages of, 103
Economics, 16, 135, 157, 220, 222–27, 262–63; centrally planned, 101, 132, 165–66; industrial, 137; market, 40, 63, 76, 94–98, 162, 165–66, 171–73; microeconomics, 30, 41–42; political, 7n. 6, 125; power of, 120n. 29; relationship between sociology and, 68–70, 94–97, 183–88, 197–99; research on, 198; of scale, 152–53, 155; of technology diffusion, 171–73, 182, 184–85. *See also* Capitalism; Development economics; Growth economics
*Economics as Culture* (Gudeman), 59, 61–62
*Economics of Underdeveloped Countries, The* (Bhagwati), 102–3
*Economic Theory of Democracy, An* (Downs), 125
Education, 163n. 8, 172, 205–6, 219, 221, 264–65
Efficiency, 15
Egalitarianism, 258–59
Elites, role of, 38, 70, 105

Emic components, 45n. 15, 47
Emissions trading, 20–22
Empirical detail, on modernization, 74. *See also* Research
Employer-labor relationship, 13
Enclosure Movement, 9, 60
Endogenous change, 3–4, 7, 143–49, 150–51, 153, 155
England. *See* Great Britain
Enlightenment, 54, 58, 108, 114n. 17, 244, 271
Entitlement, 256, 259
Entrepreneurs, 17, 35, 70, 138–39
Environment: disease, 27; institutional, 89, 138n. 4; of political system, 106–7
Environmental Defense Fund (EDF), 20
Environmental policy, 20–21
Environmental Protection Agency (EPA), 20–21
Equality, of opportunity, 122
Equilibrium models: of cultural endowments, 22–24, 64, 73; of institutional innovation, 22–24, 64; of technology diffusion, 77–78, 137–41, 144–45, 156, 186–88, 234–35. *See also* Disequilibrium
Equity, 15. *See also* Egalitarianism
Ethics, 251, 255–56, 259–61, 268. *See also* Moral obligation
Ethiopia, 261
Ethnicity, 40, 241n. 22, 242–43, 246–48
Ethnography, 42–43, 48–49, 64. *See also* Anthropology
Etic components, 45n. 15, 47
Europe, 9, 102, 254; Eastern, 123, 132, 165; nationalism in, 240, 242–43, 244; postmodernism in, 56; radical scholars in, 81; trade with North and Latin America, 80; Western, 25, 177–78, 185, 230
European Economic Community, 123
Evangelical communities, 221. *See also* Fundamentalism

Evolutionary models, of diffusion, 75–78, 188–90, 271
Exogenous change, 3–4
Exploitation, 257
Externalities, 22, 25, 146n. 15, 147, 148. *See also* Consequences, unintended

Factor analysis, 37
Falwell, Jerry, 221
Familism, 230
Family: effect of fundamentalism on, 217, 224–26; joint, 234
Farmers, 212–13. *See also* Peasants
Farm Foundation, 175
Fertilizer, 207, 214
Feudalism, 240
Financial functioning, 210–11. *See also* Banking; Economics
Flat panel display (FPD) technology, 193
Food, 66, 124. *See also* Agriculture; Corn; *and specific commodity*
Foreign Assistance Act, 107
*Formations des Jeunes Agriculteurs* (FJAs), 205–6
*Foundations of Social Theory* (Coleman), 7n. 5, 88, 90, 98
France, 56, 112–13, 123, 156, 240, 257; nationalism in, 242–43, 244
Fundamentalism, 215–26; Christian, in United States, 220–22; economic development and, 215–17, 219–20; family life and, 217, 224–26; Islamic, 222–24; society and, 217–20. *See also* Religions
Fundamentalism Project, 216n. 2, 224–26
*Fundamentalisms and the State* (Marty and Appleby), 219

Gamma system, 13–15
Geography, of technology diffusion, 171, 173
Germany, 107n. 7, 132, 242–43, 244, 245
Ghana, 262
Gift giving, 204

Globalization, 54, 87n. 24, 215, 257, 259–60
Government: autocratic, 225; postcolonial, 243; regional, 130; structure of political power of, 210, 213–14. *See also individual ideologies*
Grants, 251, 253. *See also* Assistance, development
Great Britain, 16, 26n. 25, 35, 77n. 10, 185, 194; development assistance from, 256; economic development in, 112–13; Enclosure Movement in, 9, 60; ethnic consciousness in, 243; industrial culture in, 196; study of science in, 84
Green revolution, 24, 59n. 26, 182nn. 12–13
Gross national product (GNP), 154, 156
*Groupements des Jeunes Agriculteurs* (GJAs), 206
*Groupements Naam,* 200–202, 205, 208–14
Growth, 90, 103, 123, 143, 156, 166; rate of, 140, 142, 145, 151–52
Growth economics, 135–54; classical, 136–37; convergence in, 7, 168; in developed economics and traditional societies, 135; as development economics, 154–59; endogenous, 143–49, 150–51, 153; Keynesian, 139–41, 164; neoclassical, 141–43, 147–50, 152–54, 155n. 26, 164, 167–68; new, 143–49, 154–55, 163–68, 191, 198; of political development, 7n. 6, 112–16, 117–19, 125, 131–34; Schumpeterian, 137–39. *See also* Development economics

Health, investment in, 264–65
Hermeneutic-Dialectic (H-D) tradition, 183
Hinduism, 220, 225, 232–35, 240. *See also* Religions
Human capital, 97, 146–47, 155, 264

ICRISAT, 209
Ideas, as source of economic growth, 147
Ideology: formation and impact of, 51; influence on institutional innovation, 18–19, 33; in sociological research, 78; of Westernized elite, 38. *See also individual ideologies*
Imperialism, 80, 236, 257
Incentives, 28; social capital, 91, 93
Income, 132, 143, 151–52, 159; distribution of, 257–58; influence of technology transfer on, 192, 195; in international trade, 190–91; in relationship to productivity, 198–99. *See also* Wage
Index of Civic Engagement, 93
India, 35, 48, 226–27, 248, 250; religion in, 220, 225, 232–35
Indiana University Workshop in Political Theory and Policy Analysis, 125, 127
Individualism, 26, 230, 234–35, 239, 259–60
Industrialization, 82, 102, 137, 157, 164, 195–96, 244; in newly industrializing countries, 151, 231
*Industrialization and Democracy* (de Sweinitz), 102
Information, 162–63, 188n. 21, 190n. 22, 196
*Innovation and Growth in the Global Economy* (Grossman and Helpman), 149
Institutional innovation, 76, 159, 231, 271–72; constraints to, 195–97; costs of, 3, 6–7, 16, 18–19; demand for, 3–4, 9, 184n. 5; design of, 8, 67, 105–6, 125–33, 246–50, 271; endogenous, 7, 165; of modern corporate state, 231; nationalism as, 240–41; supply of, 16–18, 70–71; theory of, 3–5, 24n. 21, 29, 240. *See also* Development
Institutions, 4, 104–5, 211, 239, 244; bureaucratic, 76; development of, 162–63, 237–38; environments of, 89, 138n. 4; equilibrium frame-

Institutions (*continued*)
  work of, 22–24, 64; public sector, 3–5, 91, 126, 131, 133, 151; reform of, 165; sociological issues of, 99; traditional, 213–14. *See also* Corporations; Nations; Organizations; States
Interactive model, of science and technological knowledge, 85–86
Interdependence, among nations, 260
Interdisciplinary project, knowledge as, 199
Interest groups, 17, 25–26
International Institute for Applied Systems Analysis (IIASA), 185, 187
International Monetary Fund (IMF), 261–62
International Potato Center (CIP), 66
International Trade Organization (ITO), 262
Invention, 136–37. *See also* Institutional innovation; Technology
Investment: in agriculture, 263–64; of gross national product (GNP), 154, 156
Iowa State Agricultural Experiment Station, 174–75
Iowa State Agricultural Extension Service, 174, 177
Iran, 222
Irrigation systems, 128n. 37, 200–201, 206–9, 212, 213
Islam, 220, 222–24. *See also* Religions
Israel, 208, 247
Italy, 33, 92, 130, 132

Japan, 34, 64, 107n. 7, 132, 193, 230, 241n. 22; cultural endowments of, 18; economic development in, 112–13; economy of, 227, 230–31; GNP in, 156; income distribution in, 159; institutional changes of, 16; modernization of, 122n. 32; property rights in, 10

Johnson, Lyndon Baines, 20
Justice, distributive, 257–61. *See also* Ethics

Keynesian growth economics, 139–41, 164
Knowledge, 144–47; accessibility of, 66; of economic development, 166; as interdisciplinary project, 199; link with power, 58–59; sociological, 69; sociology of, 82–83, 84; technical, 193
*Kombi-Naam,* 202–3, 205–6
Korea, 149, 154, 193, 227, 230–31, 262
Kuwait, 257

Labor, 9–10, 13, 14n. 14; costs in international trade, 190–93; in Philippine village, 10; physical capital as, 146, 200n. 2
Land: models of, 61–62, 137–38; reclaiming degraded, 207–9; tenure on, 9–13, 15, 160
Latin America, 79, 80, 81, 110–11; preindustrial economy of, 164; reforms in, 106; religion in, 224–25
Law, 76, 77n. 10, 105, 237, 247. *See also* Government; Reform
Leadership, 105, 115, 121, 202–3; cooperative, 206, 210, 212–13, 230; transfer of, 244–45
Liberal impulse, 267
Liberty University, Lynchburg, Virginia, 221
*Logic of Collective Action, The* (Olson), 126
Los Boquerones, Panama, 59–62
Luddites, 194, 196

Maastricht Economic Research Institute on Innovation and Policy (MERIT), 185
Macroeconomic environment, 262–63
*Making Democracy Work: Civic Traditions in Modern Italy* (Putnam), 91–92
Malaysia, 223, 253

Markets, 40, 63, 76, 94–98, 162, 165–66; as social capital, 94–96; technology diffusion, 171–73
Marshall Plan, 254
Marxism, 6, 24, 44, 46, 79–81, 110–11, 245
Materialism, 44–49, 50, 64, 65
Measurement, of political development, 124–25
Metatheory, 85, 87, 96
Middle East, 74, 222, 227, 236, 239
Military assistance, 255
Millet, 207
Model, 22n. 19. *See also* Theory
Modernization, 35, 38–39, 115, 225, 242; nation building in, 241–43; objective of, 58, 66; paradigm of political development, 108, 110–11; theory of, 72, 74–75, 78, 82. *See also* Development; Industrialization
Moral obligation, 235n. 16, 258–60. *See also* Ethics
Mossi society, Africa, 200–203, 208, 213
Multiculturalism, 115, 246
Multinational states, 215, 241n. 22, 266
Muslim society, 240. *See also* Islam
Mutual assistance groups, 203–4. *See also* Cooperation; *Naam*
Mutual Development and Security Agency, 254

*Naam*, 200–203, 205–6, 208–14
National Commission on Philanthropy and Civic Renewal, 93
Nationalism, 240–46
Nations, 107, 114, 230, 241–43, 259–60. *See also* Developed countries; Underdeveloped countries
Natural resources, 158, 256–57
Near East, 226
Neoclassical growth economics, 141–43, 147, 150, 153–54, 155n. 26, 164, 167–68
Neo-Marxism, 79
New Christian Right (NCR), 221–22

New Economic Order, 256
New Institutional Economics, 125–26, 160, 219
New York City, 94
Nigeria, 166, 241n. 22
Nixon, Richard M., 20
Nkrumah, Kwame, 112
Norms, generalized universalistic, 76
North Africa, 222
North America, 80, 230. *See also* United States
North Korea, 231
Nuclear power, 196

OECD economies, 151, 156
Ohio State University, 177
Opportunity, equality of, 122
Organic perspective, of institutional change, 8
Organization for African Unity, 243
Organizations, 5; informal and formal, 95; spontaneous and formal, 71. *See also* Corporations; Institutions
*Oriental Despotism* (Wittfogel), 24
Oxfam, 200, 208–9

Pakistan, 222, 223
Palestinian Authority, 247
Panama, 59, 61–62
Paraguay, 62n. 27, 65
*Passing of Traditional Society, The* (Lerner), 74
*Pasteur's Quadrant: Basic Science and Technological Innovation* (Stokes), 86
Pattern, 22n. 19, 272
Peasants, 59–62, 65. *See also* Farmers
Personality formation, 36–37
Peru, 66
Philippines, 10, 15n. 15, 16, 154, 166
Physical capital, 10, 14n. 14, 146, 190–93, 200n. 2
Pious foundations (*waqf*), 237
Pluralism, cultural, 247

Policy, 102, 116–17, 195–97, 267; development assistance, 96–97; economic, 262–63; public, 133, 151; studies of, 30, 56, 105–6

*Political Basis of Economic Development* (Holt and Turner), 112–13

Political development, 17, 39, 66n. 31, 100–134, 239; cycles in, 103–4; influence of, 100–102, 106–11, 132, 267; institutional design in, 105–6, 125–31; modernization paradigm of, 108, 110–11; and political power, 119–24, 239; relationship of economics to, 7n. 6, 112–16, 117–19, 125, 131–34; research on, 74; role of cultural endowments in, 33, 63, 113–16, 215, 226–32, 238–39; strategy in, 253–54; in structural-functionalist model, 124; theory of, 100, 107n. 6, 108–9, 132, 267

"Political Development and Political Decay" (Huntington), 109

Political science research, 107, 114

Poor countries, 256. *See also* Poverty; Underdeveloped countries

Population, 11, 25, 60, 129, 137–38, 205, 221. *See also* Demographics

Postmaterialism, 118

Postmodernism, 217n. 3; in anthropology, 54–57, 59; in sociology, 55, 82–83, 85, 87

Postnationalism, 245

Poverty, 107, 159. *See also* Poor countries; Underdeveloped countries

Power, 53–54, 65, 102, 120n. 29, 124–25, 131, 203; link with knowledge, 58–59; political, 119–24, 213, 239. *See also* Political development

Preference falsification, 238, 240

*Principles of Political Economy and Taxation, The* (Ricardo), 136

*Private Truths, Public Lies* (Kuran), 238

Product cycle model, 190, 192n. 25

Production, 66; functions of, 150n. 21, 151, 153; models of, 46, 136–37; subsistence system of, 59. *See also* Agriculture; Technology

Productivity, 146, 167–68; of land and labor, 9; of new technology, 194–95; relationship of income to, 192, 198–99

Profit, 138, 194

Property rights, 6, 9–10, 25, 160–62. *See also* Land

*Protestant Ethic and the Spirit of Capitalism, The* (Weber), 26

Protestantism, 26–28, 219, 224–25. *See also* Religions

Public sector, 3–5, 91, 126, 131, 133, 151

Quebec, 248

Rashtriya Swayamsevak Sangh (RSS), 225

Rate, of economic growth, 140, 142, 163, 164n. 40

Rational choice, 88–90, 131

Reagan administration, 254

Recipient, 263. *See also* Underdeveloped countries

Redistribution, of assets, 260. *See also* Distributive justice

Reform: in agriculture, 11–12, 220, 232n. 11; policy, 105–6

Relativism, 67

Religions, 18, 33, 39, 215–16, 224–25, 232–35. *See also* Christianity; Confucianism; Fundamentalism; Hinduism; Islam; Muslim society; Protestantism

Rent, land, 9, 13

Research, 19, 30, 88n. 25, 147, 158, 182, 272; agricultural, 17, 182–84, 264; anthropological, 43, 45, 48–49, 51, 53, 63–66; cross-country, 116, 150–53, 156; economics, 143, 148–49, 198; health, 265; human capital employed in, 147; policy, 102, 116–17, 195–97, 267; political development, 74, 134; political science, 107, 114;

Subject Index    339

scientific, 83–84; sociological, 78, 79, 82, 83n. 18, 87, 96–98; on technology diffusion, 171–78, 184–85. *See also* Theory; *and individual disciplines*
Resource endowments, 5–6, 22–24, 26, 44, 64, 272. *See also* Capital
Returns-to-scale hypothesis, 152
Rice production, 24, 48
Rich countries, 256. *See also* Developed countries
*Rural Sociology* (journal), 183
Russia, 94, 101, 102. *See also* Soviet Union (USSR)

Saudi Arabia, 223
Savings and credit, 138, 140, 164, 210–11, 222–23, 228
Scale, economy of, 152–53, 155
Scholars, radical, 81
Science, 83–84, 167, 218–19, 228–29. *See also* Research; Technology; Theory; *and individual disciplines*
Science Policy Research Unit (SPRU), 185
Secularization, 242. *See also* Public Sector; Religions
Security, 254–55
Self-interest, 44, 251–52, 254, 268
Self-organization, 126
Senegal, 210–13
Service economy, 157
Singapore, 115n. 20, 231
6-S (*Se Servir de la Saison Sèche en Savane et au Sahel*), 210, 212
Slovakia, 250
Social action, concept of, 99
Social capital, 90–96, 97–98, 131, 237; relationship between social development and, 36, 93; in relationship to cultural endowments, 200–202, 206, 213–14
Social development, 96–97
Socialism, 102–3, 105, 222. *See also* Government
Social science, 3–4, 248–50, 271–72
Social Science Research Council Committee on Comparative Politics (SSRC/CCP), 74, 107–8, 114, 114n. 18
Social structure, 36
Sociology, 8, 30, 36–37, 93, 215; of knowledge, 82–83, 84; postmodernism in, 55, 82–83, 85, 87; relationship between economics and, 68–70, 94–97, 183–88, 197–99; research and theory in, 77–79, 82, 83n. 18, 87, 90, 96–98; of technology diffusion, 171–75, 178–84
*Sociology of Modernization and Development, The* (Harrison), 79
Solidarity networks, 203–4. *See also* Cooperation; *Naam*
Somalia, 257
Sorghum, 207, 213
South Asia, 18, 33, 38–40, 222, 226, 232, 239
Southeast Asia, 40, 79
South Korea, 231
Soviet Union (USSR), 102–3, 123–24, 132, 165, 232, 239, 243. *See also* Russia
Spain, 133, 243, 249–50
Spatial diffusion or transfer, of technology, 177, 191–92, 198
Spontaneous order, 10
Sri Lanka, 248
Stability, vs. change, 5
States, 40, 102, 247–48; modern corporate, 230–31; relationship between culture and, 241; separation of church and, 216. *See also* Nations
Strategic interest, 253–54
Stratification, 76
Strong Program, 85, 98
Structural-functionalism, 73–75, 76, 78, 80, 87–88, 124
Structural transformation, 157
*Structure of Scientific Revolutions, The* (Kuhn), 84
*Structure of Social Action* (Parsons), 88, 90
Sugar production, 59, 60–62
Sulfur dioxide ($SO_2$) emissions, 19–22

Sweden, 177
Switzerland, 249

Taiwan, 149, 193, 231
Tax, zakat, 223
*Techniques and Civilization* (Mumford), 196
Technology, 85, 98, 142, 144, 229; agricultural, 191n. 23, 197, 210; development of, 3–4, 147–48; equilibrium framework of, 22–24, 64, 73; flat panel display (FPD), 193; relationship between science and, 167; resistance to, 192–97
Tennessee Temple, 221
Tenure, land, 10–13, 15, 160. *See also* Property rights
Thailand, 9–13, 16, 220
Theory, 8, 81; anthropological, 50, 52, 55; of change, 3–4; Coase, 10, 22, 126; contractarian, 258–60, 268; decision, 125, 154; dependency, 79, 82n. 17, 110–11; development, 40, 101n. 3; economic development, 69, 101; Enlightenment, 108; evolutionary, 75–78, 188–90, 271; general competitive equilibrium, 156; global, 54; growth, 123, 166; information, 162–63, 190n. 22; institutional innovation, 3–5, 24n. 21, 29, 240; of institutions, 71, 126; metatheory, 85, 87, 96; microeconomic, 30, 41–42, 49; modernization, 72, 74–75, 78, 82; new trade, 149, 191–93; political development, 100, 107n. 6, 108–9, 132, 267; postmodernist, 83; Rawls's, 258–60; Schumpeterian, 136–39; of self-interest, 254; of social development, 8, 36–37, 69; sociological, 77, 79, 90, 97; of underdevelopment, 79
Theory, development economics and growth economics, 135–49; classical, 136–37; developed economies and traditional societies, 135; endogenous, 143–48; Keynesian, 139–41; neoclassical, 141–43; new, 149; Schumpeterian, 137–39. *See also* Research; *and individual disciplines*
Theory, growth economics, 103, 123, 135–49, 167
*Theory of Economic Development, The* (Schumpeter), 137
Things, as productive inputs, 148
Third World, 56, 59, 80–82, 110–11. *See also* Underdeveloped countries
Title IX, Foreign Assistance Act, 107
Totalitarianism, 103. *See also* Government
Trade, 80, 149, 190–93
Traditional societies, 74–75, 135. *See also* Underdeveloped countries
*Tradition, Change, and Modernity* (Eisenstadt), 75
Transaction costs, 161–62, 240
Transfers: of leadership, 244–45; technology, 171, 192, 195, 251. *See also* Diffusion of technology
*Transforming Traditional Agriculture* (Schultz), 41
Trust, 94
Turkey, 238, 242
Twenty-first century, vision of, 266–67

Underdeveloped countries, 70, 79–82, 102–3; diffusion studies in, 177–78; growth in, 143, 156; transfers to, 192, 251. *See also* Poor countries; Third World; Traditional societies
United Nations Economic Commission for Latin America (ECLA), 80
United States, 219–22, 241n. 22; agriculture in, 4, 253; benefits of ethnographic study to, 64; development assistance efforts of, 79, 266–67; diffusion research in, 177–78; donor self-interest of, 252; income in, 151–52, 159; industrial culture of, 195–96; international trade of, 190; political develop-

ment by, 107–8; postmodernism in, 56; radical scholars in, 81; social capital in, 93
U.S. Agency for International Development (USAID), 53, 107, 177, 254
U.S. Congress, 22
U.S. Constitution, 113–14, 249, 271
U.S. State Department, 243
Universities, 221. *See also* Education
University of Minnesota, xvii, 164n. 40
Urbanization, 74
USSR. *See* Soviet Union (USSR)
Utilitarianism, 257. *See also* Religions

Vietnam, 78, 110
Visha Hindu Parishad (VHP), 225
Vision, of twenty-first century, 266–67
Voluntary agreements, 15. *See also* Cooperation; *Naam*

Wage, 136–37. *See also* Income
*Waqf* (pious foundations), 237
Wars, 242–43
Washington Consensus, 105–6, 165, 261–62
Water retention systems, 200–201, 206–9, 212, 213
*Wealth of Nations* (Smith), 136
*West African Trade* (Bauer), 94
Western Europe, 25, 177–78, 185, 230
World Bank, 53, 70, 243, 255, 261–62, 265
World Health Organization, 265
World systems, 80–82

Yatenga, Africa, 200, 202, 207–8, 210, 213
Younger, S., 209
Yugoslavia, 243

Zakat tax (Islamic), 222, 223, 226